Fundamentals of Cosmic Particle Physics

Maxim Khlopov

Viva Books

New Delhi | Mumbai | Chennai | Kolkata | Bengaluru | Hyderabad | Kochi | Guwahati

First Indian Edition 2013

VIVA BOOKS PRIVATE LIMITED

- 4737/23 Ansari Road, Daryaganj, New Delhi 110 002
  E-mail: vivadelhi@vivagroupindia.net, Tel. 42242200
- 76, Service Industries, Shirvane, Sector 1, Nerul, Navi Mumbai 400 706
  E-mail: vivamumbai@vivagroupindia.net, Tel. 27721273, 27721274
- Jamals Fazal Chambers, 26 Greams Road, Chennai 600 006
  E-mail: vivachennai@vivagroupindia.net, Tel. 28290304, 28294241
- B-103, Jindal Towers, 21/1A/3 Darga Road, Kolkata 700 017
  E-mail: vivakolkata@vivagroupindia.net, Tel. 22836381, 22816713
- 7, Sovereign Park Apartments, 56-58, K. R. Road, Basavanagudi, Bangalore 560 004
  E-mail: vivabangalore@vivagroupindia.net, Tel. 26607409, 26607410
- 101-102 Mughal Marc Apartments, 3-4-637 to 641, Narayanguda, Hyderabad 500 029
  E-mail: vivahyderabad@vivagroupindia.net, Tel. 27564481, 27564482
- Beevi Towers, First Floor, SRM Road, Kaloor, Kochi 682 018, Kerala
  E-mail: vivakochi@vivagroupindia.net, Tel: 0484-2403055, 2403056
- 232 GNB Road, Beside UCO Bank, Silpukhuri, Guwahati 781 003
  E-mail: vivaguwahati@vivagroupindia.net, Tel: 0361-2666386

www.vivagroupindia.com

Published by arrangement with
Cambridge International Science Publishing
7, Meadow Walk, Great Abington
Cambridge CB1 6AZ, UK

ISBN : 978-81-309-2295-9

Published by Vinod Vasishtha for Viva Books Private Limited, 4737/23, Ansari Road, New Delhi-110 002. Printed and bound by Syndicate Binders, New Delhi.

To my family
the fundamentals of the world,
in which we live

Maxim Yu. Khlopov

Dedicated to the memory of Y.A. Zeldovich and A.D. Sakharov (their work laid the basis of cosmoparticle physics), A.A. Anselm, D.A. Kirzhinits, Ya.I. Kogan, M.A. Markov, B.M. Pontecorvo, V.F. Shvartsman, D. Schramm, A.A. Trushevsky, all my teachers, friends and colleagues, communications and joint work with whom are reflected in this book which they unfortunately cannot read.

Acknowledgement

I am grateful to E.Kh. Akhmedov, F. Balestra, K.M. Belotsky, Z.G. Berezhiani, A.V. Berkov, G.M. Beskin, S.I. Blinnikov, N.G. Bochkarev, V.B. Braginsky, M.I. Vysotsky, S.S. Gershtein, D.Yu. and Yu.A. Golubkov, T. Damour, A.G. Doroshkevich, V.K. Dubrovich, I.G. Dymnikova, D. Fargion, A.F. Kadnikov, A.A. Klypin, Yu.Ya. Komachenko, R.V. Konoplich, Yu.L. Levitan, A.D. Linde, V.I. Maslyankin, A.G. Mayorov, R. Mignani, P.D. Naselsky, S.T. Petkov, G. Piragino, A.G. Polnarev, D.B. Pontecorvo, L.A. and S.A. Pustilnik, S.G. Rubin, M.V. Sazhin, A.S. Sakharov, E.V. Sedelnikov, I.M. Sobol, D.D. Sokolov, E.Yu.Soldatov, A.A. Starobinsky, A.L. Sudarikov, V.M. Souslin, R.A. Sunyaev, N.G. Uraltsev, S.S. Filippov, R.R. Khomeriki, V.M. Chechetkin, K.I. Shibaev, A.A. Shklyaev, all participants of joint work on the projects Cosmoparticle Physics, Astrobelix, Astrodamus, Cosmion-ETHZ, AMS-Epcos. The results of cooperation with them have greatly enriched the contents of this book.

I am also grateful to all participants of international conferences Cosmion-94, Cosmion-96, Cosmion-97, Cosmion-99, Cosmion-2001 and Cosmion-2004. Fruitful discussions with them transformed these conferences to important milestones on the road to the establishment of cosmoparticle physics as an independent scientific discipline.

The author

Contents

Preface to the English edition

The book is based on the English translation of the Russian book « Osnovy kosmomikrofiziki » (Basics of cosmoparticle physics), which is used as a textbook for the course "Introduction to cosmoparticle physics", given during the last two decades to 5th year students of National Research Nuclear University "Moscow Engineering Physics Institute" (MEPhI), specialized in physics of elementary particles. The aim of this course is to present the general trends of development of modern cosmology and particle theory that lead to foundation of cosmoparticle physics as the new field of science, studying fundamental relationship of macro- and micro-worlds.

During the last decade the development of astroparticle physics, precision cosmology and collider physics has been progressing so rapidly that it may be now meaningless to review all the directions of these studies in a comprehensive and self-consistent way. The Virtual Institute of Astroparticle Physics (VIA) created in 2007, now incorporated into the structure of APC Laboratory (Paris, France), and operating on its website http://viavca.in2p3.fr/site.html provides online the first hand information about the current progress in this field, so that the interested reader can enter this site, get acquainted with the huge library of records of previous videoconferences with VIA lectures, online VIA transmissions from various conferences, colloquiums or seminars. He/she can also participate in an interactive regime online in all such events.

On the other hand, sometimes this progress passes by important experience of earlier works and it is the aim of this book, presenting the subject in its historical development, to fill in such gaps of the common knowledge.

The ideas and methods of cosmoparticle physics, the nontrivial cross-disciplinary links, arising in its research are to be developed in the full body of this science, being the great challenge for this Millennium. I can only express the hope that fundamentals of cosmoparticle physics, here discussed, could play the role of proper cornerstones on which this great future science will be built.

I am grateful to Victor Riecansky of Cambridge Intrenational Science Publishing for cooperation in the work on the text of the manuscript.

1

Introduction

Cosmoparticle physics is a natural step in the investigation of the relationship between cosmology and microphysics, i.e., cosmology and particle physics. This relationship is reviving the tradition of natural philosophy which examined the Universe in its totality and unity.

Cosmoparticle physics is a new level in this tradition and, being an exact science, uses quantitative methods to describe the nature and structure at both micro- and macroscales.

To follow the trend leading to the formation of cosmoparticle physics, we must consider several key stages in the development of particle physics and cosmology.

The experimentally verified part of modern elementary particle physics is based on the standard model (SM) of electroweak and strong interactions, which generalizes the principles of electrodynamics on transformation of particles in processes of strong and weak nuclear interactions.

The possibility of such generalization follows from the fundamental changes in ideas about the charge and current in quantum field theory. Instead of relying on the internal constant property of eternal charged particles in classical electrodynamics, in quantum electrodynamics, we consider the current as a locally conserved bilinear combination of single-particle operators of creation and annihilation acting on the physical vacuum.

For example, an electric current of an electron is a sequence of annihilation and creation of single-electron states. Different particles in the initial and final states may differ, which corresponds to the quantum concepts of electric current.

This makes it possible to take the next logical step in addressing different particles in the initial and final states, and generalizes the notion of the currents accompanied by the transformation of the particles.

Electromagnetic currents are a source of electromagnetic fields. In quantum electrodynamics (QED), the annihilation of the electron in the initial state and its creation in the final state is a quantum process of emission (or absorption) of quantum of the electromagnetic field.

Generalizing, we can say that the annihilation of the particles in the initial state and the creation of another particle in the final state is the process of emission (or absorption) of a quantum of the field for the corresponding interaction.

Thus, the weak interaction in annihilation of neutrinos and the creation of an electron is accompanied by the creation of a W^+-boson or annihilation of the W^--boson – the quanta of the field of the weak interaction.

With this generalization we assume symmetry between the particles in the initial and final state. Mathematically, the symmetry provides a generalization of gauge invariance of electrodynamics to a local gauge invariance of the Lagrangian density of interaction under the transformations of the considered symmetry group, allowing us to enter a unified description of weak, strong and electromagnetic interactions. The Lagrangian density of these interactions contains particles of matter (fermions) in the following form

$$L_{\text{int}} = g \cdot J_\mu \cdot V_\mu, \tag{1.1}$$

where g is the gauge constant; J_μ is current and V_μ is the field of the appropriate interaction. The observed differences in the interactions of fundamental particles are attributed to the differences in the groups of gauge symmetry and also to the existence of the mechanism of spontaneous symmetry breaking.

The standard model of particle interactions based on $SU(2)\otimes U(1)$ local gauge symmetry group of the electroweak interactions and $SU(3)_c$ symmetry of quantum chromodynamics (QCD, the gauge theory of strong interactions) does not come now in direct contradiction with the experimental data.

However, the internal theoretical problems and aesthetically attractive way to unite all the fundamental forces of nature displace the theory beyond the standard model.

Theoretical and practical need to expand the SM (standard model) follows from its internal problems such as the quadratic divergence of the radiative loop corrections to the Higgs field mass or the strong violation of the CP-invariance in QCD.

The solution of the first problem is supersymmetry – symmetry between bosons and fermions, which provides compensation for the contributions from bosonic and fermionic loops in the Higgs mass due to the difference between the Bose–Einstein and Fermi–Dirac statistics.

Since we do not observe supersymmetry in the mass spectra of known fermions and bosons, then it must be broken, and the search for supersymmetric partners heavier than the corresponding particles is one of the greatest challenges for the Large Hadron Collider and/or the next generation of accelerators. But there is very small hope that, at least in the distant future, we will succeed in search for the gravitino (the supersymmetric partner of the graviton), predicted in the local supersymmetric models, because of its very weak, semi-gravitational interactions with other particles.

To solve the problem of strong violation of *CP*-invariance in QCD we assume the existence of the invisible axion, a pseudo-Goldstone boson – 'a little brother' of π^0 with super-weak interaction, which makes it virtually impossible to search directly for this particle in accelerators.

The idea of uniting all the fundamental forces of nature is the aesthetically appealing reason for the extension of the SM.

The similarity of the description of electromagnetism, strong and weak interactions, achieved in the SM, is embedded deeply in a grand unified theory (GUT), which extends the fundamental gauge symmetry group, including the symmetry group of the SM $SU(3)_c \otimes SU(2) \otimes U(1)$ into a single group of the symmetry of the Grand Unification G.

By placing a set of known particles in representations of the group G, we see that there remain 'white spots' which should be occupied so that the representations are complete.

The wider the group G, the larger is the number of additional particles and fields, corresponding to the total symmetry.

These particles and fields correspond to the 'hidden sector' of the relevant theory, since they are hidden from direct experimental verification or because of their large mass, or because of the extremely weak interaction with the known particles.

In both cases, the (super-weak interaction, or very super-large mass) verification of the predictions requires the use of indirect methods.

The same is required to explore such parts of the hidden sector as the 'invisible axion' or gravitino that arise when dealing with internal consistency of the SM.

There is very little indirect effects to study the acceptable properties of supermassive or super-weakly interacting particles in the laboratory.

These effects are the neutrino mass, *CP*-violation, nonconservation of the baryon and lepton numbers (reflected in the neutrino oscillations, neutrinoless double β-decay, proton decay and neutron–antineutron or hydrogen–antihydrogen oscillations).

Experimental identification of these rare processes is facilitated by direct violation in these processes of conservation laws contained in the SM. A relatively small number of these effects makes attractive an extension of indirect methods of checking the hidden sector of particle physics.

The problem of correct choice of the extended hidden sector is particularly acute when studying models of the 'theory of everything' (TOE) describing all four fundamental forces of nature, including gravity.

Such a description could come from a consistent expansion of the gauge symmetry, say, from a combination of local gauge models and supersymmetry, as is done in supergravity. Here the generalization follows from the expansion of internal symmetries to the symmetry of space-time.

An alternative approach is based on the generalization of space-time geometry to describe the interaction of particles. The geometric approach connects the fundamental forces of with additional dimensions of space-time, complementing the symmetry of space-time by the symmetry of elementary particles.

Both directions are combined in superstring theory, based on a fundamentally new type of fundamental objects – strings.

Heterotic string theory connects the 10-dimensional heterotic string theory with $E_g \otimes E'_8$ gauge symmetry. So that the hidden sector of this theory should, in principle, contain the entire 'zoo' of particles, fields, and new phenomena appearing in various extensions of the SM, which are difficult or impossible to find by direct experimental searches.

That is why the Universe, as a possible source of information about elementary particles, attracts the most attention of people involved in elementary particle physics.

Ya.B. Zeldovich called the Universe 'the poor man's accelerator'. But, as later noted by A.D. Linde, even the richest man cannot build an accelerator of elementary particles to energies of the GUT or TOE which are naturally realized in the early stages of cosmological evolution.

Thus, the internal development of particle physics leads to the theory of a hot expanding Universe, called Big Bang Universe, as a natural landfill of its fundamental ideas.

Modern cosmology is based on two observational facts. On the fact that the Universe is expanding, and that the modern Universe is filled with the thermal background of electromagnetic radiation. Combining these facts leads to the ideas of Big Bang expanding Universe.

The old scenario of the Big Bang Universe was an internally self-consistent picture based on the consistent application of laws of general relativity, thermodynamics, atomic and nuclear physics, well-studied experimentally in the laboratory. These well-known laws of physics are applied to the development of the Universe as a whole under the assumption that the present Universe contains only baryonic matter and electromagnetic radiation (and neutrinos) (Zeldovich and Novikov, 1975).

Under this scenario, the first three minutes of the expansion are characterized by the occurrence of nuclear reactions which create the primary chemical composition.

This pattern found qualitative confirmation when comparing the predictions of cosmological nucleosynthesis in the Big Bang Universe with the observed abundance of light elements. The pattern gave a qualitative explanation for the observed structure of inhomogeneities as a result of gravitational instability in a weakly nonhomogeneous substance.

However, the quantitative inconsistencies, which at a deeper examination became more pronounced, made this pattern now controversial. To resolve these contradictions, it has been necessary to add new fundamental elements to the basics of its theoretical constructions.

The formation of the large-scale structure of the Universe is connected in the old Big Bang scenario with the expected anisotropy of the thermal electromagnetic background which does not correspond to the observed level of its isotropy.

On the other hand, a low baryon density is required to reproduce the observed abundance of light elements as a result of cosmological nucleosynthesis (see a review by Copi and Schramm, 1996). This is not consistent with the fairly high density of matter required to explain the formation of the large-scale structure as a result of gravitational instability.

It seemed that the solution of both of these problems was found in the old scenario of the Big Bang Universe, when in 1980 there were claims by the experimental group at ITEP (Lubimov et al 1980) that the electron neutrino mass may be about 30 eV. The neutrino thermal background, with the concentration comparable to the concentration of the background of thermal photons, is a reliable prediction of the Big Bang Universe theory. Multiplying this concentration by the neutrino mass which was measured in the experiment at ITEP, we find that the current density of massive neutrinos is 1–2 orders of magnitude greater than the density of baryonic matter.

This implies a scenario of neutrino Universe in which massive neutrinos, weakly interacting with matter and radiation, form a large-scale structure of the Universe and dominate the modern cosmological density. Anisotropy of the thermal background radiation as predicted in this picture, seemed to be consistent with the observational data.

However, subsequent experimental studies did not confirmed the claims by Lubimov et al (1980) that the neutrino mass may be so high.

On the other hand, the cosmological analysis, in general, confirming that the dark matter should dominate in the Universe during the formation of the large-scale structure, has revealed serious problems in a scenario of neutrino Universe.

These problems have taken the physics of dark matter beyond the experimentally studied standard model of elementary particles.

There is a need to modify the old scenario of the Big Bang Universe, adding a new fundamental element – the dark matter, the hidden mass, the physical basis of which relate to the hidden sector of particle physics.

The problem of finding the true physical nature of the cosmological dark matter is complicated by the following circumstance.

Models of formation of the large-scale structure of hot, cold, unstable dark matter, or more complex models which use cosmic strings and hot dark matter, late phase transitions, and so on, are very different from a cosmological point of view.

Nevertheless, these different cosmological models are not alternative in terms of elementary particle physics, because their physical bases relate to different mutually complementary parts of the hidden sector of particle physics.

Thus, in general, we should consider a mixture of different forms of dark matter.

Another important initial condition for the formation of large-scale structure is the spectrum of initial fluctuations.

It is easy to verify that the statistical fluctuations are too small to form large-scale inhomogeneities in the expanding Universe. Therefore, we must assume the existence of primary small initial inhomogeneities dating back to the very early stages of cosmological evolution.

The old Big Bang scenario did not suggest the physical mechanisms for their occurrence.

Moreover, there are fundamental questions: why is the Universe expanding? Why are the initial conditions for expansion so close to a flat Universe? Why do they have so much in common in causally disconnected regions? Why does the Universe not contain an equal amount of matter and antimatter? All of these issues could not find answers in the old paradigm.

The principal answer for the first three issues was found on the basis of inflationary cosmological models (Glinner 1965, 1970; Guth 1981), suggesting that in the past there was a phase of superluminal (in the simplest case of exponential) expansion in the early Universe.

This stage could not form if matter, radiation or relativistic plasma was dominant but it could, under certain conditions, form under the effect of various cosmological implications of the theory of elementary particles, such as a strong first-order transition or slow rolling of the scalar field to its true vacuum state.

At the same time, inflationary models offer a physical mechanism for the birth of the spectrum of initial density fluctuations.

Most of these effects are associated with experimentally unverified particle physics, in particular, it relates to mechanisms of symmetry breaking at very high energy scales.

On the other hand, the various inflationary models are based on different complementary theoretical principles, and can coexist in the full cosmological scenario.

A.D. Sakharov (1967) and V.A. Kuzmin (1970) first linked the observed baryon asymmetry of the Universe with the physical mechanism of generation of an excess of baryons in the early Universe. They found that the effects of violation of *CP*-invariance in non-equilibrium hypothetical processes that do not preserve the baryon number, could create a baryon asymmetry in the early stages of evolution of the baryon-symmetric Universe.

Grand unification models, among their other predictions, provided a physical justification of these original ideas of baryosynthesis.

Other grounds for the mechanism baryosynthesis were found later in supersymmetric models. According to these models, the primary condensate of scalar quarks could have existed in the early Universe. Scalar quarks, decaying into ordinary quarks, then created an excess of baryons.

In the SM, the disappearance of baryons due to electroweak interactions takes place at very high temperatures. For electroweak baryosynthesis to be possible, the SM should contain a much larger number of Higgs bosons, or include non-conservation of the lepton number. The latter may be associated with the mechanism of generation of the neutrino mass.

Thus, the modern cosmological paradigm reflects a fundamental change in our view as to what the Big Bang cosmology is.

From self-consistent, but contradictory at its core and incomplete old scenario of the Big Bang Universe, we arrive at a picture of inflationary cosmology with baryosynthesis and multicomponent non-baryonic dark matter/energy.

Thus, directly or indirectly, the old theory of the Big Bang Universe is supplemented in the modern version by at least three additional elements – inflation, baryosynthesis and non-baryonic dark matter. All three of these essential elements of modern cosmology are based on physical laws predicted by the theory of elementary particles which, however, have not been experimentally verified.

There are many different physical mechanisms pretending to describe inflation and baryosynthesis. There are also many different candidates for the role of dark matter particles.

Unfortunately, neither the early Universe, when there were inflation and baryosynthesis, nor dark matter can be observed directly by astronomical means. It is therefore necessary to develop a system of indirect methods of correct choices of variants associated with different cosmological scenarios and models of elementary particles on which they are based.

The problem is that the space of cosmological and physical parameters is, in general, multidimensional, since the physical basis of various mechanisms of inflation and baryosynthesis, as well as various forms of dark matter are derived from various physical considerations, and not only do not contradict but complement each other.

On the other hand, the cosmological test of models of elementary particles should, generally speaking, take into account both the concrete realization of the physical basis of inflation, baryosynthesis and dark matter, and also additional modifications of the cosmological scenario, corresponding to the chosen implementation.

It can be concluded that the internal development of elementary particle physics requires cosmological verification of the principles of microworld physics. On the other hand, this approach which lies in the area inaccessible by direct modern experimental methods, was used to construct the physical principles of modern cosmology.

Detracting from the hypothetical possibility of discovery, storage and direct study of any relics of superhigh energy physics, such as primordial magnetic monopoles (t'Hooft 1974; Polyakov 1974; Zeldovich, Khlopov 1978; Preskill 1979, Rodionov 1996) we have to resort to finding new non-trivial methods for verification of this physics.

Clear understanding of the relationship between the fundamentals of macro- and microworlds and the virtual absence of direct experimental and astronomical methods of their research led to the creation of cosmoparticle physics which studies these foundations on the basis of comprehensive analysis of the effects of their indirect manifestations.

The need for such a combination of indirect methods is derived from the main issue of both cosmology and particle physics – the Ouroboros problem: **Physical basis of modern cosmology is based on predictions of the theory of elementary particles, which, in turn, look to cosmology for their test**. Therefore, being at the forefront of fundamental physics at the largest and smallest scales, nor cosmology or particle physics can examine their grounds alone, no matter how sensitive were the methods of their direct research. These grounds are in such a close relationship that they are practically inseparable. Thus, the frontiers of our knowledge of macro- and microworld converge, and a mystical 'Ouroboros' snake that eats its own tail symbolizes the cycle of problems which fundamental physics faces in its one-dimensional development.

Cosmoparticle physics, offering a non-trivial way out of this vicious circle, is based on the fundamental relationship of the basis of cosmology and particle physics, and opens the possibility in principle to explore these reasons in a complex combination of indirect cosmological, astrophysical and microphysical effects.

In elementary particle physics, cosmoparticle physics opens the way to build a 'theory of everything' (TOE), the unified theory of fundamental forces of nature the basis of which have become available, albeit indirectly, but nonetheless for real tests in the experiments and observations.

In the theory of the Universe, it opens the way to physical cosmology with a physically self-consistent description of the structure and evolution of the Universe.

The paradoxical form characteristic of the methods of cosmoparticle physics in relation to research in traditional fields of science, is illustrates by the 'pyramid in the circle' (Fig. 1.1) showing the multidimensional nature of the solution of the Ouroboros problem. The Ouroboros head, eating the tail, meets the base of the *world system*

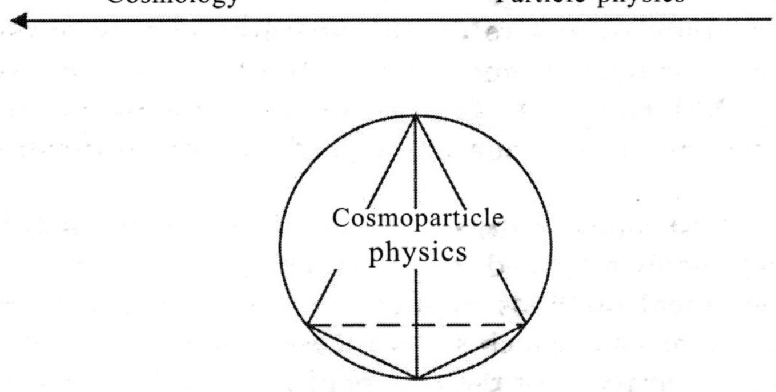

Fig 1. 'Pyramid in the circle' – a multi-dimensional solution of the Ouroboros problem.

which is available in multi-dimensional study of a combination of indirect macro- and microphysical tests.

The same pattern can serve as an illustration for astroparticle physics (astroparticle physics is the linear relationship between cosmology and particle physics corresponding to the head and tail parts of the snake) or TOE (the head eating its tail). Moreover, the contact of the body of the pyramid with the line of the snake identifies the new aspects brought by cosmoparticle physics to the traditional areas of science.

The zero- and one-dimensional nature of the most popular and widespread views on the relationship between cosmology and elementary particle physics (TOE and astroparticle physics, respectively) are not able to provide a balanced and very consistent approach to the subject as a whole. The choice of the fundamental cosmological and microphysical models is thus largely a matter of taste and fashion. And the criterion of *naturalness* is unlikely to help in this choice. As rightly pointed out by A.D. Linde, natural is of course what follows from the natural laws, the very laws of nature that we do not know and are trying to discover in both cosmology and elementary particle physics.

In fact, cosmoparticle physics studies the world as a whole in the fundamental relationship of its structure in the smallest and largest scales. It considers the *world system* in which the bases of cosmology and particle physics are so interrelated that a complete cosmological scenario based on a single particle theory and the theory of elementary particles is cosmologically viable. *The first principle* of cosmoparticle physics is that the *world system* does exists.

The *world system* establishes the correspondence between the fundamental parameters determining the processes in elementary particle physics, astrophysics and cosmology, and thus establishes a quantitative relationship between the microscopic and macroscopic effects.

The existence of a system of such links between the fundamental properties of elementary particles, astrophysical and cosmological parameters is the essence of the *second principle* of cosmoparticle physics.

Finally, the number of parameters that characterize the world system, should be less than the number of independent manifestations determined by these parameters in elementary particle physics, astrophysics and cosmology, thus providing a complete cosmoparticle test – so formulated the *third principle* of cosmoparticle physics.

Cosmoparticle physics reproduces on the largest and smallest scales the fundamental relationship between the microscopic and macroscopic descriptions, typical for theoretical physics. It offers a new level of this relationship, which, for example, takes place between thermodynamics and atomic physics, hydrodynamics and kinetics, or between the fundamental macroscopic and microscopic parameters (e.g., between the Avogadro number and the proton mass).

However, the absence of direct experiments on the cosmological scales and in superhigh energy physics on which modern cosmology is based, leads to the need to develop a system of non-trivial indirect methods in cosmoparticle physics.

Cosmoarcheology – search in the astrophysical data for traces of new physical phenomena in the Universe, is already an established direction of such indirect methods of investigation typical of cosmoparticle physics as an independent field of scientific knowledge (Sakharov 1989; Khlopov 1989; 1996b, c). The links of the cosmoarcheological chain are connected in a rather non-trivial fashion and lead to a system of astrophysical verification of the existence and study of the admissible properties of hypothetical particles, fields, objects and phenomena, predicted as cosmological implications of the theory of elementary particles.

Cosmoarcheology treats the Universe as a unique natural accelerator in which astrophysical data play the role of a particular experimental result of the *thought experiment* (*Gedanken Experiment*), carried out by *cosmoarcheology*.

As with any experiment, to achieve meaningful results, it is necessary to know the specifics of the experimental device used, as well as methods for obtaining experimental data and their analysis.

The problem is that the Universe as a 'source' of elementary particles and their 'detector' are beyond the reach of the 'experimenter'.

The astrophysical process cannot be directly reproduced in the laboratory, but no matter how complex the combination of physical effects is, theoretical astrophysics uses, as a rule, in its analysis the laws of nature, confirmed experimentally.

The difficulty lies in the fact that the basic physical laws are not known in the theoretical study of the inflationary Universe with baryosynthesis and non-baryonic dark matter.

This makes in a general case the physically self-consistent formulation of the cosmoarcheological approach model-dependent. One must take into account the relationship between the hypothetical particles or fields that are studied on the basis of astrophysical data, and the physics of inflation, baryosynthesis and non-baryonic dark matter. Since this physics depends on the model of elementary particles, it is necessary to relate cosmological consequences of the hypothesis under consideration with this scenario of cosmological evolution, which is based on the selected particle model and reproduces the basic elements of modern cosmology.

This means that the cosmological history of hypothetical particles or fields can be multi-stage and follow a very non-trivial way of cosmological evolution.

On the other hand, if the inflationary model of the baryon-asymmetric Universe with the non-baryonic dark matter is indeed the true basis of theoretical cosmology, the real picture of the cosmological evolution *should be* much more complicated than the original Gamow evolution scenario of the Big Bang Universe. We should inevitably expect that this scenario will be much more non-trivial than the simple addition to the old scenario of inflation, baryosynthesis and non-baryonic dark matter.

In fact, any physically sound theoretical scheme, which reproduces the essential elements of modern cosmology, contains a significantly large set of additional cosmologically important predictions which could be called the hidden parameters of the cosmological model. To check out these predictions, cosmoarcheology enhances observational cosmology, linking the realistic theory of the Universe with observations.

The other side of the problem of finding the physical basis of modern cosmology is the question of cosmological manifestations of a complete theory of elementary particles.

The unified description of the fundamental forces of nature is aesthetically appealing. However, the higher the level of the unified description, the more appropriate the model of elementary particle physics is and, consequently, the larger is its hidden sector. This brings into consideration more and more exotic cosmological manifestations deserving, however, considerable attention due to fundamental physical basis for their existence.

There is also a more pragmatic reason for going beyond the SM of particle interactions – removing its internal inconsistencies. These extensions of the SM also lead to the specific cosmological effects that arise as a consequence of the considered cosmological models.

The unified theories of elementary particles are characterized by the fundamental ability of using the self-consistent approach for describing the cosmological evolution and effects of new physics of particles and fields. Consideration of a separate investigation of the cosmological consequence of extension of the SM, eliminating its internal contradictions, results in the loss of such self-consistency. But such an analysis of the cosmological predictions of the theory of elementary particles is important for *cosmophenomenology of new physics* – section of cosmoarcheology, studying cosmologically significant effects of new physics in the early Universe and their possible manifestations, which can be verified in comparison with astrophysical data.

Superstring models, claiming to be the 'theory of everything' (TOE), cannot at the present level of development serve as an example of the world system in the full cosmoparticle sense. Because of the existence of a large set of their realizations it is almost impossible to analyze a fairly complete set of physical, astrophysical and cosmological predictions. It is therefore of particular interest to consider an alternative approach to the world system. The main observed properties of elementary particles and cosmological parameters, related to the physical mechanisms of inflation baryosynthesis and dark matter could be considered under a single quantitative theoretical scheme.

This approach can be illustrated by a model of horizontal unification which will be discussed in this book (see Sakharov, Khlopov 1994b; Khlopov, Sakharov 1996 and references therein). In these studies it was shown that adding the gauge symmetry $SU(3)_H$ of generations of quarks and leptons to the symmetry $SU(3)_c \otimes SU(2) \otimes U(1)$ of the standard model not only provides a realistic theoretical description of the existence of the observed similarities and differences between the three families of quarks and leptons

$$\begin{pmatrix} \nu_e \\ e \\ u \\ d \end{pmatrix}, \begin{pmatrix} \nu_\mu \\ \mu \\ c \\ s \end{pmatrix}, \begin{pmatrix} \nu_\tau \\ \tau \\ t \\ b \end{pmatrix}, \tag{1.2}$$

but also its implementation within the same theoretical scheme provides the physical basis of inflation, baryosynthesis and dark matter.

This model which offers an alternative (horizontal) way to unification, is not in any case an alternative to the widely known grand unified theory (GUT) or supersymmetric (SUSY) extension of the SM. Solution of internal problems of the minimal scheme of horizontal unification requires further SUSY and GUT extension, which is also expected to improve the agreement of its astrophysical and cosmological predictions with observations. The results of search for neutrino oscillations and double neutrinoless β-decay make consideration of these extensions unavoidable.

But even in its current form the model reflects the basic principles of cosmoparticle physics. On the basis of the local gauge model with spontaneous symmetry breaking the model implements the phenomenology of the Universe, combining the description of almost all the basic properties of known particles and the essential parameters of modern cosmology associated with the hidden sector of particle physics. Finally, the number of free parameters of the model is much smaller than the number of their corresponding manifestations in particle physics, astrophysics and cosmology, providing a certainty test of the model and its completeness.

Thus, the model demonstrates the power of the cosmoparticle approach. The fundamental scale of horizontal symmetry breaking is not known *apriori* and refers to the hidden sector of particle physics, but a comprehensive analysis of the set of physical, astrophysical and cosmological predictions can 'fix' the value of this scale in two fairly narrow ranges (around 10^6 and around 10^{10} GeV).

The second solution corresponding to a higher energy scale reproduces a widely accepted standard cosmological scenario with inflation, baryosynthesis and cold (axion) dark matter. However, the practical realization of such a scenario, which never reflects the complete physical picture, shows that even the simplest cosmological scenario contains a number of non-trivial additional items. These include a post-inflationary stage in which primordial black holes (PBH) can form, followed by evaporation of such PBHs on in the radiation-dominated (RD) stage, after primordial nucleosynthesis, the formation of the primary percolation structure of archioles, and so on.

This example shows that *in no case* are the new cosmological elements based on the hypothetical effects of particle physics, reduced *solely* to inflation, baryosynthesis and dark matter. It reflects a system of non-trivial interdisciplinary connections that should be used to build a true system of Universe creation based on methods of cosmoparticle physics.

2

Hidden sector of particle physics

1. Theory of elementary particles – the standard model

The dramatic story of finding the most fundamental constituents of matter led to the modern quark–lepton picture of the world of elementary particles and the so-called standard model (SM) of electroweak and strong interactions.

This picture was the result of the revolution which began in the 30s of the 20th century, when the common effect of quantum theory and relativity radically altered the concept of 'elementary particles', 'field' and 'charge'.

Beginning with the ancient world and up to early 20th century, the fundamental idea of the 'elementary' was essentially unchanged.

Elementary particles in the tradition of atomism were considered as eternal entities carrying the smallest possible amount of the respective properties.

Preservation of fundamental values, such as conservation of electric charge, was in this picture as the trivial reflection of the eternal nature of elementary charged particles – electrons.

The names of elementary particles were changed – the molecules were replaced by the atoms and then by elementary components of an atom – electrons and nuclei. The principle remained unchanged – all the observed changes in material objects were attributed to the redistribution of elementary particles, similar to the permutation of elements in the LEGO.

This approach proved very fruitful in chemistry, where the entire set of chemicals was reduced to a variety of combinations of about 100 different chemical elements. However, this approach led to major difficulties in analyzing the structure of atomic nuclei and nuclear transformations.

The discovery of the positron and the experimental discovery of creation and annihilation of electron–positron pairs, the problem of conservation of energy and momentum in nuclear beta decay, 'nitric catastrophe' resulting from the assumption that the presence of an electron inside the nucleus seriously contradicted the idea of an eternal electron in the scale of the atomic nucleus.

Electrons emitted in nuclear beta decay, being eternal, should have been found inside the nuclei before the decay, i.e. they should be localized in the space with the size of the nucleus

$$r_N \leqslant 10^{-13}\,\text{cm}. \tag{2.1}$$

But the uncertainty principle of quantum mechanics

$$\Delta p \cdot \Delta r \geqslant \hbar \tag{2.2}$$

predicts the uncertainty in the electron momentum

$$p_e \approx \Delta p \geqslant \frac{\hbar}{\Delta r_N} \approx 40\ \text{MeV}/c, \tag{2.3}$$

which exceeds the momentum of the electrons emitted in beta decay 5–10 times.

The uncertainty in the electron energy equal to

$$E_e \approx p_e \cdot c \geqslant 40\ \text{MeV}, \tag{2.4}$$

is the same number of times higher than the typical binding energy per nucleon in the nucleus. This binding energy is equal in magnitude to

$$E_b' \approx 7\ \text{MeV}. \tag{2.5}$$

On the other hand, if the energy of the 'nuclear electrons' was of the order of the nucleon binding energy in the nucleus, the size of the electron localization region would exceed the size of the nucleus by an order of magnitude. Moreover, taking into account the spin of an electron inside the nucleus led to the problem of the relation between spin and statistics for the nuclei, which led to the 'catastrophe' in the case of the nuclei of nitrogen ^{14}N.

The nuclear charge is

$$Q = Z \cdot |e| \tag{2.6}$$

where e is the electron charge. In the case of nitrogen, the nuclear charge is equal to $Z = 7$, and this was due to charge compensation of 14 protons by 7 electrons inside the nucleus. However, in this case the total spin of the odd (21) number of components with half-integer spin must be half-integer in units

$$\hbar = \frac{h}{2 \cdot \pi}$$

while the statistical properties of the ^{14}N nucleus correspond to the Bose-Einstein statistics of the particles with an integer spin.

Finally, the electron spectrum in beta decay is continuous, and not linear, as expected for a single particle emitted in a nuclear transition.

The hypothesis of neutrinos emitted in pairs with the electron in β-decay, provides a solution for both the 'nitric catastrophy' and for conservation of the momentum in β-decay. Additional 7 neutrinos with half-integer spin make the total number of components even, resulting in an integer spin of the nitrogen nuclei. On the other hand, neutrinos are emitted together with electrons and take away some of the energy and momentum what makes the β-spectrum continuous.

But this solution has caused even more difficulties because the light neutrinos inside the nucleus have the same momentum uncertainty as for the electron.

Even the discovery of the neutron did not solve this problem, because the problem of localization of light particles (neutrinos and electrons) within the nuclei surfaced again as the problem of localization of an electron and a neutrino in the neutron before its β-decay.

In the early twentieth century, discovery of the quantum of the electromagnetic field – the photon, which was created in emission and was annihilated in absorption, resulted in the first example of an elementary particle which could be created and annihilated.

Such an approach was not easy to use for describing the components of matter.

The idea of so-called 'Dirac sea' was proposed to preserve the concept of the eternal electron.

The 'birth' of the electron–positron pair was interpreted as the electron transition from the sea of filled states to the free state, and the appearance of a hole in the sea associated with this transition interpreted as a positron. The process of electron–positron 'annihilation' looks in this approach as filling of the 'hole' state of the sea by the free electron.

Such fermion–hole formalism has been applied in the theory of nuclear structure and the theory of semiconductors, but was forced back over the course of development of relativistic quantum field theory, essentially based on an alternative idea of creation and annihilation of particles.

In this alternative picture, the charge and its relationship with the conservation laws and symmetries become non-trivial and of fundamental importance.

Charge conservation loses the meaning of the trivial property inherent to the eternal elementary charged particle.

In the world of elementary constituents of matter that can be created and annihilated, the laws of transformation of particles and particle properties conserved in these transformations reflect the fundamental structure of the microworld, its fundamental symmetry.

In the development of quantum field theory, the symmetry of the particles and the corresponding generalization of the concept of the charge formed the basis of the set of theoretical ideas, regarded as the standard model (SM) of particles and their strong, weak and electromagnetic interactions.

Following the objectives of this book and assuming that a coherent and systematic exposition of SM is available to the reader in a number of well-known books, we focus on the aspects of the modern theory of elementary particles which are most important for the further extension of the SM and its cosmological implications.

Note that the ideas of quantum field theory, which emerged in the thirties of the twentieth century, were preceded by the idea of non-stationarity of the Universe which were put forward and received the first observational evidence in the 20s of the twentieth century. This correspondence seems far from accidental.

Beginning with ancient philosophy, the stability of elementary particles found its psychological basis in the inalterability of the world order. Changes in this order – expansion of the Universe – supported psychologically radical changes in the description of elementary particles.

1.1. Quantum electrodynamics

To follow the logic chain that leads to the ideas of the modern theory of elementary particles, we consider changing of notion of the particle, field and charge in the transition from classical to quantum-field description of electromagnetism.

At the end of 19th century, classical physical principles of electricity, magnetism and light were combined in Maxwell's electrodynamics.

The physical nature of the interaction of electric charges, magnetic interaction of electric currents, and optical phenomena have found an explanation in the existence of the electromagnetic field acting on these charges and currents and generated by them.

Mathematically, Maxwell's equations relate the spatial and temporal derivatives of the electromagnetic potential with the density of electric charge and current and their spatial and temporal derivatives.

In classical electrostatics, the electric charge density $\rho(t,\vec{x})$ is the product of the electric charge of elementary charged particles (electrons) and the number density of these particles

$$\rho(t,\vec{x}) = e \cdot n(t,\vec{x}). \tag{2.7}$$

The Lagrangian density $L_{el}(t,\vec{x})$ of electric charge interaction with the scalar potential $\varphi(t,\vec{x})$ has the form

$$L_{el} = e \cdot n(t,\vec{x}) \cdot \varphi(t,\vec{x}), \tag{2.8}$$

where the charge density and electric potential are, in general, functions of the spatial and temporal coordinates.

Taking into account the observed absence of magnetic charges, electrodynamics ascribes the magnetic effects to the motion of electric charges, so that the corresponding Lagrangian density describes the interaction of the density of electric currents

$$\vec{j} = e \cdot n(t,\vec{x}) \cdot \vec{v} \tag{2.9}$$

with the vector potential $\vec{A}(t,\vec{x})$:

$$L_{\text{mag}} = e \cdot n(t,\vec{x}) \cdot \vec{v} \cdot \vec{A}(t,\vec{x}). \tag{2.10}$$

The theory of relativity suggests the relativity of spatial and temporal variables, treating them as a single space-time event

$$x = (t,\vec{x}), \tag{2.11}$$

as well as the relativity of electricity and magnetism. Therefore, the the classical Lagrangian density of the electromagnetic interactions $L_{\text{em}}(x)$ is constructed as the interaction of the electromagnetic 4-current

$$j_\mu = (\rho \cdot c, \vec{j}) \tag{2.12}$$

with the electromagnetic 4-potential

$$A_\mu(x) = \left(\varphi(x), \vec{A}(x)\right) \tag{2.13}$$

in the form of

$$L_{\text{em}} = e \cdot j_\mu \cdot A^\mu. \tag{2.14}$$

In non-relativistic quantum mechanics, the electron number density is the density of probability of finding an electron at a fixed time t with specified coordinates $\vec{x}$, and it is associated with the electron wave function $\Psi(t,\vec{x})$ as

$$n(t,\vec{x}) = \Psi^* \cdot \Psi = |\Psi|^2. \tag{2.15}$$

Because of this modification, the Lagrangian takes the form

$$L_{\text{el}}(t,\vec{x}) = e \cdot \varphi(t,\vec{x}) \cdot \Psi^*(t,\vec{x}) \cdot \Psi(t,\vec{x}). \tag{2.16}$$

In quantum field theory, instead of the wave function which determines the probability of finding the eternal and indivisible electron in a given space-time domain, the probability of creation or annihilation of an electron or a positron is considered. The operator $\mathbf{\Psi}(x)$ and its conjugate $\mathbf{\Psi}^*(x)$ (or, more accurately, the Hermitian conjugate operator $\mathbf{\Psi}^+ = \mathbf{\Psi}^{*T}$, where superscript T denotes transposition) are superpositions of the operators of creation and annihilation of electron and positron states.

Instead of the classical electromagnetic field potential $A_\mu(x)$ we introduce the operator of the electromagnetic field $\mathbf{A}_\mu(x)$ as a superposition of creation and annihilation operators of the electromagnetic quantum. The quantum field operator of the electromagnetic interaction is given by

$$\mathbf{L}_{\text{em}} = e \cdot \mathbf{j}_\mu \cdot \mathbf{A}^\mu, \tag{2.17}$$

where the quantum field operator of the electromagnetic current has the form

$$\mathbf{j}_\mu = \overline{\Psi} \cdot \gamma_\mu \cdot \Psi, \tag{2.18}$$

where γ_β with μ = 0, 1, 2, 3 are the Dirac gamma matrices, and $\overline{\Psi} \equiv \Psi^+ \gamma_0$ is the Dirac conjugate operator.

Thus, all the processes of electromagnetic interaction in quantum electrodynamics can be reduced to a combination of elementary acts of creation and annihilation of electromagnetic quanta by the charged particles.

The difference between quantum-mechanical and quantum-field approach to the description of emission and absorption of electromagnetic quanta lies in the way they treat the charged particle emitting and absorbing these quanta.

Quantum-mechanical description of the emission or absorption of a photon by an electron uses transition currents that describe the variation of the wave function of the eternal electron in the initial and final states. As a result of emission or absorption, an electron moves from one state to another in accordance with the probability density to find it in this state at a given point in space-time.

The quantum field theory is considering such processes as the annihilation of the electron in the initial state and the creation of an electron in the final state in accordance with the creation (annihilation) of the emitted (absorbed) electromagnetic quantum.

These two approaches start to differ greatly when considering the birth or annihilation of an electron–positron pair.

The quantum mechanical approach should regard the birth of a pair as the transition of an electron from the Dirac sea to the free state, interpreting the 'hole' in the sea as a positron. Accordingly, in the annihilation of a pair the free electron fills the state of the 'hole' in the Dirac sea.

The quantum field theory considers the electron and positron states equal. This allows us to treat the creation of an electron–positron pair from the physical vacuum as the creation of an electron and positron and annihilation of the pair as the destruction of the electron and the positron.

The fundamental cause, which allows us to consider the electron in terms of its creation and annihilation, is the identity of elementary particles of the same type. Following the ideas of quantum field theory, quantum electrodynamics describes all quantum processes of interaction between charged particles and electromagnetic quanta on the basis of a single elementary act, interpreted from different points of view.

The same diagram (see Figure 2.1), viewed from different sides, describes the emission and absorption of photons by electrons (left to right) or positron (right to left), the birth of the electron–positron pair by the electromagnetic quantum (bottom to top) and single-photon annihilation of the e^+e^--pair (top to bottom). The direction of the arrow from the initial to the final state corresponds to the electron. The reverse direction represents a positron.

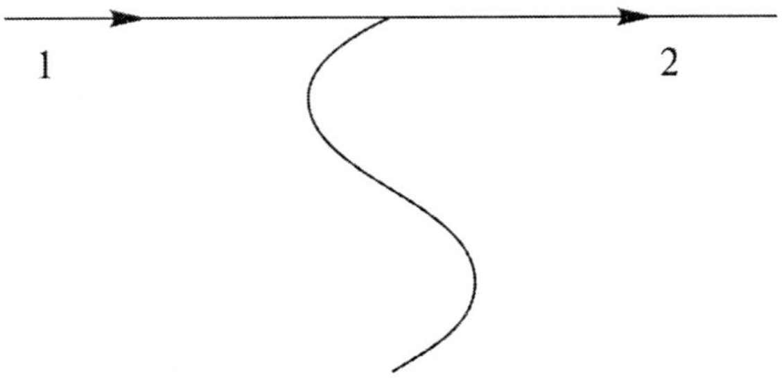

Fig. 2.1. An elementary act of quantum electrodynamics.

The elementary act of interaction in quantum electrodynamics is a sequence of processes of creation and annihilation of charged particles and electromagnetic quanta.

For example, an elementary act of photon emission by the electron (Fig. 2.1, left to right) is a process in which an electron in the initial state is annihilated, but in the final state the electromagnetic quantum and electron 2 are created.

Using an elementary diagram (Fig. 2.1), we can reduce any process of electromagnetic interaction to the relevant combinations of elementary acts, considered from different sides.

Thus, the scattering of electrons, shown in Fig. 2.2, consists of two elementary acts of interaction: 1) the annihilation of an electron 1 in the initial state and the creation of an electron 3 in the final state associated with the creation (or annihilation) of the electromagnetic quantum γ, and 2) the annihilation (or creation) of an electromagnetic quantum in the annihilation of electron 2 and the birth of electron 4.

The possibility of creation and annihilation profoundly changes our understanding of the properties of the electron. The charge of an electron and its electromagnetic field are no longer the eternal quality of the eternal particle. Conservation of the charge and the

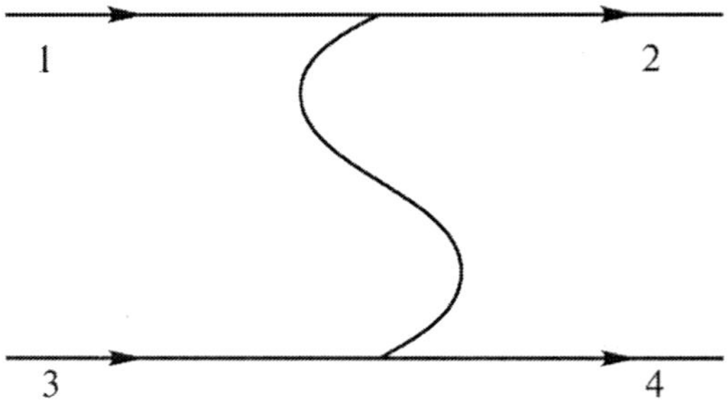

Fig. 2.2. The scattering of electrons in quantum electrodynamics.

electromagnetic field as a necessary attribute of the electric charge does not find a reason in the eternal immutability of the elementary charged particles. The electric charge, and its electromagnetic field, must find an explanation of its nature in the laws that define the creation and annihilation of charged particles, in the symmetry laws of quantum field theory.

1.2. Gauge symmetry of quantum electrodynamics

In quantum mechanics, a charged particle is described by a complex wave function. The probability density of finding an electron at a given point at a given time is determined by the square modulus of the wave function. The phase of the complex wave function is not observable. We can consider it arbitrary and any change at different points in space and at different times does not lead to observable effects.

Therefore, the quantum mechanical description of the electron must have the invariance with respect to phase transformations of the wave function

$$\Psi \to \Psi \cdot \exp\left(i \cdot e \cdot \chi(x)\right), \tag{2.19}$$

where e is an electrical charge. Such transformations with constant χ correspond to transformations of the global symmetry group $U(1)$. Local phase transformations from the coordinate-dependent function $\chi = (t, \vec{x})$ correspond to the local symmetry group $U(1)$.

In the wave equation of quantum theory, the local phase transformations add an arbitrary variable to all the space-time derivatives:

$$\partial_\mu \Psi(x) \to \partial_\mu \Psi(x) + i \cdot e \cdot \partial_\mu \chi(x). \tag{2.20}$$

These variables are proportional to the derivative of $\chi(x)$ and have no physical meaning.

To preserve the invariance of the wave function of a charged particle (electron) with respect to local phase transformations, we introduce the compensating transformations of electromagnetic potential $\boldsymbol{A}$. The local phase transformations defined by transformation (2.19) must be accompanied by a gauge (gradient) transformation of the electromagnetic potential:

$$\mathbf{A}_\mu \to \mathbf{A}_\mu + e \cdot \partial_\mu \chi(x). \tag{2.21}$$

If the derivatives of the wave functions have the form of the so-called long-derivative

$$\partial_\mu \Psi(x) \to \partial_\mu \Psi(x) + i \cdot e \cdot \mathbf{A}_\mu \Psi(x), \tag{2.22}$$

then it is invariant in relation to local phase transformations. The gauge transformations of the electromagnetic potential compensate the non-physical variables that arise in the differentiation of the wave function of a charged particle.

On the other hand, the phase transformations of the wave function compensate for the effects of the gauge transformation of the potential, making the theory gauge-invariant.

The condition of gauge invariance with the long derivative offers an aesthetic way of introducing the electromagnetic interaction into the theory of free charged particles.

The long derivative determines the form of the electromagnetic interaction of particles with different spins. For example, in the quantum theory of the electron, the derivative establishes the form of electromagnetic interaction with the electric charge and magnetic moment of the electron.

Moreover, the idea of local gauge invariance provides a way to link the fundamental forces of nature with the symmetry of the fundamental particles of matter.

According to Noether's theorem, the exact symmetry of the Lagrangian is associated with the existence of a conserved charge.

Assuming the local symmetry, we must introduce a gauge field interacting with the charge to preserve the symmetry of the Lagrangian.

Thus, in the framework of Lagrangian field theory, we can consider the symmetry between the different types of particles, related with this symmetry, the corresponding generalized charges and introduce a gauge field through which the interaction of these charges is mediated.

In the framework of the quantum field theory we consider the symmetry between the field operators, representing a superposition of creation and annihilation operators of particles and antiparticles.

The transition from the initial to the final state formally corresponds to the transformation of symmetry, produced by the generator of the appropriate symmetry group. A set of generators of this symmetry group can be associated with the set of charges corresponding to the considered symmetry, which are the sources of the gauge fields of the interaction.

Consequently, the symmetry between the different types of particles, giving a mathematical description of their interactions, may lead to an explanation of the physical nature of the processes studied by high-energy physics at accelerators.

1.3. Symmetry of the fundamental particles

Since the 30s of 20th century, the development of experimental high-energy physics has led to the discovery of a large number of different elementary particles. However, the way to their current classification has not been simple and straightforward.

In addition to the electron and nucleons of which consists the matter that surrounds us, and electron neutrinos – the inevitable participant of the beta decay, many new particles were discovered.

Almost all new particles are unstable. With respect to the fundamental interactions, they are classified into two groups: leptons and hadrons. These groups also differ with respect to the internal structure of particles.

The leptons include the electron e and the electron neutrino v_e, as well as two other pairs of charged leptons and their neutrino: muon μ^-, muon neutrino v_μ, the tau-lepton τ^- and its neutrino v_τ. The leptons also include their antiparticles. It was experimentally proved that the neutrino is not identical with the antineutrino and that electron, muon and tau-neutrinos are different particles.

In general, in the observed processes (except for the existence of neutrino oscillations and the double neutrionoless beta decay to be discussed later), the lepton number L_l is conserved for each type of lepton: the number of leptons minus the number of antileptons is constant.

The creation of an electron in beta decay is accompanied by the birth of an antilepton – electron antineutrino.

The same is true for all other reactions involving leptons (with the above reservation taken into account).

Leptons are created (annihilated) either in pairs together with their antileptons, or this process is accompanied by the annihilation (creation) of the lepton of the same type.

For example, the muon comes in a pair with the antimuon or muon antineutrino. A single muon is formed by the reaction caused by a single muon or muon neutrino in the initial state.

The charged leptons differ greatly in mass. For example, the mass of the electron is

$$m_e = 0.511 \text{ MeV}, \tag{2.23}$$

while the mass of the muon and tau-lepton is

$$m_\mu = 105.7 \text{ MeV}, \tag{2.24}$$

$$m_\tau = 1777 \text{ MeV}. \tag{2.25}$$

Due to such a large mass, the muons and tau leptons are unstable.

Currently, there are only experimental upper limits on the neutrino mass. In the table of the properties of the particles (Nakamura et al 2010), these limits are specified at the level of

$$m_{\nu_e} < 2 \text{ eV}, \tag{2.26}$$

$$m_{\nu_\mu} < 0.19 \text{ MeV}, \tag{2.27}$$

$$m_{\nu_\tau} < 18.2 \text{ MeV}. \tag{2.28}$$

As we shall see, the question of neutrino mass and instability of massive neutrinos is crucial for cosmology.

The charged leptons take part in the weak and electromagnetic interactions. Neutrinos participate only in weak interactions.

Leptons are considered as fundamental particles and have no internal structure.

Hadrons, in contrast, are more numerous because they are composite particles. The hadron structure is explained on the basis of the quark model. All hadrons are involved in strong reactions.

There are two groups of hadrons: baryons and mesons.

Baryons are fermions – they have half-integer spin. The baryons include the proton, neutron, hyperons, baryon resonances and their antiparticles called antibaryons.

In all the observed reactions, the baryon number B is conserved. This means that in any process the number of baryons minus the number of antibaryons remains unchanged. The other side of the conservation of baryons is that the proton, being the lightest baryon, is stable. In the table of the properties of the particles (Nakamura et al 2010) the lower limit of the lifetime of the proton is provided

$$\tau_p > 2.1 \cdot 10^{29} \text{ years}, \tag{2.29}$$

which was obtained in the searches for de-excitation of the residual nucleus after disappearance of the proton in ^{16}O and does not depend on the possible modes of proton decay, as well as restrictions

$$\tau_p > 10^{31} \div 5 \cdot 10^{32} \text{ years}, \tag{2.30}$$

arising from the negative results of search for specific channels for proton decay.

The mesons have integer spin, so they are bosons. Pions, K-mesons and D-mesons are examples of relatively long-lived mesons.

The most numerous group of hadrons contains the so-called

resonances. The lifetime of a resonance with the mass m is of the order of

$$\tau \sim \frac{\hbar}{m \cdot c^2} \tag{2.31}$$

which is the characteristic time for reactions proceeding through the strong interaction. The major part of the table of particle properties (Nakamura et al 2010) is occupied by the data on baryon and meson resonances.

All hadrons are classified by their internal quantum numbers: spin, isospin, strangeness, charm, parity, etc.

The combination of these properties determines the 'flavour' of the particle. The symmetry between the 'flavours' can be used for possible classification of the hadrons. The symmetry of the 'flavours' provides the arrangement of hadrons in multiplets, i.e., the identification of sets of different particles with the representation of the relevant symmetry group of 'flavours'.

Thus, the proton and the neutron form a doublet with respect to the group of the isotopic spin (or isospin) $SU(2)$, and the pions π^+, π^- and π^0 are an isotopic triplet. Strong interactions and, in particular, the nuclear interactions are invariant under the isospin symmetry. Based on isospin invariance, we can consider the various particles in the same multiplet as different quantum (isospin) states of one particle. Thus, the proton and the neutron are treated as 'isospin up' and 'isospin down' states of one particle – the nucleon N.

Nucleons and pions, together with strange particles – hyperons, mesons and their antiparticles, and the resonances were classified on the basis of the symmetry of 'flavours'.

Fundamental representations of this group (a triplet and an antitriplet) did not find equivalents among the known particles, and these ideas have been identified with the hypothetical u-, d-, s-quarks and antiquarks.

Therefore, the quark model can classify all hadrons as bound states of quarks and antiquarks. The baryons are considered as bound states of three quarks, antibaryons as bound states of three anti-quarks and mesons as bound systems of the quark and the antiquark.

Thus, the proton and the neutron are composed of u- and d-quarks and:

$$p = (uud) \tag{2.32}$$

$$n = (udd) \tag{2.33}$$

the positively charged pion

$$\pi^+ = \left(u\bar{d}\right) \tag{2.34}$$

and the antiproton is represented as the bound state of three antiquarks

$$\bar{p} = \left(\bar{u}\,\bar{u}\bar{d}\right). \tag{2.35}$$

In this approach, the symmetry is only used for the classification of particles, and the physical meaning of the quarks was not clear.

It is easy to see that in units of the elementary charge, defined as the absolute value of the electron charge, the quarks must have a fractional electric charge equal to

$$Q_u = +\frac{2}{3} \quad Q_d = -\frac{1}{3} \quad Q_s = -\frac{1}{3} \tag{2.36}$$

for u-, d- and s-quarks and

$$Q_{\bar{u}} = -\frac{2}{3} \quad Q_{\bar{d}} = +\frac{1}{3} \quad Q_{\bar{s}} = +\frac{1}{3} \tag{2.37}$$

for the antiquarks.

It is assumed that all the quarks must also have the fractional baryon number equal to

$$B_q = +\frac{1}{3} \tag{2.38}$$

for quarks and

$$B_{\bar{q}} = -\frac{1}{3} \tag{2.39}$$

for the antiquarks. Experiments were also carried out with 'charmed' hadrons (which include c-quarks with the electric charge $Q_e = +2/3$) and b-hadrons (with the b-quark, $Q_b = -1/3$). Experimental data indicate the existence of the t-quark, $Q_t = +2/3$).

Experimental searches for free quarks and antiquarks in the Earth crust failed, and an upper limit was set on the abundance of free quarks in the Universe (the ratio of their permissible concentration n_q to hydrogen concentration n_{H})

$$\frac{n_q}{n_H} < 10^{-21}. \tag{2.40}$$

On the other hand, the calculations carried out by Zeldovich et al (1965) for the primordial abundance of free quarks in a hot Universe predict that such an abundance is 10 orders of magnitude higher than the upper limit (2.40).

This contradiction leads to the idea that free quarks cannot exist in nature, and the idea that quarks exist only bounded inside the hadrons and cannot exist as free particles.

The quarks must have an additional degree of freedom called colour. The introduction of colour solves the problem of the spin-statistics for the quarks.

For example, in the quark model, the Δ^{++}-resonance which has the spin 3/2 and electric charge +2, should be a bound system of three identical quarks that are in the same spin state. If these quarks have different colours, all three u-quarks may be in the ground state with $l = 0$ which is also assumed by the quark model.

Colour plays a major role in the gauge theory of strong interactions called quantum chromodynamics (QCD).

The name itself suggests an analogy with quantum electrodynamics. Colour is the determining factor in this analogy, playing the role of the generalized charge in the dynamics of strong interactions.

Gauge colour symmetry is based on the colour group $SU(3)_c$.

Colour transitions, defined by the generators of the $SU(3)_c$ group, are sources of fields of colour interactions.

These fields glue quarks into hadrons, and quanta of these fields are called gluons g.

The symmetry between the 'flavours' of the quarks and the leptons is used to construct the gauge theory of weak interactions. The role of the charges of the weak interaction is played by transitions of the weak isospin. By analogy with electrodynamics, the currents of the weak interaction are the sources of gauge fields of the weak interaction Quanta of these fields, $W^{\pm}$- and Z^0-bosons, have a large mass

$$m_W = 80\ \text{GeV},$$ (2.41

$$m_Z = 91\ \text{GeV}$$ (2.4

and their exchange is effective only at close distances. This expla the short-range weak interactions.

Thus, the gauge symmetry between the fundamental particles matter (leptons and quarks) forms the basis of modern standard mo of electromagnetic, weak and strong interactions.

1.4. Standard model of electroweak and strong interactions

The standard model is based on the $SU(3)_c \otimes SU(2)_L \otimes U(1)$ grou gauge symmetry assumed for the leptons and quarks.

This model assumes that the quark of each 'flavour' has

colour states, i.e., it is identified with the triplet representation of the $SU(3)_c$ symmetry group of strong interaction QCD. Invariance of the Lagrangian of the free quark field with respect to the gauge symmetries includes the existence of gauge fields mediating colour interactions and called gluons.

Since the $SU(n)$ group has

$$N = n^2 - 1 \tag{2.43}$$

generators, there exists eight generators of the colour gauge group, which generate eight gauge fields of colour interaction. Thus, gauge invariance of QCD implies the existence of eight gluons.

The structure of the Lagrangian of the quark–gluon interaction has a form similar to the electrodynamics Lagrangian (QED)

$$L_{gs} = g_s \cdot \overline{q}_i \gamma_\mu q_j A^\mu_{ij}, \tag{2.44}$$

where i, j = 1,2,3 are the colour indices, and the gauge coupling constant (g_s) determines the dimensionless constant of QCD

$$\alpha_s = \frac{g_s^2}{4 \cdot \pi \cdot \hbar \cdot c}, \tag{2.45}$$

which plays in QCD the same role as the fine structure constant

$$\alpha_{em} = \frac{e^2}{4 \cdot \pi \cdot \hbar \cdot c} \tag{2.46}$$

QED.

These dimensionless constants determine the probability of the
entary interaction between the quark in QCD and the electron in
respectively.

to radiation effects, these constants are not really constant.
they are called 'running' constants. Their values depend on
between the charges, that is, according to quantum theory,
scale of the interaction.

harge renormalization due to radiative corrections leads
d problem of '0-charge'. That is, if we consider a final
short distances, then at large distances the value of the
d to zero.

mportant role in radiation-induced effects is played by
character of the colour symmetry group. The gluon
as an octet of the colour group. This means that they
our charge and are the source of the gluon field.

In contrast to the electromagnetic field tensor, defined as

$$F_{\mu\nu} = \partial_\mu A_\nu - \partial_\nu A_\mu, \tag{2.47}$$

the tensor of the gluon field has the form

$$G^a_{\mu\nu} = \partial_\mu A^a_\nu - \partial_\nu A^a_\mu + g_s f^a_{bc} A^b_\mu A^c_\nu, \tag{2.48}$$

where a, b, c = 1,...,8 are the $SU(3)_c$ octet indices, f^a_{bc} are the $SU(3)_c$ structure constants, antisymmetric in the permutation of indices b, c.

Thus, in contrast to the Lagrangian of the electromagnetic field, defined as

$$L_{emf} = -\frac{1}{4} F_{\mu\nu} F^{\mu\nu}, \tag{2.49}$$

the Lagrangian of the gluon field, which has a very similar structure

$$L_{gf} = -\frac{1}{4} G^a_{\mu\nu} G^{a\mu\nu}, \tag{2.50}$$

contains expressions with the product of three or four operators of the gluon field, which determine the three- and four-gluon couplings. These special features of the gluon interactions lead to non-trivial properties of the running coupling constant in QCD.

The QCD charge tends to zero at short distances, but increases with distance and on the confinement scale becomes infinitely large.

The dependence of the QCD constant on the energy scale can be parametrized

$$\alpha_s = \frac{4 \cdot \pi}{\beta_0 \cdot \ln\left(\dfrac{\mu^2}{\Lambda^2_{QCD}}\right)}, \tag{2.51}$$

where β_0 is the function which depends on the number and type of colour states that contribute to the radioactive effects, and Λ_{QCD} is the confinement scale.

Thus, at high energy scales corresponding to small distances, the QCD running coupling constant tends to zero. This phenomenon is called the asymptotic freedom.

Qualitative confirmation of asymptotic freedom appears in the parton picture of deep inelastic scattering of leptons and photons on nucleons. At large momentum transfers, i.e., at short distances, the leptons and photons are scattered by the point-like constituents of the nucleon – partons. The quark–parton model interprets the partons in terms of momentum distribution of quarks in the nucleon.

On the other hand, the QCD confinement means that no coloured object can be in the free state. Being born as free particles at short distances, the quarks and gluons should be 'dressed' at large distances and appear only in the form of jets of hadrons.

At high temperatures

$$T > \Lambda_{QCD} \tag{2.52}$$

according to QCD in the Universe there should be phase transitions in hadronic matter to the phase of quark–gluon plasma. This approach gives the physical meaning to the relativistic equation of state at the so-called 'hadronic phase' in the early Universe. The phase of quark-gluon plasma can be studied experimentally in high energy collisions of nuclei and nucleons at accelerators.

Electromagnetic and weak interactions are combined in the SM on the basis of the $SU(2)_L \otimes U(1)$ group of gauge symmetry. They are treated as a unified electroweak interaction. Left-handed polarized states of quarks and leptons form a doublet representation of the weak isospin group $SU(2)_L$. Their right-handed states are singlets in $SU(2)_L$ transformations. Inequivalence of the left-handed and right-handed states reflects parity violation in weak interactions.

Three generators of $SU(2)_L$ group correspond to the three W gauge bosons. The charged W^-- and W^+-bosons mediate the interaction of charged weak currents. The neutral W-boson and the B-gauge boson $U(1)$ of the gauge group are combined into a photon and a neutral Z^0-boson of the weak interaction of neutral currents.

Gauge interaction of the weak currents, carried out by intermediate W and Z bosons is reduced to an effective four-fermion interaction at momentum transfer much smaller than the mass of these bosons.

Thus, the interaction of charged currents mediated by $W^{\pm}$-bosons at low energies is reduced to an effective four-fermion Fermi interaction with the Fermi constant G_F, as expressed through the gauge constant g_W and the mass m_W

$$\frac{G_F}{\sqrt{2}} = \frac{g_W^2}{8 \cdot m_W^2}. \tag{2.53}$$

The original symmetry of electroweak interactions is broken to the symmetry of electromagnetic interaction via the Higgs mechanism of spontaneous symmetry breaking.

This mechanism suggests that in addition to the leptons, quarks and gauge bosons, there are additional scalar fields ϕ. In the minimal model of electroweak interactions we introduce the Higgs doublet of the scalar fields

$$\phi = \begin{pmatrix} \phi^+ \\ \phi^0 \end{pmatrix} \tag{2.54}$$

together with a doublet of antiparticles

$$\overline{\phi} = \begin{pmatrix} \overline{\phi}^0 \\ \phi^- \end{pmatrix}. \tag{2.55}$$

The Lagrangian of these fields formally preserves the $SU(2)_L \otimes U(1)$ symmetry, but the potential of scalar interaction of linear combination of the neutral fields ϕ is such that the vacuum state with non-zero vacuum expectation value ϑ is energetically most favourable

$$V(\phi) = \frac{1}{4} \cdot \lambda \cdot \left(|\phi|^2 - \vartheta^2 \right)^2. \tag{2.56}$$

So there is spontaneous symmetry breaking. The symmetry of the Lagrangian is broken by the asymmetry of the ground (vacuum) state.

Regarding this true asymmetric vacuum, there is only one linear combination of the neutral field ϕ which has the physical meaning of the scalar field. Another combination of the neutral field ϕ with the charged fields $\phi^{\pm}$ acquires the meaning of the third, longitudinal components of the massive *W*- and *Z*-bosons.

Yukawa couplings of the fields ϕ to the fermions, which prior to the violation of symmetry have the form

$$L_{Yuk} = h_f \phi \overline{f} f, \tag{2.57}$$

generate after the spontaneous symmetry breaking mass terms of fermions, given by the expressions

$$L_m = m_f \overline{f} f, \tag{2.58}$$

in which the fermion masses are determined by the vacuum expectation of the Higgs fields

$$m_f = h_f \vartheta \tag{2.59}$$

and index f indicates $f = e, \mu, \tau, u, d, s, b, c, t.$.

At high temperature

$$T > T_{cr} \sim \vartheta \tag{2.60}$$

because of the effect of thermal fluctuations the Higgs potential minimum is shifted to the zero vacuum expectation value

$$\langle \phi \rangle = 0 \tag{2.61}$$

and the original $SU(2)_L \otimes U(1)$ symmetry of electroweak interactions is restored. Thus, the Higgs mechanism of spontaneous symmetry breaking causes phase transitions at high temperature in the Big Bang Universe.

The phase transitions in the early Universe are an important cosmological consequence of spontaneous symmetry breaking, predicted in the theory of elementary particles.

In the SM, the fermion masses, as well as the mass of the W- and Z-bosons are determined by the non-zero vacuum expectation value of the Higgs field. Then, restoration of the symmetry of electroweak interactions at which this vacuum expectation value is zero, should lead to the vanishing of the masses of fermions and gauge bosons. This means that, according to SM, not only physical conditions but also the fundamental physical parameters (masses of elementary particles) should have changed in the early Universe.

Most of the calculations within the SM are based on perturbation theory. In the case of small dimensionless electromagnetic, weak and QCD constants α calculations on the basis of series of the perturbation theory are justified, i.e., expansions in powers of these constants, which is possible at

$$\alpha < 1. \tag{2.62}$$

However, the effects not described by perturbation theory, are important both in QCD and in the electroweak theory. Such effects lead in QCD to the θ-term which corresponds to non-vanishing and *apriori* not small effects of violation of CP-invariance in strong interaction.

Similar effects in the electroweak theory lead to transitions which do not conserve the baryon and lepton numbers.

These transitions, called sphalerons, are suppressed exponentially at low temperatures, but at the electroweak phase transition and higher temperature their probability is of the order of 1. This phenomenon has important implications for the theory of the generation of baryon excess in the initially baryon-symmetric Universe.

2. Aesthetic and practical reasons for the extension of the standard model

Although the standard model of electroweak and strong (QCD) interactions of elementary particles does not meet any kind of direct experimental contradictions, there are several reasons to consider it incomplete and to turn to its extensions.

Considerations of mathematical beauty make the unified theories of fundamental forces of nature attractive. In these theories, the symmetry of the SM is extended to a higher underlying symmetry group, and the complete symmetry requires in addition to the known particles the existence of large numbers of new particles and fields, forming a 'hidden sector' of the relevant theory. The higher the level of unification, the larger is the 'hidden sector' that is involved by the corresponding symmetry.

No matter how attractive the aesthetic arguments are, from a pragmatic point of view the expansion of SM, needed to address its internal contradictions, is even more important.

So, supersymmetry is needed to eliminate the quadratic divergence in the calculation of radiative corrections to the mass of the Higgs boson, and the axion hypothesis is required to address violations in QCD. Such extensions of SM are usually studied separately, and as they relate to different areas of particle theory they are generally independent of each other. It is worth noting that it is this type of extension of SM that is widely involved as a physical basis of the existence of hot, cold, mixed, unstable, and other forms of dark matter in the Universe.

Cosmological consequences of both aesthetic and practical approaches to the extension of SM are primarily associated with stable or sufficiently long-lived particles or objects predicted in these models.

Since the (meta) stability in the theory of elementary particles suggests that there are some (approximate) conservation laws which reflect the proper fundamental symmetry and/or mechanism of its breaking, the majority of new fundamental laws of nature, which are predicted in the extensions of SM, lead to the cosmological implications accessible to astrophysical verification. Indeed, the new symmetries that extend the symmetry of SM, mean new exactly (or at least approximately) conserved charges. In this case, the lightest particles with such a charge should be (meta) stable.

New charges may be associated with local or global, continuous or discrete symmetry group. They may be topological, arising from the topology of the corresponding symmetry group. In most cases, the masses of hypothetical particles and objects reflect the energy scale at which the symmetry with which they are associated was broken.

Let us consider in detail the physical reasons for the extensions of the SM and their cosmologically significant consequences.

2.1. New physics of Grand Unification

The similarity in the description of electromagnetic, weak and strong interactions in the SM indicates that they can be unified.

In Grand Unified Theories (GUT) models, the gauge symmetry $SU(3)_c \otimes SU(2)_L \otimes U(1)$ is included in a unified gauge group G_{GUT}

$$G_{\mathrm{GUT}} \supset SU(3)_c \otimes SU(2)_L \otimes U(1). \qquad (2.63)$$

In this case, quarks and leptons are placed in the same representation of the unification group G_{GUT}, and the presence of the generators of this group related to the lepto-quark transitions, corresponds to the existence of gauge bosons that cause processes with non-conservation of the baryon and lepton numbers.

Such processes are very important in the mechanism of generation of baryon excess in the initially baryon-symmetric Universe.

The simplest group in which the symmetry of the SM can be fully incorporated is $SU(5)$. The number of generators of this group is 24, so that even in this simple implementation of GUT 12 more new gauge bosons must appear in addition to the known 12 bosons (8 gluons – for strong interactions, 3 – $W^{\pm}$ and Z^0 – for the weak processes, and a photon for the electromagnetic interaction). These 12 new bosons of the $SU(5)$ model take part in the processes with non-conservation of the baryon and lepton numbers.

Baryosynthesis mechanisms, in the framework of the GUT, can be verified by experimental search for rare processes violating the baryon number. In particular, proton decay, predicted by the $SU(5)$ model, goes through the channel

$$p \to \pi^0 e^+ \qquad (2.64)$$

with the lifetime

$$\tau_p \leqslant 10^{30} \div 10^{31}\,\text{years}, \qquad (2.65)$$

which is totally refuted by the results of the search for proton decay.

Historically, the strongest argument in favour of the simplest scheme of GUT followed from an analysis of the energy dependence of the current constants of weak, strong and electromagnetic interactions.

The first evaluation of such constants of the weak interaction resulted in the conclusion that all three curves, describing the energy dependence of these constants, have a common intersection point at an energy of about

$$E_{\text{GUT}} \sim 10^{15}\,\text{GeV} \tag{2.66}$$

on the condition that the set of the components of the SM does not change up to this energy.

However, direct measurement of the probability of decays of $W^{\pm}$ - and Z^0-bosons gave more precise data on their interactions with fermions. This resulted in the opposite conclusions. To ensure that all three curves intersect at the same point, it is necessary to add an additional set of the particles to the theory, such as, for example, such as, for example, a set of supersymmetric partners.

In all models unifying electromagnetism with the other interactions into a single compact symmetry group, we obtain solutions such as a magnetic monopole.

These solutions follow from the topology of violation of the symmetry of GUT. The GUT monopoles are stable topological defects with the Dirac magnetic charge

$$g = \frac{\hbar \cdot c}{2 \cdot e} \tag{2.67}$$

and the mass of the order,

$$m \sim \frac{\Lambda}{e} \tag{2.68}$$

where Λ is the scale at which the $U(1)$ symmetry, corresponding to electromagnetism, is separated from other interactions.

Choosing a scale equal to the scale of the GUT, with the order of magnitude of

$$\Lambda \sim E_{\text{GUT}} \sim 10^{15}\,\text{GeV}, \tag{2.69}$$

it turns out that the mass of monopoles should be of the order

$$m \sim 10^{16}\,\text{GeV}, \tag{2.70}$$

making it impossible for the birth of such a particle in accelerator experiments, and even in the collision of cosmic rays with ultrahigh energies.

On the other hand, magnetic monopoles should arise in the early Universe as a result of the phase transition with breaking of the symmetry of GUT. The calculation of their primordial abundance showed (Zeldovich, Khlopov 1978; Khlopov 1979; Preskill 1979) that their concentration must be so high that it could lead to the serious problem of overproduction. The solution to this problem has stimulated the development of both cosmology and particle physics.

2.2. New physics in modern models of Grand Unification

The apparent failure of the simplest model of GUT, based on a minimal group $SU(5)$ requires the use of both more complex models of GUT and extension of the theoretical basis for a unified description of the fundamental forces of nature.

Some GUT models attract the non-trivial topology of breaking of the symmetry of GUT which leads to the existence of the so-called topological defects.

In the case of discrete symmetry, such a two-dimensional topological defect is a domain wall arising at the border between two degenerate vacua which correspond to different ground states related by a discrete symmetry.

The energy density per unit area of the domain wall is determined by the scale Λ of breaking of discrete symmetry

$$\rho_\sigma \sim \Lambda^3. \tag{2.71}$$

If the continuous symmetry $U(1)$ is broken, this leads to the existence of one-dimensional topological defects – cosmic strings.

The corresponding energy density per unit length of the strings is proportional to

$$\rho_\mu \sim \Lambda^2, \tag{2.72}$$

where Λ is the scale of symmetry breaking.

The combination of breaking of continuous and discrete symmetries leads to more complex types of topological defects, such as domain walls bounded by strings, and the like. In particular, the walls, bounded by the strings, arise in the theory of the invisible axion which will be discussed below.

One might expect that realistic Grand Unified Theory will naturally combine the entire set of theoretical ideas that lead to the extension of the SM on the basis of various physical reasons.

One of the most popular ideas, applying for inclusion in the GUT, is supersymmetry. The symmetry between bosons and fermions is not only an aesthetic requirement, but also plays an important practical role in removing the quadratic divergences in radiative corrections to the Higgs boson mass. Supersymmetry also provides a physical basis for the hierarchy of different scales of symmetry breaking.

Due to supersymmetry, the hierarchy of the scales of breaking of symmetries of electroweak interactions and GUT, which differ by 13 orders of magnitude, becomes stable with respect to radiative effects.

In addition, the effects of renormalization in supersymmetric GUT models can induce the Higgs mechanism of spontaneous breaking of electroweak symmetry at low energies.

Since there is no symmetry between the known bosons and fermions it is assumed in the supersymmetric models that a set of new particles exists – the supersymmetric partners of the known bosons and fermions.

Supersymmetric partners have the same charges as normal particles, but differ from them by spin so that the supersymmetric partners (superpartners) of the known fermions are bosons and the superpartners of the known bosons – fermions.

In supersymmetric models there is correspondence between the quarks and leptons and the scalar particles with zero spin, squarks (quarkinos) and sleptons, which have the same charges as the ordinary quarks and leptons, respectively.

Supersymmetric partners of gauge and Higgs bosons are fermions with spin 1/2, called gauginos and higgsinos. The photino, gluino, zino and vino as the supersymmetric partners with the 1/2 spin for photons, gluons, W- and Z-bosons, respectively, are introduced.

To explain the absence in nature of supersymmetric particles, we must assume that supersymmetry is broken, so that the superpartners have much greater mass than the normal particles.

In supersymmetric models, we introduce a new quantum number R-parity, which distinguishes normal particles and their supersymmetric partners.

The exact R-parity corresponds to the absolute stability of the lightest supersymmetric particle (LSSP). Its mass is generally associated with the scale of supersymmetry breaking. In a wide class of supersymmetric models LSSP is a linear combination of zino, photino and higgsino called the neutralino.

In some supersymmetric models, the R-parity is not a strictly conserved quantity that corresponds to the metastability and decay of the LSSP to ordinary particles. The sufficiently long-lived metastable LSSP may have important cosmological significance.

In the locally supersymmetric models, the scale of supersymmetry breaking determines the mass of the gravitino – the superpartner of the graviton. Local supersymmetry allows unification of the weak, strong and electromagnetic interactions with gravity based on supergravity.

The supergravity model is defined by the number of different sets of supersymmetric partners equal to the number of different types of gravitinos N. The case $N=1$ is the simplest case of only one type of gravitino. The maximum number of different types of gravitinos (8) is considered in $N = 8$ supergravity.

In supergravity, the gravitino has semi-gravitational interaction with other particles inversely proportional to the Planck mass $m_{\rm Pl}$. Because of the extreme weakness of their interaction, such particles are not available for research in accelerator experiments, even if the accelerator reaches the energy threshold of their production. Cosmology is a unique source of information on possible properties of the gravitino.

The GUT models which are more sophisticated than the minimal $SU(5)$ model, can include in a natural way the physics of the neutrino mass. Disregarding from the entire set of physical and cosmological manifestation of the neutrino mass, discussed below, we note here only one cosmologically interesting effect associated with the so-called see-saw mechanism for generating the neutrino mass.

The see-saw mechanism implies the existence of the state of the heavy right-handed neutrino with Majorana mass M_R determined by the scale of non-conservation of the lepton number. Mixing due to the Dirac neutrino mass of this state with the state of the normal left-handed neutrino leads to the generation of Majorana mass of ordinary left-handed neutrinos at the level of

$$m_\nu = \frac{m_D^2}{M_R}, \tag{2.73}$$

where m_D is the Dirac neutrino mass, usually associated with the mass of the corresponding charged lepton. This explains why the Majorana mass of the neutrino is small compared with the Dirac mass of the corresponding charged leptons.

The lifetime of the heavy right-handed state is determined by mixing with the ordinary left-handed neutrino ($\sim m_D/M_R$) and is inversely proportional to the mass of the light left-handed state of the neutrino. So the see-saw mechanism leads to the prediction of supermassive metastable neutrinos.

The realistic model of Grand Unification should also contain the solution of the problem of strong *CP*-violation in QCD. This solution involves the Peccei–Quinn symmetry (see details below in Section 3.3). Spontaneous breaking of this symmetry leads to the existence of the pseudo-Goldstone boson called the axion, with the mass equal to

$$m_{\alpha} = \frac{m_{\pi} \cdot f_{\pi}}{F} \tag{2.74}$$

where m_{π} and f_{π} and are the mass of the pion and the pion constant and F is the scale of Peccei–Quinn symmetry breaking. The interaction of the axion with the fermion is inversely proportional to F, and the lifetime of the axion with respect to decay to 2γ has the order of magnitude

$$\tau(\alpha \to \gamma\gamma) = \frac{64 \cdot \pi \cdot F^2}{m_{\alpha}^3} \sim F^5. \tag{2.75}$$

The realistic theory of unification should also consider the problem of equivalence of the right and left coordinate systems. Beginning with the classical work of Lee and Yang (1956), the solution to this problem assumes the existence of mirror partners of the ordinary particles.

The mirror particles should not have normal interactions and their own mirror gauge interactions should be symmetric with respect to the relevant interactions of the ordinary particles. The mirror particles, having the same mass spectrum and the same internal mirror interaction, as well as their ordinary partners, interact with the ordinary particles only by gravity. They represent a very interesting form of dark matter, which will be discussed in detail in chapter 10.

The inclusion of the mirror partners with the ordinary particles in a unified GUT model leads at this breaking of a GUT symmetry to the existence of Alice strings which are cosmic strings which alter the relative mirrority of objects passing between them.

It should be noted that the topological conditions for the existence of solutions of the type of cosmic strings as a result of breaking of non-Abelian symmetry imply the existence of a strict discrete symmetry between the subgroups which remain unbroken after the breaking of the original symmetry. The symmetry between the ordinary and mirror particles is practically the only known physically justified discrete symmetry which may be strict in particle physics. This makes the Alice strings well justified in terms of fundamental physics as a candidate for the role of cosmic strings.

The main hope for a unified theory is primarily associated with the development of models of superstrings. In this approach all the basic principles of unification models are combined: the gauge symmetry, multi-dimensional space-time, supergravity and mirror symmetry are combined in it on the basis of string (membrane) theory. The most fundamental objects here are the strings (membranes) in a multidimensional space-time, and the excitation of these strings is chatacterized by high symmetry and they reproduce the effective field theory in four-dimensional space-time.

In this approach it is argued that the fundamental constants of the quantum field theory are finite. This gives a quantitative determination to the entire fundamental physics, but the problem of such models is that they contain a too large hidden sector. Development of methods for studying this sector is an important task of cosmoparticle physics.

For example, in the widely discussed model of the heterotic string it is assumed that the initial gauge group is $E_8 \otimes E'_8$. The exact symmetry between the ordinary world (E_8) and the mirror world (E'_8) is initially assumed. The initial space-time dimension in the theory is 10 because only the 10-dimensional theory of this type is finite. The original symmetry of the ordinary and mirror particles is broken by the combined effects of compactification and breaking of gauge symmetry. So shadow matter forms. Initially mirror, partners lose discrete symmetry with the ordinary particles.

In the effective 4-dimensional theory to which the model of the heterotic string is reduced after compactification, the gauge symmetry of the ordinary particles comes from the E_8 symmetry broken to E_6, to compensate for the effects of curvature of compactificated dimensions. The world of the ordinary particles is accompanied here by the extremely diverse world of shadow particles and their interactions corresponding to broken gauge symmetry.

Thus, even in the case of the simplest realization of superstring models, we are faced with the problem of testing the E'_8 model of the shadow world. To estimate the complexity of the problem, it should be noted that the E_8 group has 248 fundamental fermions and 248 gauge bosons.

The mechanism of violation of gauge symmetry as a result of compactification on Calabi-Yau manifolds or orbifolds, used in superstring models, leads to the prediction of homotopically stable solutions with mass

$$m \sim \frac{r_c}{\alpha'} \tag{2.76}$$

where r_c is the radius of compactification and α' is the string tension. These objects are sterile with respect to gauge interactions and can affect ordinary matter only gravitationally.

Thus, even a brief discussion of current trends in the unification of fundamental forces of nature offers many examples of this 'zoo' of particles we encounter on the way to constructing a unified theory of elementary pasticles of which the Universe is constructed. GUT models take into account all these hypothetical objects, but they are all linked with new phenomena the direct experimental search of which is sometimes very difficult or even impossible.

The purpose of cosmoparticle physics is to check the indirect effect of the theory of the microworld on the basis of a cross-disciplinary approach. In many cases, the cosmological effects are important, and sometimes they are the only source of information about the possible existence of such objects.

2.3. Restrictions on baryon and lepton photons

Verification of the equivalence principle imposes very strong restrictions on the allowed parameters of the new long-range forces, including the existence of interaction associated with the baryon or lepton charge. Experimental tests of the equivalence principle imply that the gravitational mass is equal to the inertial mass, and if long-range effect depends on the baryon or lepton charge, the results of testing the equivalence principle should depend on the chemical composition of matter.

It is easy to show (Okun' 1981) that, if there are massless baryonic photons, their interaction with the baryons must be very weak: $\alpha_B < 10^{-49}$ (this should be compared with $\alpha = 1/137$ for the conventional photon). This restriction follows from the equality of gravitational and inertial masses, tested with the accuracy up to 10^{-12}. This restriction on α_B follows from the fact that the baryonic photons would create around the Earth a unique 'Coulomb' field which would repulse the baryons from the Earth. The strength of this repulsion would be proportional to the number of baryons in the sample, not its mass, and would be different, for example, for lead and copper samples with identical masses. The force acting on the i-th sample with mass M_i with A_i nucleons is

$$F = \chi \cdot \frac{M_i \cdot M}{r^2} + \alpha_B \cdot \frac{A_i \cdot A}{r^2} = \chi \cdot \frac{M_i \cdot M}{r^2} \cdot \left\{ 1 + \frac{\alpha_B}{\chi} \cdot \left(\frac{A}{M} \right) \cdot \left(\frac{A_i}{M_i} \right) \right\}$$

where M is mass of the Earth, A is the number of nucleons in the Earth ($M/A \approx m_p$); $\chi = 6 \cdot 10^{-39} m_p^{-2}$ is the gravitational constant.

A similar formula would describe in this case also the attraction of bodies toward the Sun. This obvious observation was made because the attraction of bodies toward the Sun is measured with higher accuracy than attraction to the Earth.

The nucleon mass in the nucleus of lead is about 1 MeV higher than in the core of copper

$$\frac{M_{\rm Pb}}{A_{\rm Pb}} - \frac{M_{\rm Cu}}{A_{\rm Cu}} \approx 10^{-3} \cdot m_p.$$

From experiments, it follows that

$$\frac{\alpha_B}{\chi \cdot m_p} \cdot \left(\frac{A_{\rm Cu}}{M_{\rm Cu}} - \frac{A_{\rm Pb}}{M_{\rm Pb}} \right) < 10^{-12}.$$

Consequently $\alpha_B < 10^{-9} \cdot \chi \cdot m_p^2 \leqslant 10^{-49}$.

A similar argument can be made for the constants of interaction of leptonic 'photons' with electrons

$$\frac{\alpha_L}{\chi \cdot m_p} \cdot \left(\frac{Z_{\rm Cu}}{M_{\rm Cu}} - \frac{Z_{\rm Pb}}{M_{\rm Pb}} \right) < 10^{-12}.$$

Thus, $\alpha_L \leqslant 10^{-47}$.

Due to the instability of the muon, the upper limit for the interaction constant of hypothetical muon 'photons' with the muon charge is many orders of magnitude worse (higher) than for the constant of the baryon and lepton photons.

Based on these estimates, it is natural to assume that the massless vector particles, associated with the lepton and baryon charges, apparently do not exist.

In the past three decades it has been widely discussed that the baryon and/or leptonic photons are not massless, but very light, so that their Compton wavelength is of the order of a kilometer. In this case, restrictions α_B on and/or α_L are less stringent than those discussed above. Some authors have reported the observation in experiments such as the Eotvos experiment of the so-called fifth force with a similar range. However, no reliable experimental confirmation of the existence of the fifth force is yet available. Therefore, in most theoretical models the gauge interactions are not associated with the lepton and baryon numbers.

It should be noted that the existence of strict new gauge $U(1)$ symmetries is a quite natural prediction of models of superstring phenomenology. Appropriate new long-range interactions are assigned in this case to the new particles, such as quarks and leptons of the fourth generation (Khlopov, Shibaev 2002).

If the lepton charge is not conserved, then the Goldstone boson – Majoron (see below) is associated with this violation.

3. Neutrino mass and the invisible axion

We complete the overview of the main directions of development of the physics of Grand Unification (GUT) by a more detailed analysis of the new physics associated with two widely discussed extensions of the standard model – the mass of neutrinos and the invisible axion model, considered as a solution to the problem of strong *CP*-violation in QCD.

As a result of this discussion it may seem that these two issues relate to different and unrelated to each other sections of the modern theory of elementary particles. Therefore, the cosmological scenario of hot dark matter (HDM), based on the hypothesis of the existence of massive neutrinos, and the cold dark matter scenario (CDM), based in a wide range of cases on the invisible axion model, can be considered as alternatives.

Based on cosmoparticle physics, we will see in Chapter 11 that the physics of neutrino mass and the invisible axion can be described within a unified scheme of horizontal unification which provides a physically self-consistent selection of the model of the cosmological dark matter and assumes joint examination in the cosmology of massive neutrinos and the invisible axion.

3.1. Physics of the neutrino mass

The search for the neutrino mass began immediately after W. Pauli first proposed in 1930 the idea of the existence of such particles.

The only known type of neutrino processes at that time was nuclear beta decay. The neutrino was not observed directly, and scientists tried to detect the effects of the neutrino mass indirectly through precise measurements of the beta spectrum near the maximum electron energy.

In the beta decay of nuclei with mass number A and charge Z to the nucleus with the same mass number and charge $(Z+1)$

$$(A,Z) \rightarrow (A,Z+1) + e^- + \overline{\nu}_e \tag{2.77}$$

the total maximum electron energy is equal to

$$E_{\max} = \left(M(A,Z) - M(A,Z+1) - m_\nu\right) \cdot c^2 \tag{2.78}$$

so that formally precise measurements of the maximum electron energy, together with accurate data on the masses of the initial and final nuclei might allow determination of the neutrino mass.

However, in practice, the most accurate data on the masses of the nuclei were obtained by measuring the β-spectrum, so that to determine the neutrino mass it was necessary to use more sophisticated methods.

If the neutrino mass is non-zero, then the maximum electron energy corresponds to the minimum neutrino energy at which the massive neutrino is non-relativistic. This means that near the maximum electron energy the shape of the β-spectrum must be changed due to the transition from the relativistic relation between energy and momentum

$$E_\nu \approx p_\nu c \tag{2.79}$$

to the non-relativistic relation

$$E_\nu = \sqrt{(p_\nu c)^2 + m_\nu^2 c^4} \approx m_\nu c^2 + \frac{p_\nu^2}{2m_\nu} \tag{2.80}$$

The first experimental validation of the existence of the non-zero neutrino mass, based on the change in the shape of the β-spectrum near the maximum electron energy, was made in the late 30's of the 20th century (Alichanian et al 1938; Alikhanian and Nikitin, 1938). The problem was that the values of the neutrino mass were of the order of 1 MeV and had different values in different nuclear decays.

The mystery was soon explained by Zavelski (1939) who found that the theoretical shape of the spectrum, which was used by Alikhanian Nikitin et al (Alikhanian, et al 1938; Alikhanian, Nikitin 1938) near the maximum electron energy cannot be used to describe the spectrum of β-decay in general. Correct description of the whole spectrum left unchanged its shape near the maximum electron energy, thus setting an upper limit on the neutrino mass.

It was the first but not the only episode in the dramatic history of exploration and 'discovery' of the neutrino mass (Zeldovich, Khlopov 1981b).

The latest edition of the Tables of Properties of Particles does not discuss the positive statements about the neutrino mass and give only upper limits on the mass of all three types of neutrino.

The non-zero neutrino mass corresponds to the Lagrangian mass term in the form of

$$L = m_\nu \overline{\Psi}_\nu \Psi_\nu, \tag{2.81}$$

which connects the states of neutrinos with different helicity, i.e., the left- and right-handed polarized states.

In simple words, the non-zero neutrino mass means that the velocity of neutrinos can be very close to the velocity of light but never reached this velocity so that there is always a reference system in which neutrino of the same polarization is moving in the opposite direction, i.e., has the opposite helicity.

In particular, the mass term in the Lagrangian links the left-handed neutrino state with a certain right-handed neutrino state which must also exist in nature.

Only left-handed neutrinos and right-handed antineutrinos take part in all known processes. Therefore, the neutrino mass associates the left-handed neutrino either with some new unknown neutrino state or with the known state of the right-handed antineutrino. Both options take us beyond the standard model.

In the first case, it is necessary to introduce a new state of the right-handed neutrino into particle theory.

If this state exists, then the mass term looks like the mass terms of all known fermions. This situation corresponds to the Dirac neutrino mass.

In the second case, the neutrino mass term implies a transition in which the lepton number is not conserved

$$\nu_L \rightarrow \overline{\nu}_R \tag{2.82}$$

In the Lagrangian, this neutrino mass corresponds to a term changing the lepton number

$$\Delta L = 2 \tag{2.83}$$

In this case, the neutrino and the antineutrino correspond to the different helicity states of a single particle – a truly neutral Majorana neutrino, and the neutrino mass term is called the Majorana mass term.

The experimental upper limits show that the neutrino mass, if it is non-zero, should be several orders of magnitude smaller than the mass of the corresponding charged lepton.

The theoretical idea to explain the possibility of large differences between the masses of neutral and charged leptons is called the 'see-saw' mechanism (Gell-Mann et al 1979).

This idea is based on the fact that, unlike all the other fermions (leptons and quarks), the neutrino is electrically neutral, so that the conservation of the electric charge does not prohibit the existence of the Majorana masses of the neutrino.

The see-saw mechanism implies that the Dirac neutrino mass m_D is of the same order as the mass of the corresponding charged lepton, but the right-handed state of neutrinos, corresponding to the Dirac mass, has the Majorana mass, much larger than m_D (see Fig. 2.3).

The Majorana mass term is generated by the following sequence of transitions (Fig. 2.3):

$$\nu_L \to \nu_R \to \bar{\nu}_L \to \bar{\nu}_R \tag{2.84}$$

and equals

$$m_\nu = \frac{m_D^2}{M}. \tag{2.85}$$

As a result, there are two physically different massive neutrino states: the very heavy right-handed neutrino with Majorana mass

$$m_R \approx M \tag{2.86}$$

and the normal left-handed neutrino related by the Majorana mass with the normal right-handed antineutrino

$$m_L = m_\nu = \frac{m_D}{M} \cdot m_D, \tag{2.87}$$

which is $\propto (m_D/M)$ factor less than the mass m_D of the corresponding lepton.

Thus, the see-saw mechanism leads to the mass of normal (left-handed) neutrinos by several orders of magnitude smaller than the mass of the corresponding charged lepton and the mass of the right-handed neutrino higher than the energy threshold that is accessible in experiments in modern accelerators.

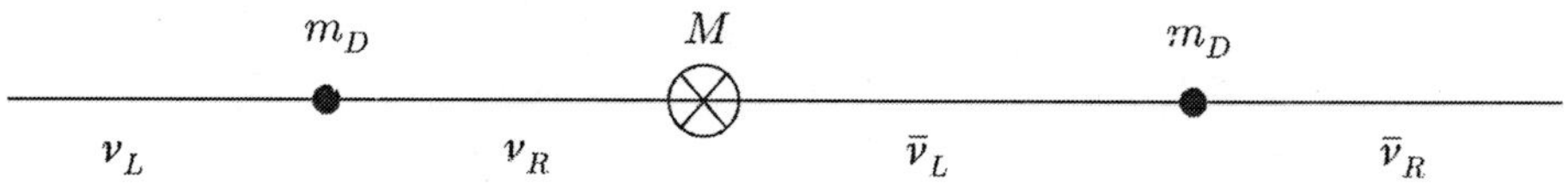

Fig. 2.3. The sequence of transitions leading to the see-saw mechanism of generation of the Majorana mass of the left-handed neutrino.

Formally, the right-handed neutrino can decay due to the small mixing (of the order $\propto(m_D/M)$) with the state of the light left-handed neutrino. We can estimate the probability of decay of the right-handed neutrino, for example, via

$$\bar{\nu}_L \to Z\nu_L. \tag{2.88}$$

This probability is of order

$$W(\bar{\nu}_L \to Z\nu_L) \propto \frac{\alpha}{\pi} \cdot \left(\frac{m_D}{M}\right)^2 \cdot M = \frac{\alpha \cdot m_\nu}{\pi}, \tag{2.89}$$

where α is the running coupling constant of weak interaction. Thus, for the small neutrino mass the heavy right-handed state is the long-lived particle.

Moreover, the smaller the mass of ordinary light neutrinos, the more massive and long-lived is the heavy right-handed partner. This makes the supermassive metastable neutrino a very interesting object for cosmological implications.

Note, however, that if the Dirac neutrino mass is generated in the Lagrangian by the Higgs mechanism which occurs in the case of the standard model, the coupling of the neutrino with the Higgs boson can lead to rapid decay of heavy neutrinos into light neutrinos and the Higgs boson.

Although in its simplest form the see-saw mechanism leads to a small Majorana mass for the usual left-handed neutrino, in more complex versions of this mechanism there may also be small Dirac masses for light neutrinos. This possibility is realized in theory by the repeated application of the see-saw mechanism and re-determination of the right-handed antineutrino states as states of the right-handed light neutrino. So, in general, the Lagrangian of the neutrino mass may contain both Dirac and Majorana mass terms.

It should be noted that the existing experimental limits do not greatly limit the strength of the interaction of the light right-handed neutrinos. The right-handed weak interactions involving right-handed neutrinos may be due to the intermediate boson being only 5–7 times heavier than the W- or Z-bosons.

The Dirac mass term implies the existence of a magnetic moment of the neutrino. In the absence of right-handed neutrino currents the magnetic moment of the neutrino is of the order

$$\mu_\nu \propto G_F m_\nu, \tag{2.90}$$

where G_F is the Fermi constant.

The existence of the right-handed currents and mixing of bosons that carry their interactions with the W- and Z-bosons can lead to the magnetic moment of the neutrino proportional to the mass of the charged lepton

$$\mu_\nu \propto G_F m_l. \tag{2.91}$$

3.2. Instability of the neutrino

In the standard model, the states of the electron, muon or tau-neutrinos are determined on the basis of conservation of the lepton number in weak interactions. Thus, the electron neutrino is the neutrino state, which is created in the final state or together with the positron, or when an electron is annihilated in the initial state.

Since the see-saw mechanism is based on non-conservation of the lepton number, we can expect that the eigenstates of the neutrino mass do not coincide with the states with a definite lepton number.

In this case, for example, the electron neutrino produced in β-decay is a superposition of several neutrino states with a definite mass.

Since the neutrino states with a definite lepton number ν_l, where l denotes e, μ, τ, are given by a superposition of states ν_i with masses m_i as follows

$$|\nu_l\rangle = \sum_i a_{li} |\nu_i\rangle, \tag{2.92}$$

where a_{li} are the mixing coefficients.

If the neutrino masses are different, then the neutrino mass states, produced with a given energy in a given initial ratio propagate with different velocities, and thus their superposition will change at some distance from the source. This means that it is possible to observe the effect of neutrino oscillations: for example, muon or tau-neutrinos will be detected at some distance from a pure source of electron neutrinos.

Neutrino oscillations are described in terms of the oscillation length which for the neutrino with energy E is equal to

$$L = \frac{E}{\delta m_{\nu}^{2}}, \tag{2.93}$$

where δm_{ν}^{2} is the difference between the squares of neutrino masses.

The second characteristic of the oscillations is their amplitude determined by the mixing parameters and defining the value of maximum admixture of neutrinos of another type in the neutrino flux of the source with a given lepton number.

The amplitude of neutrino oscillations also determines the maximum deficit of neutrinos of the given type because of the oscillations in the unobservable neutrino states.

Accordingly, there are two principal opportunities of search for neutrino oscillations.

We can register other types of neutrinos at a distance from a pure source of a given type of neutrino, or we can monitor the neutrino flux of a given type at a certain distance and find the lack of these neutrinos due to oscillations.

Propagation of neutrinos in matter can cause resonant amplification of oscillations (Mikheyev, Smirnov 1985; 1986; Wolfenstein 1978): at the resonant density the oscillation amplitude is of order 1 even for very small amplitudes in vacuum. This effect is discussed as a possible explanation for the observed solar neutrino deficit. The resonance conditions can also occur during the propagation of neutrinos in a longitudinal magnetic field (Akhmedov, Khlopov 1988; Akhmedov, Khlopov 1988).

In recent years, the collection of data of flux measurements of atmospheric neutrinos in experiments MACRO, Kamiokande and Superkamiokande, fluxes of neutrinos from the Sun in the SNO experiments, the fluxes of antineutrinos from nuclear power carriers in Japan and South Korea in the KamLAND experiments, as well as T2K experiment, in which neutrino fluxes from the KEK accelerator in the Superkamiokande installation prove the existence of oscillations $\nu_e - \nu_\tau$ and $\nu_\mu - \nu_\tau$ and possibly $\nu_\mu - \nu_e$. The additional proofs are coming from OPERA experiment, in which neutrino flux created in CERN, Switzerland, is detected in the Italian National Laboratory in Gran Sasso. Analysis of these data makes it possible to determine the mass difference and mixing parameters of the neutrino.

Another possible form of instability of the neutrino are decays of heavy neutrinos into lighter ones. It is easy to establish that in the standard model such decays of massive neutrinos as

$$\nu_H \to \nu_L \nu_L \bar{\nu}_L \tag{2.94}$$

or

$$\nu_H \to \nu_L \gamma \tag{2.95}$$

due to the creation of the virtual lepton and W-boson (see Fig. 2.4 and 2.5) and strongly suppressed.

Strong suppression of the probability of these processes is due to the fact that they correspond to the second order in the weak and electroweak interactions and their probabilities have the order of magnitude

$$W\left(\nu_H \to \nu_L \nu_L \bar{\nu}_L\right) \propto G_F^4 m_H^9 \tag{2.96}$$

and

$$W\left(\nu_H \to \nu_L \gamma\right) \propto \alpha_{em} G_F^2 m_H^5, \tag{2.97}$$

where α_{em} is the electromagnetic fine structure constant, G_F is the Fermi constant, m_H is the mass of heavy neutrinos ν_H.

Moreover, in an orthogonal neutrino mass matrix the mixing factors are normalized by the condition

$$\sum_l a_{il}^{+} a_{lj} = \delta_{ij} \tag{2.98}$$

which causes the vanishing of the amplitude of the off-diagonal transitions

$$\nu_H \to \nu_L, \tag{2.99}$$

for the processes corresponding to

$$H = i \neq j \equiv L. \tag{2.100}$$

The non-vanishing transition amplitudes (2.99) may arise because of taking into account the differences in the masses m_l of the charged leptons in the intermediate state, which leads to the additional suppression of the probability of neutrino decays (2.96) and (2.97), which finally has the form

$$W\left(\nu_H \to \nu_L \nu_L \bar{\nu}_L\right) \propto G_F^4 m_H^9 \left(\sum_l a_{Ll}^+ a_{lH} \frac{m_l^2}{m_W^2}\right)^2 \qquad (2.101)$$

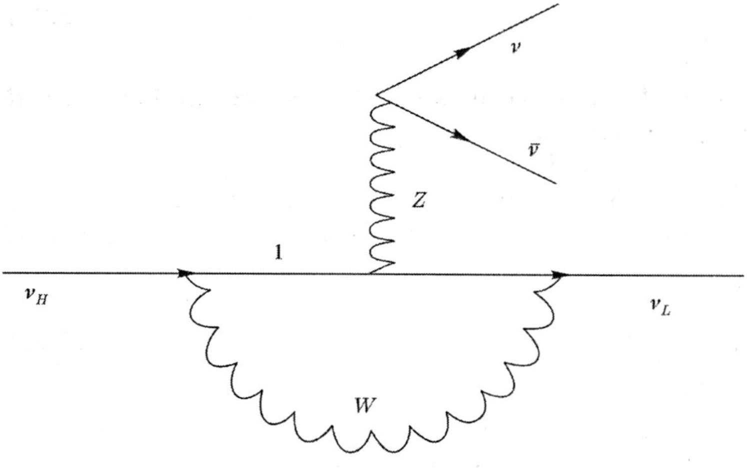

Fig. 2.4. In the standard model, the neutrino can decay into three neutrinos as a result of the creation of a virtual pair of charged leptons and the *W*-boson with the emission of *Z*-boson which decays into a neutrino pair.

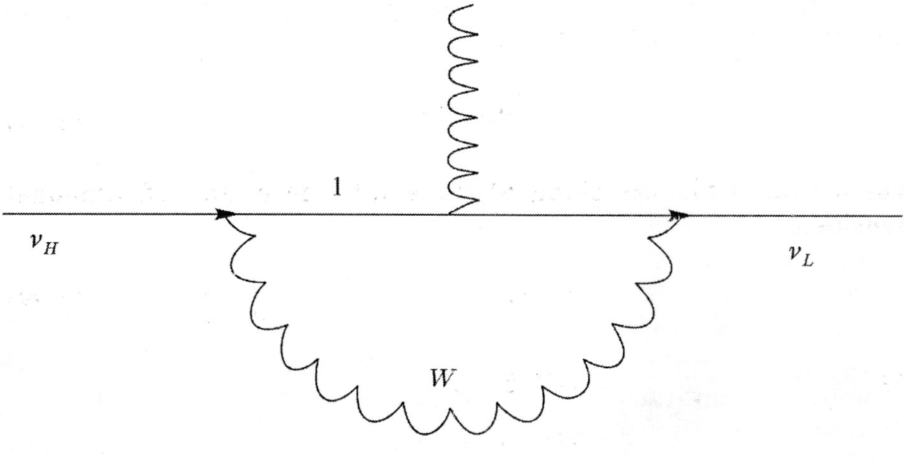

Fig. 2.5. In the standard model decay of the neutrino to a neutrino and a photon can occur due to the creation of a virtual pair of charged leptons and the *W*-boson with emission of a photon by the virtual charged particle.

and

$$W\left(\nu_H \rightarrow \nu_L \gamma\right) \propto \alpha_{em} G_F^2 m_H^5 \left(\sum_l a_{Ll}^{+} a_{lH} \frac{m_l^2}{m_W^2} \right)^2 . \tag{2.102}$$

On the other hand, we see that breaking of the symmetry of generations plays an important role in the mechanism of instability of the neutrino. Its manifestation is not confined to the difference of neutrino masses, since the difference in the masses of the charged leptons also plays an important role. The non-trivial role of breaking of the symmetry of generations is apparently manifested in all the mechanisms of instability of the neutrino.

We can expect that the nature of these mechanisms is related to the physics of the neutrino mass.

Thus, the non-conservation of the lepton number assumed by the see-saw mechanism can be realized through the spontaneous breaking of global symmetries of the lepton charge which corresponds to the existence of a Goldstone boson, called the Majoron.

The coupling of the neutrino with the Majoron M can lead to Majoron decay modes of the heavy neutrino

$$\nu_H \rightarrow \nu_L M. \tag{2.103}$$

The so-called triplet Majoron model (Gelmini, Roncandelli 1981) suggests a very low scale of interaction of the Majoron which leads to a strong inter-neutrino interaction due to exchange of the Majoron. This model is excluded by the measured width of the Z-boson to which Majoron effects should make an unacceptably large contribution.

The model of the singlet Majoron (Chikashige et al 1980) is free from this problem, because the high energy scale of interaction of the Majoron in this model leads to a very weak interaction of the Majoron with the known particles.

In the simplest singlet model of the Majoron diagonalization of the mass matrix of the neutrino at the same time leads to the diagonalization of the matrix of the coupling constant of the neutrino with the Majoron. Since the off-diagonal transition (2.103) acquires a non-zero amplitude only in the next order on the small coupling constant of the neutrino with the Majoron

$$h = \frac{m_H}{F} \tag{2.104}$$

where m_H is the neutrino mass ν_H and F is the energy scale at which the breaking of the global symmetry of the lepton number takes place. It must be greater than the scal4e of electroweak symmetry breaking, i.e.,

$$F > 10^2 \text{ GeV}. \tag{2.105}$$

Decay probability (2.103) is therefore of the order

$$W\left(\nu_H \to \nu_L M\right) \propto \frac{m_H^5}{F^4}, \tag{2.106}$$

which is not much higher than that of the photon channel (2.95) of decay of the massive neutrino in the standard model.

However, the decay probability (2.103) is strongly enhanced if different signs of the lepton number are assined to different types of neutrinos (Valle 1983). In this case, the probability of the Majoron decay of the neutrino ν_H is equal to

$$W\left(\nu_H \to \nu_L M\right) = \frac{h^2}{32 \cdot \pi} \cdot m_H \propto \frac{m_H^3}{F^2}. \tag{2.107}$$

The difference between the lepton numbers which is crucial for strengthening the Majoron neutrino decay mode is successfully implemented in models of spontaneously broken symmetry of the quark and lepton families.

Spontaneous breaking of the global symmetry of families leads to the prediction of the Goldstone boson called the familon. In case of breaking of global symmetry $SU(3)_H$ of families the octet of massless familons is predicted. There is also a model of the singlet familon (Anselm, Uraltsev 1983).

In the familon models, the massive neutrino can decay due to interaction with the familons f in the familon channel

$$\nu_H \to \nu_L f, \tag{2.108}$$

the probability of which is described by the same expression (2.107) with the scale of F which here takes on the meaning of the scale of breaking of symmetry of families.

The model of the familon is verified in experimental search for familon decay modes of the quarks and charged leptons, such as

$$\mu \to ef \tag{2.109}$$

$$\tau \to \mu f \tag{2.110}$$

or

$$s \to d + f \tag{2.111}$$

predicted in this model, together with the neutrino decay (2.108).

The physics of the neutrino mass, associated with the see-saw mechanism for generating neutrino masses, can be tested in experimental search for processes that violate conservation of the lepton number

$$\Delta L = 2, \tag{2.112}$$

such as the neutrinoless double β-decay.

Some nuclei (A, Z) are stable with respect to β-decays to the adjacent nuclei (A, Z+1) and (A, Z–1) but are unstable with respect to the double β-decay, where two protons or two neutrons in the nucleus decay at the same time, that is, the following process takes place

$$(A,Z) \to (A,Z-2) + 2e^{+} + 2\nu_e \tag{2.113}$$

or (see Fig. 2.6)

$$(A,Z) \to (A,Z+2) + 2e^{-} + 2\overline{\nu}_e. \tag{2.114}$$

Non-conservation of the lepton number, corresponding to the selection rule (2.112), leads to the neutrinoless double β-decay

$$(A,Z) \to (A,Z-2) + 2e^{+} \tag{2.115}$$

or

$$(A,Z) \to (A,Z+2) + 2e^{-}. \tag{2.116}$$

In the case of neutrinoless double β-decay determined by the Majorana electron neutrino mass, the amplitude of this process, described by the diagram in Fig. 2.7, is proportional to the mass of neutrinos (Zeldovich, Khlopov 1981a,b).

Therefore, the upper limit on the probability of a neutrinoless double β-decay restricts the Majorana mass of the electron neutrino. Model-independent analysis of experimental data, taking into account all possible uncertainties gives an upper limit for this mass of about 1 eV.

3.3. Axion solution to the problem of strong CP-violation in QCD

Instanton effects in QCD give rise to a very complicated vacuum structure, leading to the so-called θ-term in the QCD Lagrangian

$$\Delta L = \frac{g^2}{16 \cdot \pi} \cdot \theta_{\text{QCD}} G_{\mu\nu} \tilde{G}_{\mu\nu} \tag{2.117}$$

where θ_{QCD} is an arbitrary and, generally speaking, not a small constant, g is the gauge constant of QCD, $G_{\mu\nu}$ is the tensor of the gluon field, and the corresponding dual tensor

$$\tilde{G}_{\mu\nu} = \varepsilon_{\mu\nu\lambda\rho} G^{\lambda\rho}. \tag{2.118}$$

It is easy to see that the θ-term violates P- and CP-invariance.

Indeed, the corresponding product of the gluon field tensor and its dual tensor (2.118) corresponds in the electrodynamics to the well-known invariant – the scalar product of the strength of electric and magnetic fields

$$\mathbf{E} \bullet \mathbf{H} = \text{Invariant}, \tag{2.119}$$

for which

$$CP(\mathbf{E} \bullet \mathbf{H}) = -1. \tag{2.120}$$

The interference of similar CP-odd combination of the gluon fields with the CP-even term in the QCD Lagrangian leads to a strong violation of CP-invariance.

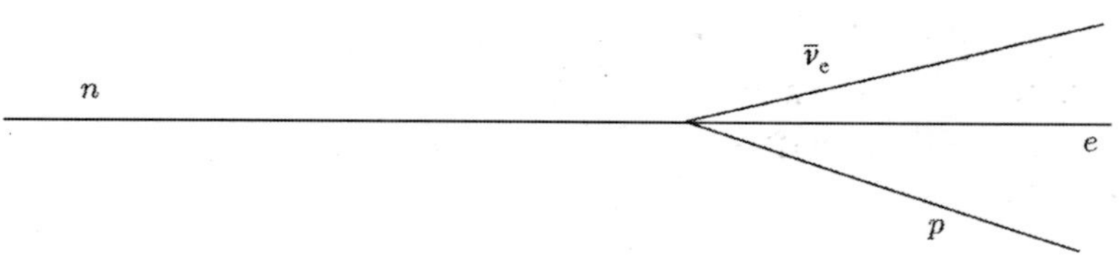

Fig. 2.6. Double β-decay is the result of the simultaneous decay of two neutrons in the β-stable nuclei.

In strong interactions, such effects are not observed. Moreover, based on negative results of searches for the electric dipole moment of the neutron, the following upper limit is obtained

$$d_n < 10^{-25} e \cdot \mathrm{cm}, \tag{2.121}$$

which implies the upper limit on the constant θ

$$\theta < 10^{-9}. \tag{2.122}$$

The main difficulty in the problem of strong CP-violation is connected with the fact that even if we assume that the initial constant is equal to zero

$$\theta_{\mathrm{QCD}} = 0, \tag{2.123}$$

the existence of the effective θ-term still can not be excluded.

This effective term is generated when the electroweak symmetry of the standard model is broken due to the appearance of complex elements in the quark mass matrix after reduction of the matrix by unitary transformations to the physical basis in which this matrix is diagonal.

Thus, the effective value of θ in the Lagrangian (2.117) is the sum of two terms with different physical nature

$$\theta = \theta_{\mathrm{QCD}} + \theta_{\mathrm{QFD}}, \tag{2.124}$$

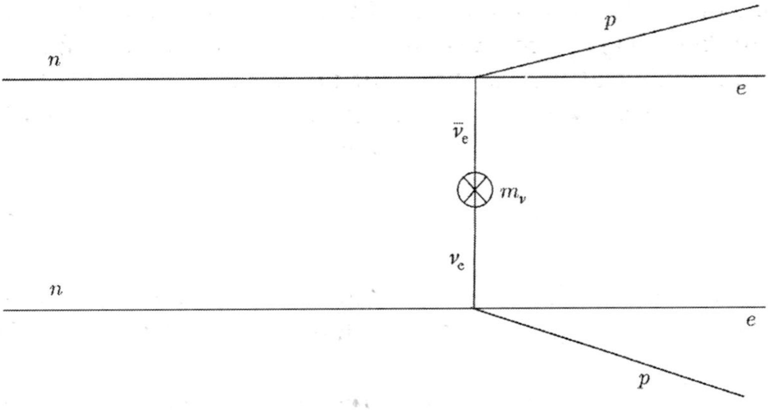

Fig. 2.7. Due to the Majorana mass, the antineutrino, emitted in the decay of one neutron, transforms into a neutrino which interacts with the second neutron.

where

$$\theta_{\text{QFD}} = \det(\hat{m}). \tag{2.125}$$

Here $\hat{m}$ denotes the total mass matrix which includes all of the coloured fermions which exist in theory. It takes into account both the known quarks and all possible hypothetical coloured fermions.

The different physical nature of the two terms in (2.124), each of which, in general, is not small, does not allow *apriori* mutual compensation corresponding to $\theta < 10^{-9}$.

A theoretical solution to these problems relates to the mechanism of natural suppression of values of θ.

In 1977, Peccei and Quinn have proposed an additional global chiral symmetry with a non-vanishing colour anomaly in order to resolve this problem.

If this symmetry remains unbroken, which corresponds to at least one massless quarks (e.g., u-quark; Wilczek 1978; Weinberg 1978), then all θ vacua are equivalent to the vacuum with

$$\theta = 0. \tag{2.126}$$

In other words, the θ-term may be expelled from the Lagrangian by the chiral phase transformations of the u-quark field.

The possible existence of the massless u-quark is theoretically not excluded (Kaplan, Manohar 1986), but it lacks the serious fundamental basis of the standard model and serious difficulties arise when comparing the results obtained on the basis of current algebra and the hypothesis of partially conserved axial current.

However, the problem of the θ-term finds its natural solution even in case of broken $U(1)_{PQ}$ symmetry. In these cases, the θ constant has the dynamic meaning of the amplitude of the pseudo-Goldstone field associated with breaking of $U(1)_{PQ}$ symmetry which is also explicitly broken by the colour anomaly. So in a vacuum the condition

$$\langle\theta\rangle_{vac} = 0 \tag{2.127}$$

is automatically satisfied thus leading to the exact mutual compensation of the terms in (2.124). Such a pseudo-Goldstone field is called the axion.

The idea of the axion solution to the problem of strong *CP*-violation can be explained as follows. The dynamic solution, found by the axion, implies that we must consider the Lagrangian (2.117) as the Lagrangian of the axion–gluon interaction. The non-vanishing colour anomaly here means that the constant of this interaction is non-zero. The zero vacuum expectation value for the axion field at the same time provides effective compensation of the θ-term in the QCD vacuum.

The most important parameter determining the properties of the axion field is the energy scale F_α at which the symmetry is broken. This scale appears in the Lagrangian of the axion–gluon interaction

$$L(\alpha gg) = \frac{g^2}{16\pi} \cdot \frac{\alpha}{F_\alpha} \cdot G_{\mu\nu} \tilde{G}_{\mu\nu}, \tag{2.128}$$

where α is the axion field.

The Lagrangian of the axion interactions with other gauge fields can be expressed in the same form with the corresponding replacement of the gauge coupling constants. Thus, the coupling constant C of the axion with the gauge bosons is defined in general by the scale F_α

$$C \propto F_\alpha^{-1}. \tag{2.129}$$

In particular, the interaction of the axion with the photon leads for small but non-zero mass of the axion to the axion decay

$$\alpha \to \gamma\gamma. \tag{2.130}$$

The lifetime of the axion with respect to this decay is expressed in complete analogy with the lifetime of the pion

$$\tau(\alpha \to \gamma\gamma) = \frac{64\pi}{m_\alpha^3 F_\alpha^2}. \tag{2.131}$$

Since the interaction of the neutral pion is similar to that of the axion, the axion mass can be expressed through the scale F_α and the corresponding parameters of the pion, so that the axion mass will be determined by this scale

$$m_\alpha = A_\alpha \frac{f_\pi}{F_\alpha} m_\pi, \tag{2.132}$$

where A_α is the constant which depends on the choice of the specific axion model.

The scale F_α also determines the constant h of the axion interaction with fermions

$$h \propto F_\alpha^{-1}. \tag{2.133}$$

The simplest variant of the axion model is the model of the Weinberg–Wilczek axion (Wilczek 1977; Weinberg 1978), in which the scale F_a coincides with the scale ϑ of electroweak symmetry breaking in the standard model. This model has been ruled out by a combination of experimental and astrophysical constraints (Vysotsky et al 1978, see review and references in Kim 1987; Cheng 1988; Raffelt 1996, 2000). Analysis of these constraints has led to a very high estimate of the lower limit for the axion scale

$$F_a \gg \vartheta. \tag{2.134}$$

Thus, the scale F_α should be linked with a new energy scale in elementary particle physics. On this scale, the axion interaction is elusive and, consequently, makes the axion invisible.

3.4. Models of the invisible axion

In all models of the invisible axion, this particle appears as a Goldstone boson associated with the phase of the complex $SU(2)\times U(1)$ singlet Higgs field. The coupling of the axion with the gauge bosons arise in these models after $U(1)_{PQ}$ symmetry breaking by the mechanism determined by the existence of the non-vanishing colour anomaly $U(1)_{PQ}$–$SU(3)_c$–$SU(3)_c$.

In the most general case, the Lagrangian of interaction of the axion with the fermions (quarks and leptons) and photons has the form

$$L = g_{\alpha\beta}\alpha\bar{f}_\alpha\left(\sin\theta_{\alpha\beta} + i\gamma_5\cos\theta_{\alpha\beta}\right)f_\beta + C_{\alpha\gamma\gamma}aF_{\mu\nu}\tilde{F}^{\mu\nu}, \tag{2.135}$$

where the indices

$$\alpha, \beta = 1,2,3$$

indicate the generation of fermions f and the constants $\theta_{\alpha\beta}$,

$$g_{\alpha\beta} \propto F_\alpha^{-1} \tag{2.136}$$

and

$$C_{\alpha\gamma\gamma} \propto F_\alpha^{-1} \tag{2.137}$$

depend on the model of the axion.

Because of the general theorem (Anselm, Uraltsev 1983), the diagonal couplings of the axion with the fermions in the Lagrangian (2.135) can only be pseudoscalar, which gives

$$\theta_{\alpha\alpha} = 0 \tag{2.138}$$

so that, neglecting higher order effects, there is no long-range interaction through axion exchange.

The off-diagonal couplings of the axion which, in general, are not excluded, can be both scalar and pseudoscalar or a combination thereof, corresponding to

$$\theta_{\alpha\beta} \neq 0, \tag{2.139}$$

depending on the model chosen.

There are three main types of invisible axion models.

Dine–Fishler–Sredintsky–Zhitnitski (DFSZ) **axion** (Dine et al 1985; Zhitnitski 1980). The model considers only the known types of fermions (quarks and leptons) and extends the SM only in the Higgs sector of the theory. The number of Higgs doublets φ increases as compared with the SM, and new singlet Higgs fields σ are also added.

In the DFSZ model, the Lagrangian of Yukawa interaction of Higgs bosons with fermions has the form

$$L_{Yuk} = g_{\alpha\beta}^{(f)} \overline{f}_{L\alpha} f_{R\beta} \varphi_f + \text{h.c.}, \tag{2.140}$$

where $f_{L(R)}$ are the left (right) components of the quark and lepton fields, $f = u, d, e,$ the indexes of the generation in the case of three generations of the quarks and leptons are equal to $\alpha, \beta = 1, 2, 3$, and φ_f are the Higgs doublets.

For a doublet

$$\varphi_u = \begin{pmatrix} \varphi_u^+ \\ \varphi_u^0 \end{pmatrix} \tag{2.141}$$

the vacuum expectation value of the neutral component is defined as

$$\left\langle \varphi_u^0 \right\rangle = \frac{\vartheta_u}{\sqrt{2}} \tag{2.142}$$

and for the doublets

$$\varphi_d = \varphi_e = \begin{pmatrix} \varphi_d^+ \\ \varphi_d^0 \end{pmatrix} \tag{2.143}$$

the corresponding vacuum expectation value is given by

$$\langle \varphi_d^0 \rangle = \vartheta_d \sqrt{2}. \tag{2.144}$$

It is believed that these vacuum expectation values are related by

$$\left(\vartheta_u^2 + \vartheta_d^2\right)^{1/2} = \vartheta = \left(\sqrt{2} G_F\right)^{-1/2}. \tag{2.145}$$

In the most general case, the doublets φ_e and φ_d may be different. In particular, we can choose

$$\varphi_e = \overline{\varphi}_u. \tag{2.146}$$

However, the choice of these doublets, described here, is determined by including the model into the framework of GUT models.

Lagrangian (2.140) is invariant in respect of global chiral transformations of the symmetry group

$$\begin{aligned} f_{L\alpha} &\to \exp(i\omega) f_{L\alpha} \\ f_{R\alpha} &\to \exp(-i\omega) f_{R\alpha} \\ \varphi_f &\to \exp(2i\omega)\varphi_f \\ \sigma &\to \exp(-2i\omega)\sigma \end{aligned} \tag{2.147}$$

which is ensured by the presence in the Higgs potential of a term given by the expression

$$L_H = \lambda \varphi_u \varphi_d \sigma^2 + \text{h.c.} \tag{2.148}$$

Under the assumption that

$$\vartheta \ll \langle \sigma \rangle = \frac{V}{\sqrt{2}} \tag{2.149}$$

the amplitude of the axion field is mainly determined by the phase of the field σ. Contribution of the phases of the fields φ_u and φ_d in this amplitude is suppressed as

$$\left\langle Arg\varphi_{u,d} \middle| \alpha \right\rangle \sim \frac{\vartheta_{u,d}}{V}. \tag{2.150}$$

The off-diagonal couplings of these axions with the fermions are absent, and other parameters in the equations (2.135)–(2.137) for the DFSZ axions are given by

$$A_a^{\mathrm{DFSZ}} = 2N_g \frac{\sqrt{z}}{(1+z)}, \tag{2.151}$$

where $N_g = 3$ is the number of quark–lepton generations.

$$F_a = V$$

$$g_f = \frac{m_f}{F_a}\left[X_f - \frac{(\delta_{fu}+\delta_{fd})N_g}{(1+z)}\right] = \frac{m_f m_a}{m_\pi f_\pi A^{\mathrm{DFSZ}}}\left[X_f - \frac{(\delta_{fu}+\delta_{fd})N_g}{(1+z)}\right]. \tag{2.152}$$

$$G_{a\gamma\gamma} = \frac{m_a \alpha_{em}}{8\pi f_\pi m_\pi}\frac{1+z}{\sqrt{z}}\left(\frac{8}{3}-\frac{2}{3}\frac{(4+z)}{(1+z)}\right) = \frac{\alpha_{em} N_g}{4\pi F_a}\left(\frac{8}{3}-\frac{2}{3}\frac{(4+z)}{(1+z)}\right)$$

Here the ratio of the masses of u- and d-quarks in accordance with the estimates of the current algebra is assumed to be

$$z = \frac{m_u}{m_d} \approx 0.56. \tag{2.153}$$

For all the up quarks (u-type quarks: u, c, t ...) parameter X_{up} is

$$X_{up} = \frac{2}{\left(x + \frac{1}{x}\right)x}, \tag{2.154}$$

and for all charged leptons and down quarks (d-type quarks: d, s, b, ...) the relevant parameters are

$$X_{lept} = X_{down} = \frac{2x}{\left(x + \frac{1}{x}\right)}, \tag{2.155}$$

where the ratio of vacuum expectation values is denoted as

$$x = \frac{\vartheta_u}{\vartheta_d}. \tag{2.156}$$

The fermionic index f of the parameters g_f and in the Kronecker symbols δ_{fu} and δ_{fd} has the values
$f = u, d, e, s, c, \mu, b, \tau, \ldots$

In equation (2.152) the Kronecker symbols δ_{fu} and δ_{fd} mean that the corresponding term takes into account only the u- and d-quarks, and α_{em} is the electromagnetic fine structure constant.

2. **Kim–Shifman–Vainshtein–Zakharov (KSVZ) hadronic axion model** (Kim 1980; Shifman et al 1980). Here, expansion of the SM takes place in both the Higgs and fermion sectors of the theory: additional Higgs fields and additional fermions are introduced. In addition to the usual quarks and leptons, the existence of new quark colour triplet Q is assumed.

The Lagrangian of the Yukawa interactions in this case has the form

$$L_{Yuk} = g\bar{Q}_L \sigma Q_R + \text{h.c.} \tag{2.157}$$

This Lagrangian is invariant in respect of global chiral transformations

$$\begin{aligned} Q_L &\to \exp(i\omega) Q_L \\ Q_R &\to \exp(-i\omega) Q_R \\ \sigma &\to \exp(-2i\omega)\sigma. \end{aligned} \tag{2.158}$$

Such an axion has no couplings with the leptons at the 'tree level', and its relationship with the light quarks only exists because of mixing with the neutral pion.

The parameters of the KSVZ axion in (2.135)–(2.137) are given by

$$A_a^{KSVZ} = \frac{\sqrt{z}}{(1+z)},$$
$$g_{u,d} = \frac{m_u}{F_a(1+z)} = \frac{m_u m_\alpha}{f_\pi m_\pi (1+z) A_a^{KSVZ}} \tag{2.159}$$

and

$$C_{\alpha\gamma\gamma} = \frac{m_\alpha \alpha_{em}}{8\pi f_\pi m_\pi} \frac{1+z}{\sqrt{z}} \left(6Q_{em}^2 - \frac{2}{3}\frac{(4+z)}{(1+z)} \right) = \frac{\alpha_{em}}{8\pi F_a} \left(6Q_{em}^2 - \frac{2}{3}\frac{(4+z)}{(1+z)} \right), \tag{2.160}$$

where Q_{em} is the electric charge of the heavy quark. All Yukawa couplings of the KSVZ axion for

$$f = e, s, c, \mu, \tau$$

are negligible.

3. **The archion model** (Berezhiani, Khlopov, 1990a, b, c; 1991; Berezhiani et al., 1992; Berezhiani, 1983; 1985) was developed within the framework of a model of horizontal unification (MHU) which will be discussed in Chapter 11. The theory naturally includes global $U(1)_H$ symmetry. Spontaneous breaking of symmetry leads to the prediction of the Goldstone boson of the invisible axion type.

The boson, which Berezhiani and Khlopov (Berazhiani, Khlopov, 1990a, b, c) call the archion, has both diagonal and off-diagonal couplings with the fermions as regards the flavour.

The global $U(1)_H$ symmetry in horizontal unification can be identified with the Peccei–Quinn symmetry owing to the existence of the triangle anomaly for the interaction of the axial currents $U(1)_H$ with the gluons.

In the simplest form of horizontal unification the gauge symmetry $SU(3)_c \otimes SU(2) \otimes U(1) \otimes SU(3)_H \otimes U(1)_H$ with a minimal set of heavy fermions the anomalies are compensated, and the archion remains almost massless like the massless Goldstone boson, called the arion (Anselm, Uraltsev, 1982). However, in contrast to the arion which can interact with the photon, the interaction of the archion with photons does not take place because of the parallel compensation of the corresponding triangular anomaly current–photon–photon.

In any realistic extension of the model of horizontal unification to the GUT symmetry, for example, when extending to the $SU(5) \otimes SU(3)_H$ symmetry, there is no compensation because of the availability of additional heavy fermions, so that the archion is similar to a hadronic axion with the strongly suppressed interaction with the leptons.

The symmetry of horizontal unification leads to the following Yukawa couplings

$$L_{Yuk} = g_f \overline{f}_{L\alpha} F_{R\alpha} \varphi^0 + g_{nf} \overline{F}_{R\alpha} F_L^\beta \xi_{\alpha\beta}^{(n)} + \mu_f \overline{F}_L^\alpha f_R^\alpha + \text{h.c.}, \qquad (2.161)$$

where μ_f are the mass parameters of the same order of magnitude for

$$f = u, d, e.$$

In the Lagrangian (2.161)

$$F = U, D, E \qquad (2.162)$$

designate the additional heavy fermions and $\xi_{\alpha\beta}^{(n)}$ are horizontal Higgs bosons that cause breaking of the symmetry of generations.

The Lagrangian (2.161) is invariant in respect of global chiral transformations of $U(1)_H$ symmetry

$$\begin{aligned} &f_L \to \exp\{i\omega\} f_L, f_R \to \exp\{-i\omega\} f_R \\ &F_L \to \exp\{i\omega\} F_L, F_R \to \exp\{-i\omega\} F_R, \\ &\xi^{(n)} \to \exp\{2i\omega\} \xi^{(n)}, \varphi \to \varphi, \end{aligned} \qquad (2.163)$$

where

$$n = 0, 1, 2.$$

In the archion model, the parameters defined in equations (2.135)–(2.137) are given by

$$\begin{aligned} &A_a = A_c \frac{\sqrt{z}}{(1+z)}, \\ &g_{33} = \frac{m_3^{(f)}}{V_1}, \ g_{22} = \frac{m_2^{(f)}}{V_1}\left(S_{23}^{(f)}\right)^2, \ g_{11} = \frac{m_2^{(f)}}{V_1}\left(S_{13}^{(f)}\right)^2 \\ &g_{23} = \frac{m_3^{(f)}}{2V_1} S_{23}^{(f)}, g_{13} = \frac{m_3^{(f)}}{2V_1} S_{13}^{(f)}, g_{12} = \frac{m_2^{(f)}}{2V_1} S_{13}^{(f)} S_{23}^{(f)} \end{aligned} \qquad (2.164)$$

$$C_{a\gamma\gamma} = \frac{m_a \alpha_{em}}{4\pi f_\pi m_\pi} \frac{z}{1+z}\left(\frac{A_{em}}{A_c} - \frac{2}{3}\frac{(4+z)}{(1+z)}\right)^{-1} = \frac{\alpha_{em} A_c z^{2/3}}{4\pi F_a (1+z)^2}\left(\frac{A_{em}}{A_c} - \frac{2}{3}\frac{(4+z)}{(1+z)}\right)^{-1}$$

where $S_{ij} = \sin\varphi_{ij}$ with $i, j = 1,2,3$ are the angular parameters of unitary rotation matrices V_f, diagonalizing the mass matrices of the fermions

$$V_{fL}^{+} m^{f} V_{fR} = \hat{m}_{diag}^{f},$$

and the phases are determined by the phases of the matrix elements V so that

$$\varphi_{12} = \arg\left(\frac{V_{23}V_{13}}{V_{12}}\right), \varphi_{13} = -\arg V_{13}, \varphi_{23} = -\arg\left(V_{12}V_{23}\right). \qquad (2.165)$$

The scale

$$V_1 = F_a$$

is the scale of $U(1)_H$ symmetry breaking, and A_c and A_{em} are the colour and electromagnetic anomalies, respectively.

In all three models of the invisible axion the Lagrangian of its interaction with nucleons has the form

$$L_{NN'} = i\bar{N}'\gamma_5\left(g^{(0)} + g^{(3)}\tau_3\right)N\alpha, \qquad (2.166)$$

where for the DFSZ axion

$$g^{(0)} = \frac{M}{2F_a}G_A^{(0)}, g^{(3)} = \frac{M}{2F_a}\cdot\frac{1-z}{1+z}G_A,$$

and for the hadronic axion and the archion

$$g^{(0)} = \frac{M}{2N_g F_a}\left(X_u + X_d - N_g\right)G_A^{(0)},$$

$$g^{(3)} = \frac{M}{2N_g F_a}\cdot\left(X_u - X_d - \frac{1-z}{1+z}\right)G_A.$$

Here M is the nucleon mass and

$$G_A = G_A^{(0)} = 1.23$$

is the isotriplet axial nucleon form factor determined by experimental measurements of the parameters of weak nucleon currents.

3

Hidden parameters of modern cosmology

1. Old scenario of Big Bang Universe

1.1. Expanding Universe

Cosmology is studying the Universe as a whole. Stars and interstellar objects, clusters of galaxies and superclusters surrounding the giant void in which the galaxies are not observed is a striking manifestation of non-homogeneity in the distribution of luminous matter in the Universe. However, on scales larger than the size of the largest cosmological structures, the distribution of matter is nearly homogeneous and isotropic.

Thus, the simplest and most reasonable assumption is the homogeneous and isotropic cosmological model. This model takes into account the presence of matter in the Universe by only one parameter – the cosmological density ρ, which is the mass density averaged over the scales exceeding the size of the observed structures.

In the early 20s of XX century, the successful application of general relativity (GR) to the Universe as a whole allowed us to prove the impossibility of stationary solutions of Einstein equations.

The solution for the stationary Universe can be obtained by simply adding to the cosmological density the cosmological term which has no physical basis in the properties of ordinary matter.

Friedman showed that the Universe without the cosmological constant must be non-stationary. Non-stationarity of the Universe means that it may contract or expand as a whole.

In the late 20s of XX century, astronomical observations showed that astronomical objects, interpreted previously as a faint nebula in our galaxy (Milky Way), are in fact distant galaxies like the Milky Way and contain billions of stars.

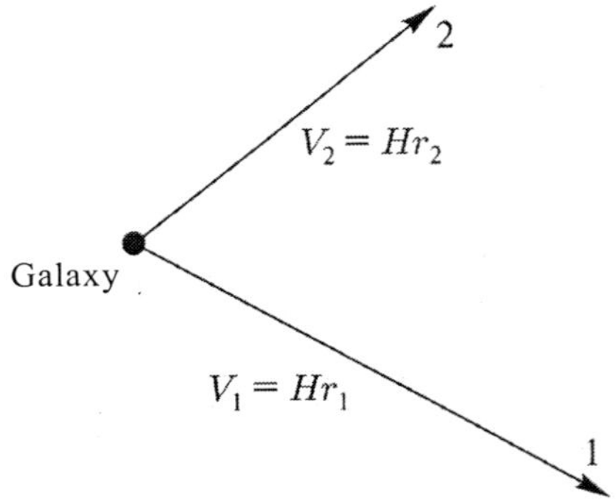

Fig. 3.1. 'Recession' of galaxies.

Discovery of other galaxies and their luminosity analysis led to the discovery of a systematic shift in their spectra to low frequencies (*red shift*). The observation of red shift in spectra of distant galaxies was interpreted as the Doppler effect associated with the 'recession' of galaxies.

The rate of 'recession' of galaxies is proportional to the distance according to Hubble's law

$$\vec{V} = H \cdot \vec{r}, \tag{3.1}$$

where the proportionality factor has the dimension of inverse time and is called the *Hubble constant.*

Although we observe the Universe in which galaxies are moving away from us (see Fig. 3.1), it is easy to show that this does not mean a dedicated position of our Galaxy in an expanding world.

Simple considerations of vector algebra (see Fig. 3.2) shows that for an observer from another galaxy, our Galaxy, like the rest of the galaxies, from his point of view, moved away from his galaxy at a velocity proportional to the distance according to Hubble's law.

Thus, the effect of 'recession' of galaxies is considered as an argument in favour of homogeneous and isotropic expanding models. The Hubble constant is the second important parameter of this model.

The Hubble constant is constant in space but not in time. Having the dimension of inverse time, it determines the time scale of the cosmological expansion.

The quantity inversely proportional to the Hubble constant, $1/H$, determines the time elapsed from the start of the expansion of the Universe t_U, called the age of the Universe. For example, for the stage of matter dominance, this age is

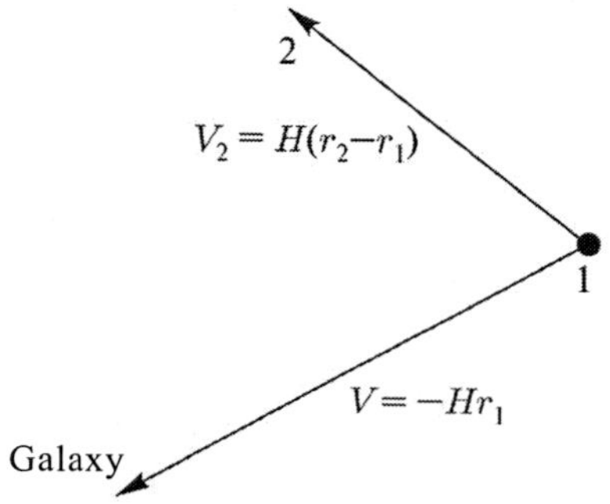

Fig. 3.2. From a remote galaxy we can observe also the same pattern of 'recession' of galaxies as from our Galaxy.

$$t_U = \frac{2}{3} \cdot \frac{1}{H}. \tag{3.2}$$

Description of the model of the expanding Universe, based on the general relativity theory, coincides exactly in some case with the results of classical treatment of the isotropic and homogeneous sphere of radius R and density ρ, expanding in accordance with the Hubble law.

The test body with mass m in this case has the total energy consisting of the kinetic energy of expansion and the potential energy of gravitation,

$$E_{tot} = E_{kin} + E_{pot} = \frac{m \cdot v^2}{2} + \left(-\frac{G \cdot m \cdot M}{R} \right). \tag{3.3}$$

Given the fact that $v/R = H$ we find that the total energy is

$$E_{tot} = \frac{4\pi GmR^3}{3}\left(\frac{3v^2}{8\pi GR^2} - \rho\right) = \frac{4\pi GmR^3}{3}\left(\frac{3H^2}{8\pi G} - \rho\right) \tag{3.4}$$

so that the sign of the total energy, which determines the future of expansion, is given by the relation between the density ρ and combination

$$\rho_c = \frac{8 \cdot H^2}{3 \cdot \pi \cdot G}, \tag{3.5}$$

which has the dimension of density and is called the critical density.

If the cosmological density exceeds the critical density,

$$\rho > \rho_c \tag{3.6}$$

the total energy is negative, and the expansion must inevitably give way to a contraction.

For the cosmological density lower the the critical density

$$\rho < \rho_c \tag{3.7}$$

expansion will never stop.

The first case corresponds to the model of 'closed world', the latter to the model of 'open world'. The Universe in which the cosmological density is exactly equal to the critical value is called a flat Universe.

According to general relativity (GR) theory, closed, flat and open models correspond to models with positive

$$k = +1, \tag{3.8}$$

zero

$$k = 0 \tag{3.9}$$

and negative curvature of space-time

$$k = -1. \tag{3.10}$$

In a non-stationary Universe, all changes in the distances are proportional to the scale factor $a(t)$.

In a closed Universe, the scale factor has the meaning of the radius of our world.

In flat and open cosmological models, the scale factor just determines the dependence of the scales on time.

A comoving volume is proportional to $a^3(t)$, so that the density of the substance will be inversely proportional to the companion volume, and consequently, will decrease with the expansion as

$$\rho_b \propto a(t)^{-3}. \tag{3.11}$$

Time evolution of the scale factor follows from Einstein's equations which in the case of dominance of matter in a flat Universe are given by

$$\left(\frac{\dot{a}}{a}\right)^2 = \frac{8\pi G}{3} \cdot \rho \tag{3.12}$$

$$\frac{\ddot{a}}{a} = -\frac{4\pi G}{3} \cdot \rho. \tag{3.13}$$

Solving these equations gives the time dependence of the scale factor at the stage of the dominance of matter in the Universe

$$a(t) \propto t^{2/3}. \tag{3.14}$$

It is easy to find the correspondence between the exact form of these equations following from general relativity, and the discussed simple picture of the homogeneous and isotropically expanding sphere, considered on the basis of Newtonian physics.

Given the fact that

$$\frac{\dot{a}}{a} = H, \tag{3.15}$$

we can reduce the first equation to determination of the critical density, putting

$$\rho = \rho_{cr}. \tag{3.16}$$

The second equation comes from Newton's equation for radial motion of the test body

$$m\ddot{\vec{R}} = -\frac{GmM}{R^3}\vec{R}. \tag{3.17}$$

Astronomical measurements of both cosmological density and the Hubble constant contain uncertainties which do not allow experiments to determine the choice of the cosmological model.

In fact, the cosmological density in the Einstein equations corresponds, in general, to any form of matter that exists in the modern Universe.

Its estimates for the luminosity of galaxies provide only a lower limit on the total density which corresponds to the contribution of visible matter consisting of electrons and atomic nuclei only. The mass of atomic nuclei, i.e., baryons dominate this contribution. Therefore, the estimates of luminous matter give a lower limit on the baryonic density of the Universe. It was as a lower limit, since non-luminous baryonic matter could also exist, but is not included in these estimates.

Dynamic effects offer a more versatile way to estimate the cosmological density, regardless of its physical nature. Estimates based on the effect of gravitational lensing can also be attributed to these generic methods.

These estimates provide a density an order of magnitude higher than the density of luminous matter. They take into account the existence of dark matter in the Universe. The physical nature of dark matter is one of the most important problems in cosmoparticle physics.

Critical density is determined by the Hubble constant. Complexities involved in experimental determination of H are associated with the correct account of systematic errors in measuring distances to galaxies and their velocities.

Nearby galaxies, the distances to which can be estimated more accurately, do not give a correct idea of the Hubble expansion rate as the effects of their movement in the local supercluster of galaxies are dominant in the measurement of velocities.

The speeds of 'recession' of distant galaxies provide more reliable information about the Hubble expansion rate, but it is necessary to use complex multi-step procedures for measuring the distances to them.

The change of an almost an order of magnitude in the measured value of the Hubble constant (which took place from early 30s of XX century) is a vivid illustration of the above-mentioned difficulties.

Moreover, even a decade ago the values, measured by various groups, differed by a factor of 2.

The measurements were performed with rather small statistical errors, so the differences in these values were due to systematic errors

$$H = (50 \div 100)\frac{\text{km}}{\text{s} \cdot \text{Mpc}}. \tag{3.18}$$

In further discussions we take for definiteness

$$H = 70\frac{\text{km}}{\text{s} \cdot \text{Mpc}}. \tag{3.19}$$

The critical density for the chosen value of H is

$$\rho_{cr} = 1 \cdot 10^{-29}\frac{\text{g}}{\text{cm}^3}. \tag{3.20}$$

Dimensionless cosmological density Ω, which is widely used in cosmology is defined as

$$\Omega = \frac{\rho}{\rho_{cr}}. \tag{3.21}$$

Given the existence of various forms of matter in the Universe, we introduce the dimensionless density for some of its forms. Thus, the dimensionless density of luminous matter is of the order of magnitude

$$\Omega_{lum} = \frac{\rho_{lum}}{\rho_{cr}} \approx 0.01, \tag{3.22}$$

and this density is significantly less than the total relative density of baryons

$$\Omega_b = \frac{\rho_b}{\rho_{cr}}. \tag{3.23}$$

Comparison of the observed abundance of light elements predicted by the theory of Big Bang nucleosynthesis (see Chapter 5) is widely used to estimate the total baryon density, which can be found in both luminous and in non-luminous objects. These estimates give

$$\Omega_b \approx 0.05. \tag{3.24}$$

1.2. Thermal electromagnetic background – trace of hot Universe

Turning to the past of the expanding Universe, we come to the inescapable conclusion that the conditions in the Universe were to change over time. The modern picture of the Universe, full of galaxies and stars, was supposed to form as a result of the expansion of dense almost homogeneous and isotropic plasma. Such a plasma could be in a state of cold degenerate matter, as it is in white dwarfs and neutron stars.

An alternative idea to G. Gamow that the primary plasma was hot found no observational confirmation for more than 30 years because this concept predicted the existence in the modern Universe of a thermal electromagnetic background, which, according to this model, should have been preserved from the early hot stages of cosmological evolution.

Discovery of the thermal electromagnetic background in 1965 (Penzias & Wilson, 1965) confirmed the correctness of this approach.

The existence of blackbody background radiation with a Planck power spectrum, $F_{em}(\nu)$,

$$F_{em}(\nu) = \frac{2 \cdot h \cdot \nu^3}{c^2} \cdot \frac{1}{\exp\left(\frac{h\nu}{kT}\right) - 1} \tag{3.25}$$

with temperature

$$T = 2.7K \tag{3.26}$$

does not lead to any kind of significant contribution to the modern cosmological density.

The density of this background radiation, which is defined as

$$\rho_\gamma = \frac{\varepsilon_\gamma}{c^2}, \tag{3.27}$$

where ε_γ is the energy density of blackbody background radiation, which is equal to

$$\varepsilon_\gamma = \sigma \cdot T^4 \tag{3.28}$$

with the Stefan–Boltzmann constant

$$\sigma = 7.57 \cdot 10^{-15} \frac{\text{erg}}{\text{cm}^3 \cdot \text{K}^4}, \tag{3.29}$$

in the modern Universe is three orders of magnitude smaller than the density of luminous matter and thus plays no role in the dynamics of modern cosmological expansion.

The observed isotropy of the thermal background (the measured deviatons from isotropy are approximately 10^{-5}) provides the strongest confirmation of the homogeneity and isotropy of the cosmological model.

The quantity

$$s = \frac{\sigma \cdot T^3}{k\rho}, \tag{3.30}$$

where σ is the Stefan–Boltzmann constant, and k is the Boltzmann constant, determines the specific entropy of the Universe, while the dimensionless ratio of the photon and baryon number densities defines the 'entropy per baryon'

$$S = \frac{n_\gamma}{n_b} = 7 \cdot 10^7 \Omega_b^{-1}. \tag{3.31}$$

Here the number density of photons at temperature T is equal to

$$n_\gamma \approx 20 \left(\frac{T}{1\,\text{K}} \right)^3 \text{cm}^{-3}, \tag{3.32}$$

which corresponds to

$$n_\gamma = 400 \text{ cm}^{-3} \tag{3.33}$$

at

$$T = 2.7 \text{ K}$$

and the number density of baryons is expressed through the relative baryon density

$$n_b = \frac{\rho_b}{m_N} = \frac{\rho_{cr}}{m_N} \Omega_b \text{cm}^{-3} = 6 \cdot 10^{-6} \Omega_b \text{cm}^{-3}, \tag{3.34}$$

where the nucleon mass $m_N = 1.6 \cdot 10^{-24}$ g and the critical density is $\rho_{cf} = 1 \cdot 10^{-29}$ g/cm^3 for the value adopted for the Hubble constant H = 70 km/(s · Mpc)

The quantity reciprocal with respect to S is the ratio of the number densities of the baryons and photons

$$r_b = S^{-1} = \frac{n_b}{n_\gamma} = 1.4 \cdot 10^{-8} \Omega_b. \tag{3.35}$$

It is regarded as the third most important cosmological parameter that determines the baryon asymmetry in the Universe.

The high specific entropy of the modern Universe reflects the existence of a hot period in the early stages of cosmological evolution.

Moreover, in the early Universe the role of the thermal background dominated in the dynamics of expansion.

The reason for this lies in both the high ratio of the number of photons and baryons, and in the different dependence on the scale factor of the densities of matter and radiation.

Matter density ρ_m is inversely proportional to the cube of the scale factor, whereas in the case of radiation we must also take into account the red shift in the mean photon energy.

The frequency of the photon emitted in the early Universe undergoes a red shift z which is defined as

$$z = \frac{\nu_{em}}{\nu_{abs}} - 1, \tag{3.36}$$

where ν_{em} is the frequency of the emitted photon, and ν_{abs} is the frequency of the absorbed photon.

For the same reasons, the blackbody spectrum undergoes a red shift so that the temperature $T(t)$ of the radiation at earlier time t is related to its current value T_0 as follows,

$$T(t) = (1 + z(t)) \cdot T_0 = \frac{a(t = t_U)}{a(t)} \cdot T_0, \tag{3.37}$$

where the redshift $z(t)$ is determined by the ratio of the scale factors given in our time, when

$$t = t_U,$$

and in the period t.

Given the decrease in temperature during the expansion, we obtain

$$\rho_\gamma \propto \varepsilon_\gamma(t) \propto T(t)^4 \propto a(t)^{-4}. \tag{3.38}$$

Turning back in time (to large red shifts and, consequently, small-scale factors), we inevitably come to the period of radiation dominance when the contribution of the radiation energy density to the total energy density was dominant and, therefore, played a decisive role in the dynamics of cosmological expansion.

In the stage of radiation dominance the density of matter is negligible compared with the density of radiation, i.e.

$$\rho_\gamma \gg \rho_m. \tag{3.39}$$

Thermodynamics of radiation determines the radiation pressure

$$p = \frac{1}{3}\varepsilon_\gamma = \frac{1}{3}\rho_\gamma c^2. \tag{3.40}$$

Thus, the equation of state of the Universe at the stage of dominance of radiation has the form

$$p = \frac{1}{3}\varepsilon \tag{3.41}$$

and the pressure should be taken into account in the Einstein equations.

At the stage of radiation dominance these equations have the form

$$\left(\frac{\dot{a}}{a}\right)^2 = \frac{8\pi G}{3}\rho, \tag{3.42}$$

$$\frac{\ddot{a}}{a} = -\frac{4\pi G}{3}\cdot\left(\rho + 3\cdot\frac{p}{c^2}\right). \tag{3.43}$$

In view of the relation between pressure and density, given by an equation of state, these equations are used to determine the dependence of the scale factor on time at the radiation-dominated stage (RD stage)

$$a(t) \propto t^{1/2}. \tag{3.44}$$

Note that pressure is not taken into account in Einstein's equations for the phase of matter dominance (MD-stage), which formally corresponds to the cosmological equation of state

$$p = 0. \tag{3.45}$$

Of course, this does not mean that matter has no pressure. The dust equation of state of the Universe (3.45) means that the non-zero pressure of non-relativistic matter has no cosmological significance, since in the expansion equation the pressure is negligible compared with the energy density of matter.

Indeed, consider an ideal non-relativistic gas of particles with mass m and the number density of these particles n at a temperature

$$T \ll mc^2. \tag{3.46}$$

Its equation of state is

$$p = \frac{3}{2} nT. \tag{3.47}$$

Given that the number density of the particles is related to their mass density by the ratio

$$n = \frac{\rho}{m} \tag{3.48}$$

and that temperature T is associated with thermal velocity

$$v_T \ll c \tag{3.49}$$

as

$$T = \frac{mv_T^2}{2} \tag{3.50}$$

we have

$$p = \frac{3}{2} nT = \frac{3}{2} \frac{\rho}{m} \frac{mv_T^2}{2} = \frac{3}{4} \rho v_T^2 \ll \rho c^2 = \varepsilon. \tag{3.51}$$

1.3. Physical cosmology of the old Big Bang scenario

The simplest physically based approach to the history of the Big Bang Universe was associated with a physically self-consistent picture of the cosmological evolution of radiation and baryonic matter.

In such a picture in the modern Universe we consider only the known types of particles, and their early history is determined by the laws of thermodynamics, operating in a hot expanding Universe.

Based on the known laws of atomic and nuclear physics, it is easy to show that the baryonic matter in the early Universe was in a state of the plasma which was in equilibrium with radiation. Therefore, in the early Universe, the temperature of the thermal electromagnetic background was equal to the temperature of all particles that were in equilibrium.

Expanding Universe can be regarded as a closed system with the thermodynamic parameters that vary during the expansion.

For a two-component mixture of matter and radiation we can easily verify that the expansion will be adiabatic. The expansion rate

$$\Gamma \sim t^{-1}, \tag{3.52}$$

where t is the cosmological time, was in the early Universe much slower than the rate of processes that establish equilibrium between matter and radiation.

This means that the following conditions applies to the radiation-dominated (RD) stage

$$n(\sigma v)t \gg 1, \tag{3.53}$$

where n is the number density of electrons and nuclei, (σv) is the rate of their interaction with radiation. Therefore, thermal equilibrium is established in the RD stage.

These thermodynamic conditions are met until the very early stages of expansion, when the formally cosmological solution of general relativity leads to a singularity.

However, there are physical reasons that limit the applicability of the classical relativistic cosmological solution for time intervals less than

$$t \sim t_{Pl}. \tag{3.54}$$

Planck time

$$t_{\mathrm{Pl}} = \left(\frac{G\hbar}{c^5}\right)^{\frac{1}{2}} = 5 \cdot 10^{-44}\,\mathrm{s}, \tag{3.55}$$

being a dimensional combination of the speed of light, Planck's constant and the gravitational constant, determines the time scale on which the description of space-time requires the inclusion of quantum effects.

At the Planck time the size of the cosmological horizon, i.e. the area in which the light signal can establish a causal link, is equal to the Planck length

$$l_{\rm Pl} = \left(\frac{G\hbar}{c^3}\right)^{\frac{1}{2}} = 1.5 \cdot 10^{-33}\ \text{cm}. \quad (3.56)$$

The Planck time and length are a natural constraint on the interpolation of the classical picture of expansion, given by general relativity.

Thus, the classical description of the history of the expanding Universe is only valid for the period

$$t > t_{\rm Pl} \quad (3.57)$$

and at distances

$$l > l_{\rm Pl}. \quad (3.58)$$

In the hot expanding Universe, Planck time (3.55) corresponds to Planck temperature equal to

$$T_{\rm Pl} = m_{\rm Pl}c^2 \sim 10^{19}\ \text{GeV}. \quad (3.59)$$

Here $m_{\rm Pl}$ is the Planck mass which is the third important dimensional combination of the fundamental physical constants

$$m_{\rm Pl} = \left(\frac{\hbar c}{G}\right)^{\frac{1}{2}} = 2 \cdot 10^{-5}\ \text{g}. \quad (3.60)$$

In the old Bing Bang cosmology the Planck scale meant a physically reasonable boundary of the 'smooth' interpolation of the observed expansion to its beginning.

The period

$$t_{\rm Pl} < t < 1\ \text{s} \quad (3.61)$$

refers to the period of the very early Universe.

Physical processes in this period were determined by high-energy physics. In the absence of information about the properties of particles at superhigh energies, the old scenario simply postulated the dominance of relativistic particles in the very early Universe and the general outline of cosmological evolution given by the dependence of density on time

$$\rho = \frac{3}{32\pi G t^3} = 4.5 \cdot 10^{-5} \left(\frac{1\ \text{s}}{t}\right)^2 \frac{\text{g}}{\text{cm}^3}. \tag{3.62}$$

In units

$$\hbar = c = 1 \tag{3.63}$$

the time dependence of cosmological density (3.62) has the form

$$\rho = \frac{3}{32\pi} \cdot \frac{m_{\text{Pl}}^2}{t^2}. \tag{3.64}$$

The simplest assumption is that at very high temperatures

$$T \gg m \tag{3.65}$$

all particles are relativistic and in equilibrium with the plasma and radiation.

Thermodynamics of a relativistic gas leads to the following relations between the energy density ε in the Universe and its temperature T

$$\varepsilon = \rho \cdot c^2 = \kappa \cdot \sigma \cdot T^4, \tag{3.66}$$

where the Stefan–Boltzmann constant is equal to

$$\sigma = \frac{\pi^2 \cdot k^4}{15 \cdot \hbar^3 \cdot c^3} = 7.57 \cdot 10^{-15} \frac{\text{erg}}{\text{cm}^3 \cdot \text{K}^4}, \tag{3.67}$$

The Boltzmann constant k is equal to

$$k = 1.38 \cdot 10^{-16} \frac{\text{erg}}{\text{K}}$$

and κ is the effective number of types of relativistic particles with regard to their statistical properties.

From the relations (3.62) (3.66) and (3.67) we obtain the equality

$$(kT)^4 = \frac{1}{\kappa} \cdot \frac{45}{32 \cdot \pi^3} \cdot \frac{\hbar^3 c^5}{G t^2}, \tag{3.68}$$

which leads to the dependence of temperature on time

$$T = 1.5 \cdot 10^{10}\,\mathrm{K} \cdot \kappa^{-\frac{1}{4}} \left(\frac{1\,\mathrm{s}}{t} \right)^{\frac{1}{2}}. \tag{3.69}$$

For the temperature, expressed in MeV, we obtain

$$T = 1.3\ \mathrm{MeV} \cdot \kappa^{-\frac{1}{4}} \left(\frac{1\,\mathrm{s}}{t} \right)^{\frac{1}{2}} \tag{3.70}$$

and in units $\hbar = c = k = 1$ this relation has the form

$$T = \frac{1}{\kappa^{\frac{1}{4}}} \cdot \left(\frac{45}{32 \cdot \pi^3} \right)^{\frac{1}{4}} \cdot \left(\frac{m_{\mathrm{Pl}}}{t} \right)^{\frac{1}{2}}. \tag{3.71}$$

The same ratio can be used to express the cosmological time through the temperature

$$t = 2.25 \cdot 10^{20}\,\mathrm{s} \cdot \left(\frac{1\mathrm{K}}{T} \right)^2 \cdot \kappa^{-\frac{1}{2}} \tag{3.72}$$

or the temperature in MeV,

$$t = 1.7\,\mathrm{s} \left(\frac{1\ \mathrm{MeV}}{T} \right)^2 \kappa^{-\frac{1}{2}}. \tag{3.73}$$

In units $\hbar = c = k = 1$ this expression has the form

$$t = \kappa^{-\frac{1}{2}} \cdot \left(\frac{45}{32 \cdot \pi^3} \right)^{\frac{1}{4}} \cdot \frac{m_{\mathrm{Pl}}}{T^2}. \tag{3.74}$$

The total number density of all types of relativistic particles, assuming the average energy of particles

$$\langle E \rangle \sim 3kT, \tag{3.75}$$

is

$$n_r = \frac{\varepsilon}{3kT} = 0.01 \cdot \kappa^{\frac{1}{4}} \cdot \left(\frac{c}{\hbar G} \right)^{\frac{3}{4}} t^{-\frac{3}{2}} = 5 \cdot 10^{31}\,\mathrm{cm}^3 \kappa^{\frac{1}{4}} \left(\frac{1\,\mathrm{s}}{t} \right)^{\frac{3}{2}}. \tag{3.76}$$

In any system, thermodynamic equilibrium is established if the rate of processes maintaining the equilibrium exceeds the rate of change of system parameters such as density, temperature and so on.

In the expanding Universe, the rate of this change coincides with the rate of expansion.

If the characteristic time of the physical process is longer than the cosmological time, the system goes out of equilibrium with respect to this process.

The question of exit from equilibrium is one of the most important issues of physical cosmology, and we will discuss it in more detail.

If the system contains several different types of particles, and there are transitions between particles of different types, then under the condition of equilibrium an equilibrium distribution is established which then does not further change with time.

If the system parameters change more rapidly than the fastest process that leads to creation or annihilation of particles of this type, the particles of this type come out of equilibrium with other particles.

Let us consider the special case of the creation and annihilation of particle-antiparticle pairs.

For a particle with a charge there is an antiparticle. The particle and the antiparticle can annihilate. The pairs of the particles and antiparticles can be created in the collision of other particles. Unpaired creation or annihilation of the particles is also possible.

In particular, if the fastest process among these processes is the annihilation of the particle–antiparticle pairs, the particles exit from equilibrium when the characteristic time for annihilation τ is longer than the cosmological time t, i.e. when

$$\tau > t. \tag{3.77}$$

For particles and antiparticles with the concentration

$$n_p = n_a = n \tag{3.78}$$

and the relative velocity v the process of their annihilation with the cross section σ has a characteristic time

$$\tau = (n\sigma v)^{-1}. \tag{3.79}$$

If $\tau > t$, further changes in the concentration of particles are determined only by the expansion of the Universe. In this case, the particles become frozen out.

Decoupling of weakly interacting particles is a special case of freezing out of the particles. Here τ is the characteristic time of interaction with other particles, so that n is the concentration of other particles, and (σv) is the rate of their interaction with the discussed weakly interacting particles.

In the process of decoupling at

$$t \sim \tau \tag{3.80}$$

the equilibrium distribution of the weakly interacting particles does not change, and in subsequent adiabatic expansion the gas of the decoupled particles preserves the equilibrium distribution as long as the particles do not appear to be non-relativistic at

$$T \sim m, \tag{3.81}$$

where m is the mass of the particles.

In the old Big Bang scenario the physical motivation for a simple thermodynamic equilibrium pattern was based on the fact that among the known particles only nucleons, electrons, photons and neutrinos are sufficiently long-lived to have cosmological significance. Hundreds of other types of elementary particles, given in the tables of particle properties, are unstable.

In elementary particle physics, the unstable particles with mass m (e.g., resonances) have lifetimes

$$\tau \sim \left(\frac{\hbar}{mc^2} \right) \tag{3.82}$$

and even the so-called 'metastable' particles with a very long lifetime on the scale of particle physics

$$\tau \gg \left(\frac{\hbar}{mc^2} \right) \tag{3.83}$$

are very short-lived on the cosmological scale which for particles with mass m corresponds to the cosmological time $T \sim m$, and is of the order of magnitude

$$t_c \sim \left(\frac{m_{Pl}}{m} \right) \cdot \left(\frac{\hbar}{mc^2} \right). \tag{3.84}$$

Therefore, it seemed reasonable to consider in cosmology only the known stable particles which are virtually confined to a set of photons, electrons and positrons, nucleons, and antinucleons, neutrinos and antineutrinos.

The cosmological effect of all other known particles is reduced to a small quantitative change in the relativistic equation of state of the very early Universe. Only an exponentially large set of such particles could significantly affect the cosmological evolution, for example, leading to the existence of the Hagedorn maximum temperature. However, the development of the quark model and quantum chromodynamics has not left any physical basis for these features.

By the end of the first second of the expansion the entire result of the evolution of the very early Universe was reduced in the old Big Bang scenario to the conditions of equilibrium of the gas of photons, electron–positron pairs, neutrino–antineutrino pairs, and a small but crucially important admixture of nucleons with the concentration of the order

$$n_b \sim 10^{-9} n_\gamma. \tag{3.85}$$

The period of the radiation-dominated stage

$$t > 1 \text{ s} \tag{3.86}$$

corresponds to the temperature

$$T < 1 \text{ MeV}, \tag{3.87}$$

at which the nuclear and atomic processes, well-studied in the laboratory, took place.

The obvious difference between laboratory and cosmological terms for these processes is the very low density of matter in the Big Bang Universe.

From equations (3.32) (3.35) and (3.76) we find that the number density of baryons, equal to

$$n_b = 7.6 \cdot 10^{23} \text{ cm}^{-3} \Omega_b \left(\frac{1 \text{ s}}{t} \right)^{\frac{3}{2}}, \tag{3.88}$$

practically corresponds to the conditions of laboratory vacuum

$$n_b < 7.6 \cdot 10^{16} \text{ cm}^{-3} \tag{3.89}$$

at

$$t > 10^4 \text{ s} \tag{3.90}$$

and

$$\Omega_b \sim 0.1. \tag{3.91}$$

This explains the specific features of the processes of elementary particle physics, nuclear and atomic physics in the early Universe.

In the period

$$t \sim 0.1 \div 1 \text{ s}, \tag{3.92}$$

corresponding to the temperature

$$T \sim 3 \div 1 \text{ MeV}, \tag{3.93}$$

the characteristic time of the reactions of the weak interaction begins to exceed the cosmological time, so that neutrinos become decoupled from other particles. This leads to the existence of the thermal background of neutrinos with the concentration determined by the equilibrium at the time of decoupling.

At temperature T the number density of the relativistic neutrinos and their antineutrinos is

$$n_{\nu\bar{\nu}} = 2\int \frac{1}{\exp\left(\frac{pc}{kT}\right)+1} \cdot \frac{d^3p}{(2\pi\hbar)^3}. \tag{3.94}$$

Under the same conditions the number density of the photons is determined by Planck's distribution and equals

$$n_\gamma = 2\int \frac{1}{\exp\left(\frac{pc}{kT}\right)-1} \cdot \frac{d^3p}{(2\pi\hbar)^3}. \tag{3.95}$$

There is a simple analytical relation between the equilibrium number density of the neutrinos (and their antineutrinos) and photons, using the following mathematical procedure (Okun 1981).

Thermodynamic functions of the relativistic gas of fermion-antifermion pairs are given by integrals of the Fermi–Dirac distribution.

The number density of the fermions is equal to

$$n_f = g_f \int \frac{1}{\exp\left(\frac{pc}{kT}\right)+1} \cdot \frac{d^3 p}{(2\pi\hbar)^3}, \tag{3.96}$$

where g_f is the statistical weight, and the energy density of the fermion-antifermion gas in the relativistic case ($E \approx pc$) is

$$\varepsilon_f = g_f \int pc \cdot \frac{1}{\exp\left(\frac{pc}{kT}\right)+1} \cdot \frac{d^3 p}{(2\pi\hbar)^3}. \tag{3.97}$$

We can express these equations in terms of integrals of the functions of the Fermi–Dirac distribution as

$$n_f = \frac{g_f}{2\pi^2} \cdot \left(\frac{kT}{\hbar c}\right)^3 \cdot I_{2f} \tag{3.98}$$

and

$$\varepsilon_f = \frac{g_f}{2\pi^2} \cdot \left(\frac{kT}{\hbar c}\right)^3 \cdot (kT) \cdot I_{3f}, \tag{3.99}$$

where the integrals are defined as

$$I_{nf} = \int_0^\infty \frac{x^n}{\exp(x)+1} \cdot dx, \tag{3.100}$$

with $x = pc/kT$.

In the same way, the corresponding quantities for the relativistic gas of bosons can be expressed in terms of integrals of the Bose–Einstein distributions

$$n_b = \frac{g_b}{2\pi^2} \cdot \left(\frac{kT}{\hbar c}\right)^3 \cdot I_{2b} \tag{3.101}$$

and

$$\varepsilon_b = \frac{g_b}{2\pi^2} \cdot \left(\frac{kT}{\hbar c}\right)^3 \cdot (kT) \cdot I_{3b}, \tag{3.102}$$

where g_b is the statistical weight of the considered bosons and integrals I_{nb} are equal

$$I_{nb} = \int_0^\infty \frac{x^n}{\exp(x)-1} \cdot dx. \tag{3.103}$$

All these integrals are tabulated, but it turns out that the difference between them has a very simple analytical form

$$I_{nb} - I_{nf} = \int_0^\infty \frac{x^n}{\exp(x)-1} \cdot dx - \int_0^\infty \frac{x^n}{\exp(x)+1} \cdot dx = \frac{1}{2^n} \cdot I_{nb}. \tag{3.104}$$

The final analytical formula that relates the integrals for the relativistic fermions and bosons, is given by

$$I_{nf} = \left(1 - \frac{1}{2^n}\right) \cdot I_{nb}. \tag{3.105}$$

Using this relation, the equilibrium number density of the photons and left-handed neutrinos (with their right-handed antineutrinos) can be linked as follows

$$n_{\nu\bar{\nu}} = \left(1 - \frac{1}{2^2}\right) \cdot n_\gamma = \frac{3}{4} n_\gamma \tag{3.106}$$

and for the equilibrium energy densities

$$\varepsilon_{\nu\bar{\nu}} = \left(1 - \frac{1}{2^3}\right) \cdot \varepsilon_\gamma = \frac{7}{8} \varepsilon_\gamma. \tag{3.107}$$

In such a relationship between the photons and the relativistic electron–positron pairs it should be noted that the number of degrees of freedom in the gas of photons (left- and right-handed states) is half the value of the same quantity for the gas of the electron–positron pairs (left- and right-handed electrons and positrons).

Thus, the equilibrium number density of the electron–positron pairs is

$$n_{e^-e^+} = 2 \cdot \left(1 - \frac{1}{2^2}\right) \cdot n_\gamma = \frac{3}{2} n_\gamma \tag{3.108}$$

and their equilibrium energy density is

$$\varepsilon_{e^-e^+} = 2\cdot\left(1-\frac{1}{2^3}\right)\cdot\varepsilon_\gamma = \frac{7}{4}\varepsilon_\gamma. \quad (3.109)$$

During the decoupling of neutrinos, their number density was related to the density of photons by the equilibrium relation (3.106).

After decoupling of the neutrinos at

$$t > 100\text{ s} \quad (3.110)$$

when in the course of the expansion the temperature dropped to

$$T < 100\text{ keV} \quad (3.111)$$

the equilibrium electron–positron pairs annihilated thereby increasing the number density of photons without any significant impact on the number density of neutrinos. The corresponding change in the concentration ratio of neutrinos and photons can be calculated on the basis of conservation of entropy.

At a temperature

$$T = T_1 \quad (3.112)$$

immediately before the decoupling of the neutrino the specific entropy of the equilibrium mixture of photons and the electron–positron pair was

$$S_\gamma + S_{e^-e^+} = \frac{4}{3}\cdot\frac{\sigma T_1^3 + \frac{7}{4}\sigma T_1^3}{\rho_1}, \quad (3.113)$$

and that of the pair of neutrinos and antineutrinos

$$S_{\nu\bar{\nu}} = \frac{4}{3}\cdot\frac{7}{8}\frac{\sigma T_1^3}{\rho_1}. \quad (3.114)$$

At some lower temperature

$$T = T_0 \quad (3.115)$$

after decoupling of neutrinos and electron–positron annihilation the specific entropy of the photons was

$$S_\gamma = \frac{4}{3}\cdot\frac{\sigma T_{0\gamma}^3}{\rho_0} = S_\gamma + S_{e^-e^+} = \frac{4}{3}\cdot\frac{11}{4}\frac{\sigma T_1^3}{\rho_1} \tag{3.116}$$

and the neutrino entropy

$$S_{\nu\bar{\nu}} = \frac{4}{3}\cdot\frac{7}{8}\frac{\sigma T_{0\nu}^3}{\rho_0} = \frac{4}{3}\cdot\frac{7}{8}\frac{\sigma T_1^3}{\rho_1}, \tag{3.117}$$

which uses the separate conservation of entropy in the mixture of photons, electrons and positrons and in the decoupled gas of the neutrino-antineutrino pairs in the course of adiabatic expansion.

Thus the ratio of the neutrinos to photons decreases and finally we obtain a prediction for the current concentration of primordial gas of the left-handed neutrinos and right-handed antineutrinos of one species

$$n_{\nu\bar{\nu}} = \frac{3}{4}\cdot\frac{4}{11}\cdot n_\gamma = \frac{3}{11}n_\gamma, \tag{3.118}$$

where the factor 4/11 takes into account the increase in photon number density due to the annihilation of electrons and positrons.

Another important consequence of decoupling of neutrinos lies in the fact that when the characteristic time scale of weak interaction exceeds the cosmological time, the β-processes in which the protons are converted into neutrons and neutrons into protons,

$$\begin{aligned} &\nu_e n \to ep \\ &\bar{\nu}_e p \to e^+ n \\ &ep \to \nu_e n \\ &e^+ n \to \bar{\nu}_e p \end{aligned} \tag{3.119}$$

come out of equilibrium.

At a temperature

$$kT < \Delta m_{np}c^2, \tag{3.120}$$

where the mass difference between the proton and the neutron is

$$\Delta m_{np}c^2 = m_n c^2 - m_p c^2 = 1.28\ \text{MeV}, \tag{3.121}$$

these processes supported by the equilibrium concentration ratio of neutrons and protons which is equal to

$$\frac{n}{p} = \exp\left(-\frac{\Delta m_{np}c^2}{kT}\right). \quad (3.122)$$

When the β-processes come out of equilibrium, this relation is 'frozen out' and cannot be changed until

$$t \sim \tau_n \sim 10^3 \text{s} \quad (3.123)$$

when a neutron decays.

However, most of the neutrons do not decay because at

$$t \sim 100 \text{ s}, \quad (3.124)$$

corresponding to the temperature

$$T \sim T_D \sim 10^2 \text{ keV} \quad (3.125)$$

they formed deuterium in the reaction with protons

$$n + p \rightarrow D + \gamma. \quad (3.126)$$

At higher temperatures

$$T > T_D \quad (3.127)$$

in the process of photodisintegration

$$D + \gamma \rightarrow n + p \quad (3.128)$$

all the deuterium formed is destroyed, but at

$$T < T_D \quad (3.129)$$

photodisintegration is not effective and the deuterium formed takes part in subsequent thermonuclear reactions

$$\begin{aligned} D + D &\rightarrow T + p \Rightarrow T + D \rightarrow {}^4He + n \\ D + D &\rightarrow {}^3He + n \Rightarrow {}^3He + n \rightarrow T + p. \end{aligned} \quad (3.130)$$

These reactions result in the so-called 'primordial chemical composition'. Its main components are hydrogen and ^{4}He, which correspond to the weight fraction of about 75% and 25%. A small

($\sim 10^{-5} \div 10^{-4}$) fraction of deuterium and ^{3}He and a much smaller fraction of ^{7}Li also form.

The absence of stable nuclear states with atomic number $A = 5$ creates an almost impenetrable barrier to further nuclear transformations in primordial nucleosynthesis. Primordial abundance of light elements plays a very important role in our further considerations.

At a temperature

$$T_{\mathrm{rec}} \sim 4 \cdot 10^3\,\mathrm{K}, \tag{3.131}$$

corresponding to the period of recombination of hydrogen

$$t_{\mathrm{rec}} \sim 10^{13}\,\mathrm{s}, \tag{3.132}$$

when the electrons and protons form neutral hydrogen atoms, the photons cease to interact with matter.

At a temperature

$$T_m \sim \frac{1}{3} \cdot \frac{n_b}{n_\gamma} \sim \frac{1}{3} \cdot r_b \cdot m_p \tag{3.133}$$

the baryonic density of matter

$$r_b = m_p n_b \tag{3.134}$$

exceeds the radiation density

$$\rho_\gamma \sim 3 T_\gamma n_\gamma. \tag{3.135}$$

Moment t_m when matter begins to dominate

$$\rho_b > \rho_\gamma \tag{3.136}$$

is determined by

$$r_b = \frac{n_b}{n_\gamma} = 1.4 \cdot 10^{-8} \Omega_b \tag{3.137}$$

and is fairly close to the period of recombination.

So, the modern stage of dominance of matter starts at the time

$$t > t_{RD} \sim t_m. \tag{3.138}$$

The gravitational instability of the neutral gas developed at this stage. Small fluctuations in matter density become stronger, forming the structure of the observed inhomogeneities to which the galaxies their clusters and superclusters, clusters of stars, individual stars and so on also belong.

The growth of density fluctuations takes considerable time. For a long time after t_{RD} the matter expands almost uniformly, and the growth of inhomogeneities is reduced to increase of the density contrast of different areas.

Only relatively recently, when

$$t_s \geqslant 10^{16}\,\mathrm{s} \tag{3.139}$$

the first inhomogeneities, separated from the general cosmological expansion, formed. Their sequential evolution led to the formation of galaxies. Thermonuclear reactions in stars born at this stage lead to the formation of heavy elements. Explosions of stars enrich the interstellar space with heavy elements. Therefore, heavy elements so important for bodies around us and our existence formed as a result of stellar nucleosynthesis.

Based on the physical laws well-proven in the laboratory, the old Big Bang scenario is physically self-consistent. Moreover, its principal prediction was confirmed by qualitative observations. Discovery of blackbody radiation and measurements of the abundance of light elements seemed to confirm this picture.

However, the development of the theory of elementary particles and attempts to apply its predictions to cosmology initiated a critical analysis of the old Big Bang model.

This process began in the late 70s of the twentieth century with the optimistic hope that the Grand Unification models give an explanation of the observed baryon asymmetry of the Universe. It was expected that the baryosynthesis mechanism based on the simple $SU(5)$ model of Grand Unification, would find experimental confirmation in the search for proton decay.

In the early 80's of 20th century, experimental evidence for the existence of a non-zero electron neutrino mass stimulated the development of the theory of large-scale structure formation in the Universe with massive neutrinos. And this trend seemed to find a physical basis in the measurements of neutrino mass.

And although either fast proton decay or the tens of eV mass of the electron neutrinos have not found experimental confirmation, theoretical analysis of the cosmological consequences of the theory of elementary particles has revealed the internal inconsitency of the Big Bang model.

A number of problems, unsolved in the old cosmological picture, found a solution in this analysis, paving the way for the new picture of inflationary models with baryosynthesis and non-baryonic dark matter/energy.

Physical justification of these models lost the ability for direct experimental verification and requires special methods for their study. Detailed development of physically self-consistent cosmological models and a comprehensive check of their physical basis is a major task of cosmoparticle physics.

2. Inflationary cosmology with baryosynthesis and dark matter/energy

From the old Big Bang scenario it follows that the early cosmological evolution would occur at very high temperatures. This provision essentially means the dependence of the conditions in the very early Universe on the laws of ultrahigh energy physics.

Our present knowledge of these laws is based on the prediction of particle theory. Thus, the studies of particle theory have a direct impact on the picture of the early Universe. In principle, all cosmologically significant phenomena, predicted by particle theory, could find their place in the history of the Universe.

Since this theory of ultrahigh energy physics has not yet been developed, we can only start from reasonable assumptions about events that should take place in cosmology.

The development of particle theory and its applications in the Big Bang model of the Universe revealed at least three phenomena that are widely recognized as necessary additions to the Big Bang scenario, with both aesthetic and practical point of view.

This leads to a change in the cosmological picture. The modern model of the Big Bang Universe usually implies an inflationary scenario with baryosynthesis and dark matter/energy.

The development of this scenario offers an excellent opportunity to explain the basic cosmological parameters based on physical mechanisms. So, the choice of an open, closed or flat cosmological model is associated with the mechanism of inflation. In the simplest case, it leads to the prediction of a flat Universe.

The observed ratio of baryons to photons is regarded as the result of baryosynthesis which defined the modern density of baryons.

The difference between the total and baryon density is attributed to non-baryonic dark matter/energy. In the simplest case, the density of dark matter is determined by mass and concentration of the frozen-out weakly interacting particles.

An attractive way to determine the cosmological parameters through the parameters of particles and fields is not the only advantage of the new approach. Inflation, baryosynthesis and dark matter/energy eliminate the internal contradictions of the old Big Bang cosmology.

However, the price for this are unknown physical foundations and a vast array of possible implementations based on different approaches. The hope to resolve these ambiguities is related to cosmoparticle physics.

Assuming that the inflationary baryon-asymmetric cosmology with non-baryonic dark matter/energy is closer to reality than the Gamow scenario of Big Bang, we are faced with the problem of observational manifestations of determining the choice of inflationary model, the baryosynthesis mechanism and the physical nature of non-baryonic dark matter and energy. We refer to this problem as a problem of hidden parameters of modern cosmology.

In cosmoarcheology, the hidden parameters determine the properties of the Universe as a natural accelerator. It is necessary to include all the details of the considered cosmological model so that the conclusions about the allowable properties of hypothetical particles and fields would be internally self-consistent.

Of course, we can ignore the hidden parameters but then the cosmological conclusions should be amended to the extent of their uncertainty.

We now turn to the aesthetic and practical reasons for the extension of the Big Bang model, leading to the modern cosmological picture. Given the extensive literature (see for example, books by Linde 1990; Kolb, Turner 1990; Khlopov, Rubin 2004; Mukhanov 2005; Gorbunov, Rubakov 2011a, 2011b), we can only briefly review these ideas, paying more attention to their physical significance and non-trivial ways of detailed verification.

2.1. Magnetic monopoles in the old Big Bang model

The question of the existence of isolated magnetic poles – magnetic monopoles – arises almost in all phases of the fundamental theory of magnetism.

Coulomb developed magnetostatics based on the strict symmetry between electric and magnetic forces. He believed that there should be long-range ‘Coulomb’ force between isolated magnetic charges.

Classical electrodynamics has broken this symmetry, attributing magnetism to the motion of electric charges. But in the next step of research – in the framework of quantum mechanics – the symmetry between electricity and magnetism was again included in the studies.

Paul Dirac (Dirac 1931, 1948) showed that the existence of magnetic monopoles with magnetic charge

$$g = \frac{\hbar c}{2e} \tag{3.140}$$

may explain the quantization of the electric charge in quantum theory.

Similar ideas were proposed by Schwinger and Mandelstam in the context of quantum field theory. In this case, the elementary magnetic charge is twice the quantum-mechanical value (3.140) obtained by Dirac.

In 1974, t'Hooft and Polyakov showed that the magnetic monopole can appear in gauge models of electroweak interactions. However, the existence of a magnetic monopole-type solution requires a compact group with single symmetry which includes the electromagnetic symmetry $U(1)$ that has no place in the framework of the standard model of electroweak interactions.

A common feature of these approaches to the problem of magnetic monopoles was their exotic status in the theory of elementary particles. Magnetic monopoles were possible, attractive, but not a necessary element of the theory of electromagnetism.

It seemed that the experiments that established more stringent upper limits on the concentration of monopoles in the Earth and the lunar soil and cosmic rays, as well as more stringent lower limits on the mass of a monopole in the experimental search at accelerators, put the idea of the existence of magnetic monopoles in serious doubt.

The status of magnetic monopoles has changed radically in terms of models of Grand Unification. In these models, the $U(1)$-symmetry of electromagnetic interactions were included in the compact group of symmetry of GUT and, in doing so, the existence of magnetic monopoles with the Dirac magnetic charge (3.140) became the inevitable topological consequence of violation of the symmetry of GUT. The order of magnitude of the predicted mass of these monopoles was

$$m \sim \frac{\Lambda}{e}, \tag{3.141}$$

where Λ is the scale of GUT symmetry breaking.

For

$$\Lambda \sim 10^{15}\,\text{GeV} \tag{3.142}$$

the monopole mass is of the order

$$m \sim 10^{16}\,\text{GeV}, \tag{3.143}$$

which explains the negative results of searches of monopoles using accelerators.

Moreover, no artificial or natural source of energetic particles in the Universe can provide the conditions for the production of particles with such enormous mass. It would appear that only extrapolation to the earliest stages of cosmological expansion could lead to the natural conditions for the creation of magnetic monopoles.

According to the Big Bang model, the particles with the mass m of any kind should be in equilibrium at temperature

$$T > m, \tag{3.144}$$

if the interaction of these particles is strong enough to satisfy the conditions of equilibrium with plasma and radiation. This means that the reaction rate (σv) is high enough to satisfy the condition (cf. (3.53))

$$n(T)(\sigma v) > \Gamma, \tag{3.145}$$

where $n(T)$ is the number density of the particles at temperature T and (cf. (3.52) and (3.74)) $\Gamma \sim (T^2/m_{\text{Pl}})$ is the rate of cosmological expansion.

When the temperature drops to

$$T < m, \tag{3.146}$$

leaving just the equilibrium condition (3.145), the number density of particles is determined by equilibrium

$$n(T) = \left(\frac{2}{\pi^3}\right)^{\frac{1}{2}} (mT)^{\frac{3}{2}} \exp\left(-\frac{m}{T}\right). \tag{3.147}$$

At a temperature

$$T \sim T_f \tag{3.148}$$

when the rate of particle interaction becomes less than the rate of cosmological expansion

$$n(T_f)(\sigma v) < \Gamma(T_f) \tag{3.149}$$

the particles leave equilibrium. Their concentration begins to differ from the equilibrium value (3.147). The result is freezing-out of the particles and their relative concentration no longer changes.

Consider, for example, stable charged particles that are at equilibrium with their antiparticles at high temperature. Their equilibrium concentration maintains the balance between the creation and annihilation of pairs.

When the temperature drops during expansion, the creation of pairs is suppressed and the equilibrium concentration decreases exponentially in accordance with (3.147). Accordingly, the annihilation rate decreases. When this rate becomes smaller than the rate of expansion the particles and antiparticles freeze-out.

Based on this picture, it is possible to simply estimate the relic concentration of any type of particles. In the case of the monopoles the magnetic charge conservation law implies their absolute stability. For the same reasons, the birth of a monopole must be accompanied by the birth of an antimonopole with the opposite magnetic charge.

This approach can be applied to the monopoles in the early Universe, believing that at high temperatures the monopole–antimonopole pairs were in equilibrium. As the temperature decreases the rate of expansion should exceed the rate of annihilation of monopole–antimonopole pairs that would lead to freezing-out of the concentration of relic magnetic monopoles.

Assuming that the monopoles are point particles with mass m and magnetic charge g, Domogatsky and Zheleznykh (1969) calculated their primordial abundance and obtained the restrictions on these parameters based on the observational constraints on the concentration of monopoles in the Universe.

However, in the case of magnetic monopoles in GUT a serious correction should be made in the the calculation of their relic abundance. The reason for this is that the monopoles of GUT are not the ordinary particles, i.e. quanta of the respective fields, but emerge as a topological defect as a result of breaking of GUT symmetry. A simple example of the Higgs triplet will be used to illustrate the birth of monopoles in a phase transition in GUT (Kibble 1976; Khlopov 1988).

Consider a triplet scalar field

$$\vec{\Phi} = \{\Phi_1, \Phi_2, \Phi_3\} \tag{3.150}$$

with the Higgs potential depending on the

$$\left|\vec{\Phi}\right| = \Phi_1^2 + \Phi_2^2 + \Phi_3^2 \tag{3.151}$$

and having a minimum at zero temperature for a non-zero vacuum mean field

$$\left\langle \left|\vec{\Phi}\right|^2 \right\rangle = \Lambda^2. \tag{3.152}$$

At high temperature, thermal fluctuations change the shape of this potential, so that at

$$T > \Lambda \tag{3.153}$$

its minimum corresponds to the zero vacuum expectation value

$$\left\langle \left|\vec{\Phi}\right|^2 \right\rangle = 0 \tag{3.154}$$

and the symmetry is restored. During expansion of the Universe the temperature drops to a critical value

$$T \sim T_{cr} \sim \Lambda, \tag{3.155}$$

where a phase with the vacuum expectation value (3.152), corresponding to the broken symmetry, appears during the phase transition process.

But the asymmetric vacuum is degenerate, since any triplet field satisfying

$$\left\langle \left|\vec{\Phi}\right|^2 \right\rangle = \Phi_1^2 + \Phi_2^2 + \Phi_3^2 = \Lambda^2 \tag{3.156}$$

corresponds to the minimum of the Higgs potential. In the space of the components of the field, the minimum is reached on the sphere

$$\Phi_1^2 + \Phi_2^2 + \Phi_3^2 = \Lambda^2. \tag{3.157}$$

The coordinates on this sphere can be defined by two variables, φ, θ, corresponding to polar and azimuthal angles. Any values of φ and θ correspond to the minimum of the Higgs potential.

The process of phase transition is accompanied by the rearrangement of vacuum in ordinary space. At each spatial point the ground state, which is the physical vacuum, transforms into an asymmetric state defined by (3.152) and (3.156). However, values φ and θ are arbitrary and their values may be different in different spatial locations.

For the terms with derivatives in the Lagrangian field

$$L = \left(\partial_\mu \vec{\Phi}\right)^2 - V\left(\left|\vec{\Phi}\right|^2, T\right) \tag{3.158}$$

to be minimal, $\varphi(x)$ and $\theta(x)$ should vary slowly over distances smaller than the correlation length

$$r_c \sim \frac{1}{e\Lambda}. \tag{3.159}$$

However, the changes of $\varphi(x)$ and $\theta(x)$ along the path, considerably longer than r_c, can take place. If they change by 2π along a closed contour, the contraction of this contour leads to the formation of a singular point at which φ and θ have not been determined.

From the continuity of the triplet field Φ we obtain the zero value of the vacuum expectation value of this field at a singular point and a significant deviation of the Higgs potential from its minimum in the region with the geometric size of the order of approximately r_c around this point. Thus, a high energy density concentrates in the vicinity of a singular point, which leads to the formation of a massive object – the monopole.

This object is also called the 'hedgehog' because of the similarity of sticking needles in all directions with gradients $\varphi(x)$ and $\theta(x)$ which determine the strength of its magnetic field corresponding to the magnetic charge $+g$. It is easy to show that the same arguments lead to the formation of a neighboring 'hedgehog' with the opposite sign of the magnetic charge $-g$. So, the non-local production of a monopole–antimonopole pair takes place.

The change of $\varphi(x)$ and $\theta(x)$ corresponds to the change in the space of definition of electrically charged and neutral states. 'Combing the hedgehog', that is, choosing in the space a uniform definition of the charged and neutral states, we get (Dirac 1931, 1948) a solution of the type of magnetic monopole with the magnetic flux of the type of singular string emanating from the monopole to infinity.

Inside the GUT monopoles the symmetry of GUT is restored. In particular, the baryon charge is not conserved inside the monopole. It turns out a proton decay is induced in the singular field of a magnetic monopole

$$p \to e^+ e^- e^+ \tag{3.160}$$

(Rubakov, 1981; Callan, 1982) with the cross section defined by the proton size

$$\sigma \sim 10^{-28}\,\text{cm}^2. \tag{3.161}$$

The specific properties of GUT monopoles lead to a nontrivial pattern of their cosmological evolution in the old Big Bang scenario.

The massive magnetic monopoles simple did not exist before the phase transition in GUT because the symmetry was restored. They formed in the Universe after the phase transition both as point singularities described above and by means of a pair production.

The latter could occur in the second-order phase transition. The mass of the monopoles is determined by the vacuum expectation value of the Higgs field. At a lower temperature but close to the phase transition temperature

$$T \to T_{\text{cr}} \sim \Lambda \tag{3.162}$$

the vacuum expectation value, being the order parameter, should follow the Curie–Weiss dependence on temperature. This leads to the same temperature dependence for the magnetic monopole mass near the critical temperature (Zeldovich, Khlopov 1978)

$$m(T) \propto \sqrt{T_{\text{cr}}^2 - T^2}. \tag{3.163}$$

However, evaluation of the concentration of frozen-out monopoles was found to be independent of all these details as well as of the exact value of the initial concentration of monopoles. If it is large enough, say, as great as it can be estimated for the topological mechanism of non-local formation of the monopoles (Preskill 1979)

$$n_m \sim \frac{1}{r_c^3} \sim e^3 \Lambda^3 \tag{3.164}$$

the subsequent diffusion of monopoles to antimonopoles and their annihilation ‘washes away’ information on the initial concentration of monopoles (Zeldovich, Khlopov 1978).

We now consider, following (Zeldovich, Khlopov 1978), the theory of the annihilation of monopoles and antimonopoles in the early Universe.

The annihilation cross section is determined by the Coulomb attraction of magnetic charges. At temperature T Coulomb attraction is strong at the distance

$$r \leqslant r_0 \equiv \frac{g^2}{T}. \tag{3.165}$$

If the mean free path of the monopoles relative to scattering in plasma exceeds r_0

$$\lambda \gg r_0 \tag{3.166}$$

we can consider free annihilation of the monopoles–antimonopoles. In the opposite case

$$\lambda \ll r_0 \tag{3.167}$$

annihilation should be calculated in the diffusion approximation.

The initial concentration of the monopoles is several orders of magnitude smaller than the concentration of relativistic particles in the plasma, and the mean free path of the monopoles is determined by their scattering on the charged particles

$$\lambda = \frac{1}{n_{ch}\sigma}. \tag{3.168}$$

The cross section for multiple scattering by 90° (Zeldovich, Khlopov 1978; Preskill 1979)

$$\sigma \sim \frac{(ge)^2}{Tm} \tag{3.169}$$

and we find that condition (3.167), in which the diffusion approximation is valid, is satisfied at

$$t < t_1 \sim \frac{m_{\mathrm{Pl}}}{\alpha^2 m^2} \tag{3.170}$$

which corresponds to the temperature

$$T > \alpha m \sim e\Lambda \sim eT_{\mathrm{cr}}, \tag{3.171}$$

where α is the fine structure constant, and m is the mass of the monopole.

The rate of annihilation of a monopole and an antimonopole can be found in the diffusion approximation when considering the diffusion of particles with the magnetic charge $-g$ to the absorbing sphere with radius $a \leqslant r_0$ and with the magnetic charge $+g$.

The spherically symmetric diffusion equation has the form

$$\frac{\partial n(r,t)}{\partial t} = D \cdot \frac{1}{r^2} \cdot \frac{\partial}{\partial r} r^2 \left(\frac{\partial n(r,t)}{\partial r} + \frac{g^2}{T \cdot r^2} \cdot n(r,t) \right) \quad (3.172)$$

where the diffusion coefficient D

$$D = \frac{1}{3} \lambda v. \quad (3.173)$$

The solution of equation (3.172) for the stationary distribution $\partial n / \partial t = 0$ of diffusing particles with the boundary conditions $n(\infty) = n_0 \equiv n_m$ and $n(a) = 0$ is

$$n(r) = \begin{cases} 0, & r \leqslant a \\ n_0 \dfrac{1 - \exp\left\{ \dfrac{r_0}{r} - \dfrac{r_0}{a} \right\}}{1 - \exp\left\{ -\dfrac{r_0}{a} \right\}}, & r \geqslant a. \end{cases} \quad (3.174)$$

Then, the diffusion flux is

$$q = 4\pi r^2 D \frac{\partial n(r)}{\partial r} \approx 4\pi D r_0 n_m \quad (3.175)$$

and the rate of annihilation of monopoles in the diffusion approximation is

$$\left(\frac{dn_m}{dt} \right)_{\text{ann}} = -n_m^2 4\pi D r_0, \quad (3.176)$$

and not

$$\left(\frac{dn_m}{dt} \right)_{\text{ann}} = -n_m^2 4\pi a^2 v, \quad (3.177)$$

as it would be in the case of free monopoles.

Taking into account (3.176), the equation for the relative concentration of the monopoles $\nu = n_m/n_r$ where n_r is the concentration of relativistic particles, has the form

$$\frac{d\nu}{dt} = -4\pi D r_0 n_r \nu^2 = -A\theta^{\frac{1}{2}}\nu^2, \tag{3.178}$$

where

$$A = \frac{4\pi}{3}\frac{g}{e}m \tag{3.179}$$

and the dimensionless temperature is defined as

$$\theta = \frac{T}{m}. \tag{3.180}$$

Introducing the dimensionless variable

$$\tau = \frac{t}{t_0} \tag{3.181}$$

and taking into account

$$\theta(t) = \theta(t_0)\tau^{-\frac{1}{2}} \equiv \theta_0\tau^{-\frac{1}{2}} \tag{3.182}$$

we obtain a solution for (3.178)

$$\nu(t) = \frac{\nu(t_0)}{1 + \frac{4}{3}A\theta_0^{1/2}t_0\left(\tau^{3/4} - 1\right)\nu(t_0)} \tag{3.183}$$

at

$$t_0 < t < t_1. \tag{3.184}$$

Diffusion approximation becomes invalid after

$$t > t_1 \approx \frac{g^8 e^4 m_{Pl}}{m^2}, \tag{3.185}$$

when the rate of annihilation of free monopoles is given by (3.177).

According to (Zeldovich, Khlopov 1978) the value a, which determines the annihilation cross section of free monopoles, is the maximum impact parameter for which the motion of the monopole is finite due to bremsstrahlung. It was found that the annihilation of free monopoles has virtually no effect on the results of diffusion annihilation and the concentration of primordial monopoles is given by $\nu(t \gg t_1) = \nu(t_1)$.

Thus, if the initial concentration of the monopole satisfies

$$\nu_0 \gg \frac{3}{4A\theta_1^{\frac{1}{2}} t_1}, \tag{3.186}$$

the concentration of relic magnetic monopoles does not depend on the initial concentration and is given by

$$\nu_m = \frac{n_m}{n_r} = \frac{m}{g^5 (eg) m_{\text{Pl}}} \approx 10^{-9} \left(\frac{m}{10^{16} \text{ GeV}} \right). \tag{3.187}$$

Equation (3.187) shows that for the mass of the magnetic monopole

$$m = 10^{16} \text{ GeV} \tag{3.188}$$

the concentration of the relic monopoles exceeds the baryon density

$$\frac{n_m}{n_\gamma} > \frac{n_b}{n_\gamma} \tag{3.189}$$

so that the density of magnetic monopoles

$$\rho_m = m \cdot n_m \tag{3.190}$$

turns out to be 16 orders of magnitude larger than the baryon density. To solve the problem of overproduction of monopoles an inflationary cosmological model was proposed in which the initial concentration of the monopoles is strongly suppressed (see the next Section).

Note that the question of the origin and evolution of the structure of magnetic fields in the Universe in the presence of monopoles becomes particularly interesting.

Indeed, in the mechanism of topological formation of monopoles in the phase transition the monopole field is the first to form in space and this is followed by the formation of a localized singularity of the field. It is clear that the local field, produced in the phase transition, has no orientation. Thus, the formation of monopoles and antimonopoles can be accompanied by the formation of loops of the magnetic field, forming the primordial structure of magnetic fields in the Universe (Khlopov 1988). Self-consistent evolution of these primordial fields with plasma of monopoles and antimonopoles has not yet been studied.

2.2. Inflationary recovery of the old Big Bang model

Typically, inflation is considered as an essential element of the cosmological picture. The inflationary model explains why the Universe is expanding. It provides solutions of the problem of the horizon, flatness, magnetic monopoles and other problems (Guth 1981; Linde 1984). This decision is based on the superluminal expansion which occurs when the cosmological equation of state is

$$p < -\frac{1}{3}\varepsilon \tag{3.191}$$

Under the condition (3.191), the acceleration in the cosmological equations (cf. (3.43))

$$\frac{\ddot{a}}{a} = -\frac{4\pi G}{3c^2}\cdot(\varepsilon + 3p) \tag{3.192}$$

is positive and the formal solution for the expansion law is given as

$$a(t) \propto \exp\left(\int H dt\right), \tag{3.193}$$

which is determined from (3.42)

$$\left(\frac{\dot{a}}{a}\right) = H = \sqrt{\frac{8\pi G\varepsilon}{3c^2}}. \tag{3.194}$$

Nor matter or radiation can provide negative pressure in the equation of state. We must assume some hypothetical phenomenon which took place at a very early stage of development of the Universe and led to a transient stage with negative pressure.

The simplest possibility, pointed out in 1965 by Sakharov and Gliner and first considered in the studies (Gliner 1970) and (Gliner, Dymnikova 1975), suggests that the initial state of the cosmological expansion has a maximum symmetry of space-time. This corresponded to the de Sitter vacuum equation of state

$$p = -\varepsilon = -\Lambda = \text{const.} \tag{3.195}$$

In this case, the dependence of the scale factor on time is given by the exponential law

$$a(t) \propto \exp(Ht) \tag{3.196}$$

with the Hubble constant H defined as

$$H = \sqrt{\frac{8\pi G\Lambda}{3c^2}}. \tag{3.197}$$

Bugrii and Trushevsky (1976) indicated the possibility of an exponential stage of expansion at the hadronic stage, but the conditions for implementing such an expansion regime were incompatible with the hadron physics of QCD.

In 1980, Starobinsky found that conditions similar to the equation of state (3.195) occurred in the early Universe due to the R^2 effect induced by true polarization of the vacuum of the classical gravitational field (Gurovich, Starobinsky 1979, and references in Zeldovich 1980).

However, overall clear understanding of the need for inflation in both cosmology and particle physics came only with the studies of Guth (1981). Stimulated by the necessity of solving the cosmological problems of overproduction of GUT monopoles, Guth's study identified in a completely transparent manner the list of internal inconsistencies of the old Big Bang model which were eliminated by inflation.

The old inflationary scenario (Guth 1981) suggested a strong first order phase transition at breaking of GUT, leading to strong supercooling of the Universe in a phase of restored symmetry of GUT.

At low temperatures, the Higgs potential is given as

$$V(\Phi) = \frac{1}{4}\lambda\left(|\Phi|^2 - v^2\right)^2.$$

The symmetric phase corresponds to the zero vacuum expectation value of the Higgs field. The Higgs potential is then reduced to a constant and its dominance in the energy density leads to an equation of state (3.195). Due to the time dependence of the scale factor described by (3.196), the correlation radius increases exponentially, which leads to exponential suppression of the topological formation of magnetic monopoles.

As a result of subsequent heating of the Universe the temperature becomes much lower than the temperature of phase transitions in GUT, so that the concentration of pairs of monopoles and antimonopoles, formed by pair production, is also exponentially small.

Thus, the inflationary scenario leads to suppression of the initial concentration of magnetic monopoles, which eliminates the problem of their overproduction.

The exponential growth of the scale factor in the inflationary stage resolves several problems in the initial conditions inherited from the old model.

The horizon problem is related to the remarkable homogeneity of the Universe in the area within the modern horizon, covering a large number of different regions causally disconnected at the early stage of the Universe.

To illustrate the fantastic fine tuning of the initial conditions required in the old Big Bang model, we can estimate the initial size of the region contained within the modern cosmological horizon, the size of which is currently

$$l_h = ct_U \sim 10^{28}\,\text{cm}. \qquad (3.198)$$

Since in the old Big Bang model the scale factor varies inversely with temperature, we find that the Planck time t_{Pl}, when the temperature was equal to Planck temperature T_{Pl}, this area should be of the order

$$l_h(t_{\text{Pl}}) = l_h(t_U)\cdot\frac{a(t_{\text{Pl}})}{a(t_U)} = l_h(t_U)\cdot\frac{T_0}{T_{\text{Pl}}} \approx 2.5\cdot 10^{-4}\ \text{cm}, \qquad (3.199)$$

where modern temperature of the thermal background radiation is equal $T_0 = 2.7$ K.

On the other hand, the size of the causally connected region at the Planck time is (cf. (3.56)) equal to the Planck length

$$l_{\text{Pl}} = ct_{\text{Pl}} = 1.5\cdot 10^{-33}\,\text{cm}. \qquad (3.200)$$

Thus, the observed homogeneity within the modern horizon should be provided by the fine tuning of the initial conditions in

$$N \sim \left(\frac{l(t_{\text{Pl}})}{l_{\text{Pl}}}\right)^3 \sim 3\cdot 10^{87} \qquad (3.201)$$

causally disconnected areas

In the case of a non-flat Universe, the horizon problem is compounded by the *flatness problem*.

The existing uncertainty in the estimation of the total cosmological density allows for models with both positive and negative curvature with the total density close to critical.

This uncertainty in the estimated density corresponds to a very fine tuning of the initial deviations from the flat Universe.

Deviations of cosmological density from critical density in a non-flat Universe are inversely proportional to the square of the scale factor. This means that even the uncertainty in the order of magnitude of the modern cosmological density (from 0.3 to 3 critical which in reality is much too high) corresponds to the initial deviation from the critical density at the Planck time of the order of

$$\Omega(t_{\mathrm{Pl}})-1 \leqslant \left(\frac{a(t_{\mathrm{Pl}})}{a(t_U)}\right)^2 \cdot O(10) \leqslant 10^{-63}. \tag{3.202}$$

Thus, analysis of the global properties of present-day Universe (Guth 1981) reveals a very fine tuning of the initial conditions for the non-flat Universe. It should be close to a flat Universe with accuracy (3.202) in each of more than 10^{87} causally disconnected areas.

It is easy to see that the exponential growth of the scale factor naturally solves both these problems.

If the so-called e-folding, defined as the exponent in (3.193) or (3.196), exceeds 72, then the size of the causally connected area at the Planck time increases so much that it includes the size of the present-day cosmological horizon. This explains the similarity of the initial conditions of the expansion in all areas of the visible part of the Universe.

For the non-flat Universe such e-folding provides an exponential growth of the scale factor, making the effect of curvature in the deviation from the flat model compatible with the existing uncertainties in the cosmological density.

It may be assumed (see, for example, Zeldovich, Khlopov 1984) that inflation explains the cosmological expansion as such. Acceleration at the inflationary stage (positive sign of the second time derivative in (3.192)) gives the initial impetus to the expansion, i.e. inflation leads to the Big Bang that marked the beginning of the hot Universe.

However, no matter how attractive the idea of inflation is, it causes serious problems in cosmology and physics.

In the 'old inflationary model' (Guth 1981) a strong first order phase transition, causing inflation, leads to the 'boiling' of the Universe. Dominance of the false vacuum is over when bubbles of true vacuum form in it, expand and collide.

The energy released in the collision of bubbles leads to heating of the Universe. However, thermalization of the energy released during the collision of bubbles can occur if the bubbles themselves are small. This corresponds to rapid bubble nucleation.

On the other hand, the e-folding, which is needed to explain the global properties of the present-day Universe, involves much slower bubble nucleation and, consequently, their large size. The large size of the bubbles in this case means that all the energy released during the phase transition is converted into the kinetic energy of the bubble walls. This leads to what is inside the bubble of true vacuum is empty, and is unlikely to be filled and heated after the collision of the bubble walls. The Universe is highly inhomogeneous in such a scenario.

In the new inflationary scenario (Linde 1982a; Albrecht, Steinhardt 1982; see the review in Linde 1984, 1990), the transition from the inflationary phase is due to slow rolling of the effective potential to the true vacuum. This eliminates the problem of inhomogeneities arising due to bubble nucleation. However, the fluctuations of the scalar field produce density fluctuations. To ensure a sufficiently low amplitude of these fluctuations, the effective potential must be very flat at $\langle\Phi\rangle \to 0$.

This implies a very slow rolling of the field from the symmetric phase to the true vacuum of the asymmetric phase.

The same conditions of slow roll scenarios arise in the chaotic inflation scenario (Linde 1984), which can be simply illustrated by the classical scalar field (Zeldovich, Khlopov 1984).

Consider the Lagrangian of the theory of the classical scalar field

$$L = \frac{1}{2}\left(\partial_\mu \phi\right)^2 - V(\phi) \tag{3.203}$$

and set for simplicity that the potential has the form

$$V(\phi) = \frac{1}{2} m^2 \phi^2 \tag{3.204}$$

and that the non-interacting massive scalar field is homogeneous

$$\phi = \Phi \cos mt. \tag{3.205}$$

In the Lagrangian field theory the energy-momentum tensor is

$$T^{\mu\nu} = \left(\partial^{\mu}\phi\right)\left(\partial^{\nu}\phi\right) - g^{\mu\nu}L \tag{3.206}$$

which in the case of the field (3.205) leads to the energy density equal to

$$\varepsilon = \frac{1}{2}m^2\Phi^2 \tag{3.207}$$

while the pressure is

$$p = -\frac{1}{2}m^2\Phi^2 \cos 2mt. \tag{3.208}$$

From (3.207) and (3.208) we easily obtain that at

$$t \ll \frac{1}{m} \tag{3.209}$$

the scalar field equation has the form

$$p \approx -\varepsilon. \tag{3.210}$$

The equation of state (3.210) holds at an early stage of cosmological evolution of any scalar field, if the terms with derivatives in its Lagrangian are negligible compared to the potential. A variety of options for the inflaton, i.e. for a field that causes inflation, arises on the basis of the chaotic inflation model.

To make the right choice between these possibilities, or at least set a limit for the selection, it is necessary to apply additional, model-dependent manifestations of the inflation mechanism. We list here some of the possible manifestations (Khlopov 1996), leaving more detailed consideration of these manifestations to the following chapters.

Fluctuations at the inflationary stage generate the spectrum of initial density fluctuations, leading to the formation of galaxies and large-scale inhomogeneities on the appropriate scale. The amplitude of these fluctuations is constrained by the observed isotropy of the thermal electromagnetic background.

This eliminates all inflation models with the high amplitude of the predicted fluctuations, in particular, most scenarios of GUT phase transitions.

The simplest models with the equation of state at the inflationary stage close to (3.210) predict the flat shape of the Harrison–Zeldovich spectrum. The estimate of the amplitude of initial fluctuations on the present scale of the cosmological large-scale structure then gives some information about the inflaton, i.e., the form and parameters of the potential of the scalar field.

For more complicated inflationary models, for example, multicomponent inflation, the predicted spectrum of fluctuations can differ from simple flat shape.

The phase transition at the inflation stage leads to specific peaks or plateaux in the spectrum with their position and amplitude determined by the parameters of the model. It is also necessary to take into account the effects of phase transitions after the general inflationary stage, which may modify the initial spectrum of fluctuations.

The inflationary scenario, due to either R^2-inflation or the slow rolling of the scalar field (e.g., chaotic inflation), includes at the end of inflation a long post-inflationary dust stage caused by the coherent oscillations of the inflaton field.

For example, in the case of a massive homogeneous non-interacting scalar field, discussed above, it is easy to establish that at the time t exceeding the period of field oscillations

$$t \gg \frac{1}{m} \tag{3.211}$$

the pressure, averaged over time much longer than the oscillation period will be equal to

$$\langle p \rangle_{tm\gg 1} = \left\langle -\frac{1}{2} m^2 \phi^2 \cos(2mt) \right\rangle_{tm\gg 1} = 0. \tag{3.212}$$

The duration of these stages determines the maximum temperature of the Universe after heating when the radiation-dominated stage starts. It also determines the specific entropy of the Universe after heating[1].

Initial density fluctuations grew at the post-inflationary dust stage. This growth is governed by the general law of development of gravitational instability at the stage of dominance of matter (MD-stage) in an expanding Universe and is defined as

$$\frac{\delta\rho}{\rho} \propto t^{\frac{2}{3}}. \tag{3.213}$$

If the ratio of the cosmological time of the end t_1 and beginning t_0

[1]The widely use term 'reheating' does not seem to be correct, since it assumes return to the hot Universe, what is not the case in the most of the modern inflationary models. Now the term 'preheating' is used to characterize the period after the end of inflation and before the heating of the Universe.

of the dust stage exceeds

$$\frac{t_1}{t_0} > \delta^{-\frac{3}{2}}, \tag{3.214}$$

where δ is the amplitude of the fluctuations, then corresponding inhomogeneities form. The evolution of such inhomogeneities may lead to the formation of primordial black holes (PBH).

The PBH spectrum reflects the scales on which inhomogeneities were formed, as well as the mechanism of formation of the PBHs.

The minimum probability of PBH formation corresponds to the direct formation of PBHs in the contraction of a very small fraction of specially isotropic and homogeneous configurations.

Accounting for PBH formation in the evolution of the main fraction of the inhomogeneities significantly increases the number of expected PBHs.

The peaks in the spectrum of density fluctuations, produced at the inflationary stage, may also form a PBH in the radiation-dominated stage (RD stage) with the probability depending exponentially on the amplitude of the fluctuations.

This makes the PBH an important theoretical tool for studying the physics of inflation which will be discussed in detail in Chapter 4.

Thus, there are various implementations for the inflationary stage.

Internal constraints are imposed on them by the condition of 'a graceful exit from inflation to the Friedmann stage of expansion'. With a strong first-order phase transition such exit should lead to collision of bubbles, the slow roll will result in the birth of particles by coherent oscillations of the scalar field. Such exit can be considered as the evaporation of the condensate of physical particles of the scalar field (Dymnikova, Kravchuk 1995; Dymnikova, Khlopov 1998, 2000).

Another important issue is the physical nature of inflation. The development of inflationary models causes the sector of particle theory, responsible for inflation, to be better and better hidden. The condition of the physical consistency of the physics of inflation, baryosynthesis and dark matter may be thus very important. Physically self-consistent initial conditions for inflationary cosmology (Dymnikova, Khlopov 1998) can be formulated.

We note in conclusion that the various physical mechanisms of inflation do not actually oppose each other and can be combined in the true cosmology of the early Universe. However, one important property is apparently inherent in all inflationary models.

The scale corresponding to the maximum e-folding given by the inflation sets the upper limit on the domain of homogeneity of the present-day Universe.

The flip side of the inflationary nature of the homogeneity of the visible Universe is the inhomogeneity of the Universe far beyond the borders of the present-day cosmological horizon. In studying the Universe as a whole, the physical mechanism proposed as the basis for a homogeneous and isotropic cosmological model inevitably leads to a much more complex cosmological picture.

This is even more true for the current pictures of large extra dimensions and brane cosmology or for the ideas of multiverse or landscape, making our Universe a small bubble in the foam of multi-dimensional space times of different worlds.

2.3 Baryosynthesis

The generally accepted basis of the baryon asymmetry of the Universe is a lack of antimatter in the macroscopic scales up to the scale of galaxy clusters. In the baryon-asymmetric Universe, the observed baryonic matter forms from the initial baryon excess which remains after local annihilation of nucleons and antinucleons in the first millisecond of cosmological evolution.

It is assumed that the baryon excess was formed in the process of baryosynthesis (Sakharov 1967, Kuzmin 1970) which leads to baryon asymmetry of the initially baryon-symmetric Universe.

It turns out that almost all the existing baryosynthesis mechanisms may, under certain conditions, lead to an inhomogeneous baryosynthesis and even to the formation in some places of antibaryon excess. Therefore, inhomogeneity in the distribution of the baryon excess and even existence of domains of antimatter in the Universe can provide a test for mechanisms for baryosynthesis.

In the original Sakharov's scenario (1967) of baryosynthesis the baryon excess is due to CP-violating effects in the out of equilibrium processes with baryon number non-conservation.

To illustrate Sakharov's ideas, we consider the baryon-symmetric Universe in which some particles X and an equal number of their antiparticles are out of equilibrium and decay.

The fact that the baryon number is not conserved in these decays means that the decay products of these particles in different channels have different baryon numbers. For definiteness, we can assume that there are two such different channels, namely the channel

$$X \rightarrow qq \tag{3.215}$$

and the channel

$$X \to \overline{q}l \tag{3.216}$$

where q denotes the quark (e.g., d-quark) and l the charged lepton. The corresponding channels for the antiparticles are

$$\overline{X} \to \overline{qq} \tag{3.217}$$

and

$$\overline{X} \to q\overline{l}. \tag{3.218}$$

Because of the *CPT* invariance, the full widths of the particles and antiparticles will be strictly equal. But due to *CP*-violation the relative probabilities for partial decay modes will not be equal for particles and antiparticles.

The total probability of the decay is assumed to be 1, and for the relative probability of decay (3.215) denoted as

$$Br(X \to qq) = r, \tag{3.219}$$

the relative probability of (3.216) equals

$$Br(X \to \overline{q}l) = 1 - r. \tag{3.220}$$

Then, the relative probability of decay through the channel (3.217) for an antiparticle will be

$$Br(\overline{X} \to \overline{qq}) = \overline{r} \tag{3.221}$$

and for the channel (3.218)

$$Br(\overline{X} \to q\overline{l}) = 1 - \overline{r}. \tag{3.222}$$

Taking into account the baryonic charge of quarks and antiquarks equal to

$$B_q = +\frac{1}{3}, B_{\overline{q}} = -\frac{1}{3} \tag{3.223}$$

we find that as a result of non-equilibrium decay (3.215)–(3.218) in the baryon-symmetric Universe with equal concentrations of particles and antiparticles

$$n_X = n_{\bar{X}} \tag{3.224}$$

an excess of baryons is created equal to

$$n_b = (r - \bar{r}) \cdot n_X. \tag{3.225}$$

The magnitude of the baryon excess is determined by the concentration of the decaying particles as well as the difference between the relative probabilities of the decay modes of particles and antiparticles. This difference is determined by the magnitude and sign of the phase ϕ of *CP* violation.

If the magnitude and even the sign of $\phi(x)$ changes in the space, the same processes with nonconservation of the baryon number will lead to the spatial dependence of the baryon density and even to an excess of antibaryons

$$B(x) < 0 \tag{3.226}$$

in areas where the sign of the *CP*-violating phase changes.

The spatial dependence is predicted in models of spontaneous *CP*-violation or in models where the *CP*-violating phase is associated with the amplitude of the invisible axion. The size and amount of antimatter in the domains formed in this case is determined by the parameters of models of *CP*-violation and/or invisible axion (see the review by Chechetkin et al 1982; Khlopov, Chechetkin 1987; Khlopov 1996).

Supersymmetric extensions (SUSY) of GUT models offer another possible physical origin of the baryon asymmetry. We recall (see Chapter 2) that supersymmetry implies the existence of partners of ordinary quarks with spin 0. These scalar quarks are bosons and can form a Bose condensate. It has been established (Affleck, Dine 1983; Linde 1983) that the superpotential is flat in respect of the existence of scalar quark condensates.

This condensate, formed with the positive sign of the baryon number ($B > 0$), produces a baryon asymmetry after decay of the scalar quarks into quarks and gluinos. However, this mechanism does not fix the magnitude and sign of the baryon charge of the condensate, opening opportunities for the inhomogeneous distribution of the baryon charge as well as antimatter domains.

In Chapter 2 we have already noted that at high temperature the standard model of electroweak interactions (SM) predicts the non-conservation of baryon and lepton charges. This effect imposes severe constraints on the baryosynthesis mechanisms determined by the GUT models and supersymmetric GUT models.

Processes of electroweak non-conservation of baryon and lepton numbers which obey the selection rule

$$B + L = 0, \tag{3.227}$$

wash out any initial baryon asymmetry corresponding to the selection rule

$$B - L = 0. \tag{3.228}$$

In particular, the electroweak non-conservation of the baryon charge cannot create the observed baryon asymmetry in the processes predicted in the GUT model with $SU(5)$ symmetry, since they satisfy the selection rule (3.228).

On the other hand, electroweak non-conservation of the baryon charge opens a new approach to baryosynthesis because it can be used to generate the observed baryon asymmetry.

Shaposhnikov (1996) showed that the observed excess of baryons cannot be produced within the minimal SM. Implementation of the baryosynthesis mechanism associated with electroweak non-conservation of the baryon number requires an extension of SM, which confirms the general relationship of physical fundamentals of baryosynthesis with the hidden sector of the theory of elementary particles.

Among the possible extensions of the SM required to implement baryosynthesis based on electroweak non-conservation of the baryon charge, interesting features are related to the physics of neutrino mass. The Majorana neutrino mass arises from the violation of the lepton number with the selection rule

$$\Delta L = 2. \tag{3.229}$$

CP-violation in non-equilibrium cosmological processes with non-conservation of the lepton number associated with the physics of Majorana neutrino mass, can lead in the early Universe to an excess of leptons which is converted to an excess of baryons due to the electroweak non-conservation of the baryon charge. Such a scenario of 'leptosynthesis' can be realized in a model of horizontal unification which will be discussed further in Chapter 11.

Based on the extension of the SM, the mechanisms of electroweak baryon charge non-conservation can also allow for the possibility of formation of antimatter domains, e.g., in the case of spontaneous *CP*-violation (Comelli et al 1994).

Such domains of antimatter may arise in the baryon-asymmetric Universe, and may be associated with virtually all the baryosynthesis mechanisms, as well as with the mechanisms of *CP*-violation and possible mechanisms of inhomogeneity of the primordial baryon charge.

The domain size depends on the details of the appropriate mechanisms (see review by Dolgov 2002a). With the account for inflation, the size of domains can be up to the present-day horizon, which takes place in a model of 'island Universe' (Dolgov et al 1987) with a very large scale inhomogeneity in the distribution of the baryon charge.

The average effect of the domain structure is determined by the relative amount of antimatter

$$\Omega_a = \frac{\rho_a}{\rho_{cr}}, \tag{3.230}$$

where ρ_a is the cosmological density of antimatter averaged over the large scale, and

$$\rho_{cr} = \frac{3H^2}{8\pi G}$$

is the critical density, spatial distribution and the characteristic size of domains, l or (for small domains) characteristic time, t_{ann} of their annihilation in the surrounding matter.

Antimatter domains can be regarded as a vivid manifestation of the physical nature of the origin of matter in the baryon-asymmetric Universe.

2.4. Non-baryonic dark matter and dark energy

The main arguments in favour of non-baryonic nature of dark matter in the Universe are primordial nucleosynthesis in inflationary cosmology and the formation of the large-scale structure of the Universe at the observed isotropy of relic radiation.

Since both directions will be discussed in detail in the following chapters, we briefly consider here the problems of the old Big Bang model which allows us to solve the hypothesis about the non-baryonic dark matter and dark energy.

The first line of the argument is based on the predictions of the theory of Big Bang nucleosynthesis. These predictions are consistent with the observed abundance of light elements at the baryon density

$$\Omega_b \leq 0.1 \div 0.2, \qquad (3.231)$$

while the total density, predicted by the simplest version of inflationary cosmology, corresponds to

$$\Omega_{tot} = 1. \qquad (3.232)$$

Estimate (3.232) for the total cosmological density is confirmed by the results of BOOMERANG and WMAP experiments.

The difference between the total and baryon densities can be attributed to non-baryonic dark matter and dark energy.

The second line of argument is that the formation of the large-scale structure of the Universe is compatible with the observed isotropy of the thermal electromagnetic background only if some form of weakly interacting substances causes the formation of structures with the minimum effect in the angular distribution of relic radiation.

There were proposed several scenarios of structure formation by the hot (HDM), cold (CDM), warm (WDM), unstable (UDM), mixed (H + CDM), hierarchically-decay (HDS) and other forms of dark matter.

These scenarios are physically different by the sequence in which elements of the structure form and the number of model parameters. But, given the independence of the physical basis for the various candidates for the role of dark matter, in terms of elementary particle physics hot, warm, cold, unstable, and other forms of dark matter do not contradict but complement each other, and should be considered jointly, taking into account the totality of the physical bases.

Indeed, the possible neutrino mass 1÷10 eV was regarded for a long time as a natural base model of hot dark matter. But, as we saw in Chapter 2, the massive neutralinos, predicted in supersymmetric models, or invisible axions associated with the solution to the problem of strong *CP*-violation in QCD lead us to a model of cold dark matter, resulting from physical grounds that in no way are the alternative to the physics of neutrino mass.

Therefore, a mixed scenario of hot + cold dark matter appeared to be more substantiated from the physical point of view than the simple one-parameter model of HDM or CDM.

However, these reasons are not related (at least at first glance) to the problem of quark–lepton generations and the existence of three types of neutrinos.

Physical mechanisms of breaking of symmetry of generations lead to new interactions because of which the massive neutrinos become unstable with respect to decay into lighter neutrinos and a light

Goldstone boson, familon or singlet Majoron. The instability of the neutrino, which is closely associated with breaking of the symmetry of generations, gave the physical basis of the model of unstable dark matter (Doroshkevich, Khlopov 1984; Gelmini 1984; Turner et al 1984).

Through additional parameters, namely, the lifetime of unstable particles, the UDM model eliminated the conflict between the data on the total density within the current inhomogeneities, estimated as

$$\Omega_{\text{inhom}} < 1, \tag{3.233}$$

and the predictions of inflationary cosmology (3.232). In these models, the difference was attributed to the homogeneous background of the weakly interacting decay products of unstable particles of dark matter.

In Chapter 9, we discuss how the UDM models eliminated the disadvantages of scenarios HDM, associated with too rapid evolution of the structure after its formation.

Due to the instability of neutrinos the large-scale structure formed at a redshift consistent with the observed distant objects, survived after the decay of the main fraction of dark matter that formed the structure. Combined with the cosmological constant such scenarios can be of interest for the analysis of the current data of precision cosmology.

Actual multicomponent dark matter may be even more diverse if we take into account the hypothesis of mirror and shadow matter. As stated in Chapter 2, this hypothesis restores the equivalence of left and right coordinate system in the Kaluza–Klein and superstring models.

For example, in heterotic string models $E_8 \otimes E'_8$ it is necessary to take into account the entire set of 248 matter fields and their 248 interactions that occur in the shadow E'_8 sector of the model.

Even the above list of possibilities which is far from complete presents a serious problem of the right choice of a true combination of the different candidates for dark matter in physically based multicomponent dark matter scenarios.

Thus, since the physical basis for all candidates for the role of non-baryonic dark matter is outside the framework of the standard model and loses experimentally tested foundations, we must either consider all possible ways to extend the standard model, considering all the candidates independently, or to seek quantitative methods for their unified description and evaluation of their relative contribution.

Along with dark matter, cosmology today also deals with dark energy of the Universe with the equation of state $p = w\varepsilon$, $w < 0$. At $w = -1$ the dark energy corresponds to the cosmological term. The so-called quintessence energy – an environment with the variable

$-1 \leqslant w < 0$ and hypothetical forms of 'ghost' matter with $w < -1$ are also discussed. Dark energy is assumed to be uniformly distributed and, as we shall see in Chapter 9, in this sense is similar to the decay products of unstable dark matter. Therefore, indirect arguments in favour of dark energy, arising from the comparison of estimates of the total cosmological density and total density of matter in galaxies, their clusters and superclusters, could, in fact, merely confirm the existence of uniform dark matter predicted in UDM models. Direct observational evidence in favour of an accelerated mode of expansion of the Universe, corresponding to the existence of dark energy with $w < -1/3$, are based on observations of distant supernovae of type I, measurements of the Hubble constant and the age of the Universe, as well as on the results of the combined analysis of the data on the large–scale structure and CMB. The existence of dark energy presents to cosmoparticle physics the thorny issue of its physical nature. Unlike dark matter, the physical nature of which may be associated with (meta) stable (super) weakly interacting particles predicted by the extensions of SM, the physics of dark energy can reflect a very non-trivial property of vacuum in the theory of elementary particles.

4

Cosmoarcheology of early Universe

In the new cosmological picture, the very early Universe is a period when there were inflation, baryosynthesis and freezing-out of the particles of dark matter. Uncertainties in the description of these three phenomena lead to ambiguity in the early history of cosmological evolution.

To reproduce the picture of the early Universe, it is necessary to consider all the details needed to identify the fundamental laws that govern physical processes in this period. To do this, the observational data must be associated with effects of ultrahigh-energy physics, which determine the physical conditions in the early Universe.

The full range of problems associated with the detailed development of such methods is the subject of cosmoarcheology.

Cosmoarcheology considers these astrophysical observations as a set of data obtained as a result of *'Gedanken Experiments'* undertaken to test models of cosmological evolution and models of elementary particles, on which they are based.

In the cosmoarcheology of early Universe the complexity of cosmoarcheological analysis is further aggravated by the dependence of the pattern of cosmological evolution on the model of elementary particles which determines this evolution.

To calculate the abundance of the hypothetical particles or analyze the formation of hypothetical objects in the very early Universe, we must make assumptions about the physical conditions in this period. Both the physical processes of creation of the studied particles or objects, and the cosmological context in which these processes take place are determined by the hidden sector of particle physics and should be treated self-consistently. This means that the cosmological consequences of particle theory should be considered in the framework of such a cosmological model which is based on the same theory.

In light of the inflationary model with baryosynthesis and non-baryonic dark matter/energy, the calculations of freezing-out of hypothetical particles or the formation of topological defects in cosmological phase transitions are model-dependent, and such calculations based on the old Big Bang model lose their reliability.

However, in the previous chapter we discussed ways of obtaining information about the early Universe which are less model-dependent, and are suitable for a wide range of different mechanisms for inflation and baryosynthesis. These theoretical tools of cosmoarcheological analysis are mainly primordial black holes.

1. Primordial black holes as a tool for cosmoarcheology

1.1. Primordial black holes

According to the theory of gravity any object with mass M can form a black hole, if this mass is concentrated within its gravitational radius

$$r_g = \frac{2GM}{c^2}. \tag{4.1}$$

The existence of black holes, which follows from general relativity, has historically been viewed within the framework of Newtonian gravity.

Back in the early XIX century, Laplace pointed out that the parabolic velocity on the surface of a supermassive star can reach the speed of light, so that such objects may exist but may not be observable.

Laplace derived the above expression assuming that the the sum of the kinetic energy of light particles with mass m equal to

$$E_k = \frac{mc^2}{2}, \tag{4.2}$$

and its potential energy in the gravitational field of a massive body

$$E_{\text{pot}} = -\frac{GMm}{r} \tag{4.3}$$

equals zero, that is, he found the value of the gravitational radius from the same equality

$$\frac{mc^2}{2} - \frac{GMm}{r_g} = 0, \tag{4.4}$$

which is used to calculate the escape velocity of material bodies.

Although the Laplace's argument is based on corpuscular theory of light, a quantitative answer turned out to be valid even in general relativity.

Theoretical astrophysics considers black holes as the final stage of the evolution of stars with masses exceeding several solar masses. For smaller objects the natural physical conditions of their collapse into a black hole can hardly take place in the present-day Universe.

However, as indicated by Zeldovich and Novikov (1966), black holes with any mass exceeding the Planck mass could, in principle, be formed in the early Universe, because the mass within the cosmological horizon can naturally form a black hole, if the expansion stops in this area. Such a black hole is called the primordial black hole (PBH).

The principal possibility to form a PBH is, however, quite very low in a uniformly expanding Universe, since it involves the metric perturbation of order 1. In the case of metric perturbations with a Gaussian distribution and dispersion

$$\left\langle \delta^2 \right\rangle \ll 1 \tag{4.5}$$

probability of fluctuation of order 1 will be determined by an exponentially small tail of high-amplitude outliers of this distribution.

In a Universe with the equation of state

$$p = \gamma \varepsilon, \tag{4.6}$$

where the numerical factor is in the range

$$0 < \gamma < 1, \tag{4.7}$$

the probability of formation of a black hole from a perturbation inside the cosmological horizon is

$$W_{\text{PBH}} \approx \exp\left\{ -\frac{\gamma^2}{2\left\langle \delta^2 \right\rangle} \right\}. \tag{4.8}$$

Thus, the spectrum of primordial black holes has an exponentially strong sensitivity to perturbation amplitude.

In inflationary cosmology, the spectrum of initial perturbations is fixed by the theory of the large-scale structure and the data on the isotropy of relic radiation. This corresponds to the scale of the fluctuations, starting with the large-scale structure and ranging up to the scale of the present-day horizon. The PBH spectrum reflects the initial inhomogeneity at smaller scales, suggesting, therefore, to check for the far ultraviolet range of the spectrum of initial fluctuations.

The case

$$\gamma \ll 1 \tag{4.9}$$

corresponds to the dust stage of cosmological evolution with an equation of state

$$p = 0, \tag{4.10}$$

and the estimation given in (4.8) cease to be valid.

Using (4.8), we formally find that in the limit $\gamma \to 0$ the probability tends to unity, which is wrong.

Nevertheless, a detailed analysis of the probability of PBH formation in the dust stage shows a significant increase in this probability, compared with the case of the radiation-dominated stage. This makes the PBH spectrum a sensitive indicator of the existence of early dust stages in the early Universe and the physical mechanisms based on the theory of elementary particles and leading to such a stage.

Another source of strong inhomogeneity in the early Universe, which can also lead to the formation of black holes, is related to the phase transition in the early Universe.

Expansion in the false vacuum of bubbles of true vacuum creates a strongly inhomogeneous structure of the bubble walls, so that the energy that is released during the transition from false to true vacuum is converted into kinetic energy of these walls. Collision of bubbles can concentrate this energy within its gravitational radius, creating a black hole. This makes the spectrum of PBH a sensitive indicator of the nature of the phase transition in the early Universe.

The sequence of phase transitions with symmetry breaking can lead in an inflationary Universe to the appearance of topological defects, such as closed walls whose size is determined by the parameters of transitions. Collapse of the walls leads to the formation of the structure of black holes. At certain parameters the structure of these massive black holes could become the seeds of the structure of present-day galaxies.

An important property of black holes is their independence from the forms of matter from which they were born. Therefore, the parameters of superweakly interacting forms of matter (such as shadow matter), forming PBHs, are also available for studies of the spectrum of primordial black holes.

1.2. Primordial black holes as a manifestation of dust stages in the early Universe

In a dust stage, regardless of its origin, gravitational instability develops within the cosmological horizon. In complete analogy with the formation of the modern cosmological structure the growth of small initial fluctuations in such a stage in the early Universe should have led to the formation of inhomogeneities.

Although the size of the inhomogeneities in the early Universe cannot exceed the atomic or even nuclear scale, their evolution can be analysed by the classical theory of gravitational instability.

The mass distribution and other properties of gravitationally bound objects, formed in the dust stage, depend on the initial spectrum of perturbations on the considered scale and also on the nature of the non-relativistic matter, dominant in the dust stage.

The fundamental property of any gravitationally bound objects is that they are unstable against gravitational collapse and will inevitably have to be compressed into a black hole. However, the characteristic time of instability with respect to gravitational collapse is strongly dependent on the rate of evolution of the object, which is determined by the mechanism of energy dissipation in this object.

We can distinguish two fundamentally different types of objects: 'galaxies' and 'stars', which could be formed in the dust stage, and the evolution of which could lead to the formation of black holes.

In the present-day Universe, both stars and galaxies formed from baryonic matter. The baryonic matter of the stars is characterized by the loss of energy due to radiation. This leads to a very fast time scale of the cosmological evolution of stars. The existing stellar black hole candidates are considered as the result of the evolution of stars with masses exceeding several solar masses.

Stars in galaxies represent a collisionless gas. The time scale of the collision of stars is much larger than the age of the Universe and the evolution of galaxies requires much more time than the evolution of stars.

Active galactic nuclei (AGN) are the second example of modern black-hole candidates with masses of order

$$M \sim 10^6 \div 10^9 M_\odot. \tag{4.11}$$

These analogies may be useful in the analysis of possible forms of inhomogeneities and their evolution in the dust stages in the early stages of the Universe.

Consider, for simplicity and certainty, only one type of non-relativistic matter in the early Universe. When its density ρ_m exceeds the density of relativistic particles ρ_γ

$$\rho_m > \rho_\gamma, \qquad (4.12)$$

this indicates the start of the dominance of the non-relativistic matter (MD stage) with the cosmological equation of state $p = 0$.

From the beginning of the MD stage, when

$$t = t_0 \qquad (4.13)$$

the density fluctuations in this non-relativistic matter grow within the cosmological horizon as

$$\frac{\delta\rho}{\rho} \propto t^{\frac{2}{3}}. \qquad (4.14)$$

If the initial amplitude of density perturbations is

$$\frac{\delta\rho}{\rho}(t_0) = \delta, \qquad (4.15)$$

then at time

$$t \sim t_f = t_0 \delta^{-\frac{3}{2}} \qquad (4.16)$$

density perturbations grow to

$$\frac{\delta\rho}{\rho} \sim 1 \qquad (4.17)$$

and form inhomogeneities, separated from the general cosmological expansion. These inhomogeneities are gravitationally bound systems of the considered non-relativistic matter.

The subsequent evolution of gravitationally bound systems is determined by the properties of the non-relativistic matter which forms the inhomogeneities.

The massive weakly interacting particles form gravitationally bound systems of a collisionless gas which by the nature of the subsequent evolution are similar to those of galaxies consisting of a collisionless gas of stars.

Energy dissipation in the gravitationally bound system of weakly interacting massive particles is a very slow process (Zeldovich, Podurets 1965). Basically, it is determined by a lengthy process of evaporation of particles whose velocity exceeds the parabolic velocity of the system.

In the case of binary collisions of gas particles the rough estimate of the characteristic time of evolution of such systems is

$$t_{\text{ev}} = \frac{N}{\ln(N)} t_{ff} \tag{4.18}$$

for gravitationally bound systems consisting of N particles where the free-fall time of the system at density ρ is

$$t_{ff} \sim (4\pi G\rho)^{-\frac{1}{2}}. \tag{4.19}$$

As shown by (Gurzadian and Savvidi 1987), the time evolution of the system can be significantly reduced due to collective effects in a gas of collisionless particles, so that for large N this time will be about

$$t_{\text{ev}} \sim N^{2/3} t_{ff}. \tag{4.20}$$

But even in the latter case at a time when as a result of the growth of small initial perturbations in the stage of dominance of massive weakly interacting particles in the early Universe their gravitationally bound systems form, separated from expansion, the time of evolution of such systems in black holes far exceeds the cosmological time.

This time with the order of magnitude

$$t_f \sim (G\rho)^{-\frac{1}{2}}, \tag{4.21}$$

where ρ is the cosmological density in the period in which the density fluctuations grow on this scale to $\partial\rho/\rho \sim 1$, coincides in the order of magnitude with the free-fall time within the gravitationally bound system, generated by this fluctuation

$$t_f \sim t_{ff}. \tag{4.22}$$

Since

$$t_{\text{ev}} \gg t_f \sim t_{ff}, \tag{4.23}$$

the dust stage should be sufficiently long and ends at

$$t_f > t_{\text{ev}} \gg t_{ff}, \tag{4.24}$$

so that the gravitationally bound system of weakly interacting particles has enough time to form a black hole as a result of its evolution.

Non-relativistic matter interacting with the relativistic particles and radiation forms a gravitationally bound system whose evolution is determined by the energy loss through radiation, as in the case of ordinary stars formed from ordinary matter.

Using this analogy one can naturally conclude that the time evolution of such systems is comparable with the cosmological time or is even shorter, and that the PBHs are produced abundantly as a result of this evolution, even in relatively short periods of dominance of the non-relativistic matter in the early Universe, if the inequality

$$t_e > t_f \geqslant t_{\text{ev}} \tag{4.25}$$

is valid. It turns out that the minimum estimate of the probability of PBH formation in the early dust stages can be obtained, regardless of the form of non-relativistic matter dominant in the Universe.

1.3. Direct formation of PBH in dust stages

The idea of direct PBH formation in the dust stage, first proposed in (Khlopov, Polnarev 1980) (see the review in Polnarev, Khlopov 1985; Chechetkin, Khlopov, Sapozhnikov 1982; Khlopov, Chechetkin 1987, Khlopov 2010) is as follows.

As we mentioned earlier, the mass inside the cosmological horizon would be located just within its gravitational radius, if there were no expansion. The idea of Zeldovich and Novikov (1966) was to stop the relativistic expansion which corresponds to the exponentially small probability in a homogeneous and isotropic Universe.

Another possibility is to study conditions in a dust stage in which the growth of fluctuations leads to the formation of such homogeneous and isotropic configurations that separated from the expansion are contracted within their gravitational radius.

Direct formation of primordial black holes means that once the density fluctuation grows to a value of the order of 1 and has been separated from the general cosmological expansion, the configuration contracts under its own gravitational radius.

By the time t_1, when the contraction starts, the configuration can be characterized by the following quantities:

1) the average density ρ_1 equal by the order of magnitude to the mean cosmological density at the time t_1;

2) size of the configuration r_1;

3) deviation from sphericity s, defined as

$$s = \max\{|\gamma_1 - \gamma_2|, |\gamma_1 - \gamma_3|, |\gamma_2 - \gamma_3|\}, \tag{4.26}$$

where γ_1, γ_2, γ_3 determine the deformation of the configuration along three main orthogonal axes;

4) inhomogeneity u of the density distribution within a configuration, defined as

$$u \sim \frac{\delta\rho_1}{\rho_1}. \tag{4.27}$$

Formation of a black hole as a result of contraction corresponds to the average density

$$\rho_{\text{BH}} \sim \frac{M}{r_g^3} \sim \frac{\rho_1}{x^3}, \tag{4.28}$$

where

$$x = \frac{r_g}{r_1}, \tag{4.29}$$

and

$$r_g = \frac{2GM}{c^2}$$

is the gravitational radius of the configuration with mass M.

On the other hand, the maximum density that can be achieved by compression of non-spherical configurations is

$$\rho_{\max} \sim \frac{\rho_1}{s^3}. \tag{4.30}$$

Equation (4.30) follows from the minimum size of the configuration that can be achieved as a result of contraction

$$r_{\min} \sim \max\{t_c \cdot \Delta v, \Delta r\}, \tag{4.31}$$

where t_c is contraction time,

$$\Delta v \leqslant s \cdot \frac{r_1}{t_c} \tag{4.32}$$

and characterizes the difference in the initial rate of contraction on different axes and

$$\Delta r \leqslant s \cdot r_1 \tag{4.33}$$

is a characteristic of the initial 'flatness' of the configuration.

Equations (4.28) and (4.30) imply that the configuration for the formation of a black hole configuration should be very close to spherically symmetric

$$s \leqslant x \ll 1. \tag{4.34}$$

Contraction of the main fraction of configurations with $s > x$ leads to the formation of objects with a small gravitational potential. As discussed in the previous section, the evolution of these objects depends on the properties of the particles of which they consist.

When the density in the contraction process at time t_{BH} approaches ρ_{BH}, the equation of state within the configuration can be relativistic: $p = 1/3\varepsilon$. To ensure that a sufficient condition for the formation of a black hole exists, the pressure gradient should not exceed the gravitational forces. This limits the inhomogeneity of the configuration at the time t_{BH}

$$\frac{\delta\rho_{\mathrm{BH}}}{\rho_{\mathrm{BH}}} < 1. \tag{4.35}$$

If massive particles are a collisionless gas, the equation of state cannot be changed when the density approaches ρ_{BH}. In this case, to form a black hole it is enough to compress the configuration under the gravitational radius to the moment t_{caus} where the caustics forms in the center of the configuration. The corresponding constraint is given by

$$t_{\mathrm{caus}} > t_{\mathrm{BH}}. \tag{4.36}$$

Indeed, if the condition (4.36) is not satisfied, the collisionless particles, having passed a caustic, can get out of the area inside the gravitational radius, so that the PBH does not form.

If at the time t_{BH} the configuration is fragmented into weakly interacting 'pieces' of the mass $\Delta M \ll M$, the account for the velocity dispersion of these 'pieces' will lead to the condition of PBH formation determined by the same condition (4.36).

The contracting almost spherical dust configuration is described by the Tolman solution. Analysis (Khlopov, Polnarev 1980; Polnarev, Khlopov 1985) of the Tolman solutions shows that the conditions (4.35) and (4.36) are reduced to the same constraint on the inhomogeneity of the configuration at the time t_{BH}

$$u \leq x^{\frac{3}{2}}. \tag{4.37}$$

The conditions of high sphericity and homogeneity, given by (4.34) and (4.37), determine sufficient conditions for the direct formation of primordial black holes.

To estimate the probability that the configuration meets these conditions, we can use the following arguments (Khlopov, Polnarev 1980, Polnarev, Khlopov 1981, 1982, 1985; Khlopov 1999):

1) For the normal law of distribution of configurations on inhomogeneities (u) with a dispersion of 1, the probability of configurations with the anomalously low inhomogeneity

$$u \leqslant x^{3/2} \ll 1$$

is determined by the phase volume corresponding to such configurations, so that the probability of the configuration to be sufficiently homogeneous is

$$W_u \sim u \sim x^{\frac{3}{2}}. \tag{4.38}$$

2) In general (before reduction to the main axes), six independent components of the stress tensor should be considered as random variables. The condition of sphericity (4.34) implies that 5 independent conditions must be simultaneously satisfied.

Namely, we must impose three conditions that the off-diagonal elements are very small and the two conditions that the three diagonal elements are very close to each other.

Assuming that the probability of formation of spherically symmetric configurations satisfying (4.34) is also determined by the phase volume corresponding to the configurations with $s < x$, we find that this probability should be proportional to the fifth power of x

$$W_s \sim x^5. \tag{4.39}$$

Equations (4.38) and (4.39) show that the direct formation of black holes takes place in the dust stage, with the probability of not less than

$$W_{\text{BH}} \geqslant W_s \cdot W_u \sim x^{\frac{13}{2}}. \tag{4.40}$$

Equation (4.40) defines the minimum fraction of matter that forms the black holes with mass M in the dust stage during the formation of primordial black holes.

The mass spectrum of primordial black holes formed by the direct mechanism in the early dust stages can be associated with the spectrum of density fluctuations which are formed in particular in the inflationary stage.

We can show (Khlopov, Polnarev 1980) that the formation of primordial black holes due to the direct mechanism is strongly suppressed for fluctuations within the cosmological horizon before the dust stage, as well as for fluctuations that do not have time to grow to unity before the end of this phase. Thus, the direct mechanism is effective only in the following range of PBH masses

$$M_0 < M < M_{\max}. \tag{4.41}$$

Minimum mass M_0 is the mass within the cosmological horizon at the beginning of a dust stage t_0, which is equal

$$M_0 = \frac{4\pi}{3} \cdot \rho(t_0) \cdot t_0^3 \sim m_{\text{Pl}} \cdot \frac{t_0}{t_{\text{Pl}}}. \tag{4.42}$$

The maximum mass is determined implicitly from the condition that the amplitude of perturbation of mass M, 'coming from the horizon' with the initial amplitude $\delta(M)$ reaches 1 directly at the end of the dust stage t_e. This condition has the form

$$t_e \sim t(M_{\max}) \cdot [\delta(M_{\max})]^{-\frac{3}{2}}. \tag{4.43}$$

In the interval (4.41) the probability of direct formation of primordial black holes is no less than

$$W_{\text{PBH}}(M) \geqslant [\delta(M)]^{\frac{13}{2}}. \tag{4.44}$$

It should be noted that the mechanism of direct formation of the black hole is universal because it does not depend on the form of non-relativistic matter and the period of its dominance in the Universe.

However, its formal application to the modern Universe leads to a very low minimum probability of formation of black holes with masses of order of the mass of superclusters of galaxies.

On the other hand, this mechanism provides a universal model-independent check for inhomogeneities in the dust stage in the very early Universe. The sensitivity of this test, based on astrophysical data, greatly increases when analyzing possible effects of PBH evaporation.

1.3. Evaporation of PBHs

The possibility of evaporation of a black hole, first pointed out by Hawking (1975), is an essential property of primordial black holes related to their possible astrophysical effects in the Universe after the first second of expansion.

Black holes with mass M have a gravitational radius (4.1) which can be rewritten as

$$r_g = 2GM = 2\frac{M}{m_{\text{Pl}}^2} \tag{4.45}$$

in a system of units where $\hbar = c = 1$.

In a strong gravitational field particles could be produced near the black holes (Zeldovich, Starobinsky 1976).

Therefore, due to quantum effects, particles can be emitted from the surface of the black hole (Hawking 1975). Radiation of a black hole is described as radiation from the surface of a black body with temperature T_{PBH} (Hawking 1975; Page 1981)

$$T_{\text{PBH}} = \frac{1}{4\pi r_g} = \frac{m_{\text{Pl}}^2}{8\pi M}. \tag{4.46}$$

The luminosity of the black hole is of the order of

$$\frac{dE}{dt} \sim T^4 r_r^2 \sim r_g^{-2} \sim \frac{m_{\text{Pl}}^4}{M^2}. \tag{4.47}$$

The energy loss (4.47) implies that the black hole loses its mass at a rate of

$$\frac{dM}{dt} = -\frac{dE}{dt} \sim \frac{m_{\rm Pl}^4}{M^2}. \tag{4.48}$$

From equation (4.48) it is easy to find that over the time

$$t_e \sim \left(\frac{M}{m_{\rm Pl}}\right)^3 t_{\rm Pl} \tag{4.49}$$

the black hole with mass M would lose all its mass. This means that the black hole evaporates.

With all the numerical factors taken into account, the evaporation time of a black hole is equal to (Hawking 1975; Novikov et al 1979; Hawking, 1976)

$$t_e = 10^{-27}\,{\rm s}\left(\frac{M}{1{\rm g}}\right)^3. \tag{4.50}$$

According to the theory of evolution of stars, black holes must be formed from stars with the mass greater than at least $M > 2M_\odot$.

The evaporation time of such black holes is of the order

$$t_e \sim 10^{66} \text{ years}$$

and their evaporation can be neglected. But the masses of primordial black holes (PBH) may be much smaller than the masses of stars. The PBH mass may be practically any value down the Planck mass $m_{\rm Pl}$ (Markov 1965, 1966) and even less than the Planck mass (Zeldovich, 1981).

For a PBH with the mass less than

$$M < M_e \sim 10^{15} \text{ g} \tag{4.51}$$

the evaporation time is less than the age of the Universe. Formed at the early stages of evolution of the Universe, these black holes should have evaporated completely by now.

But the effects of their evaporation can lead to observable consequences, thus providing a validation of their existence in the past. A detailed discussion of PBH and constraints on their possible concentration in the early Universe can be found in articles by Carr (1975), Hawking (1974, 1975), Zeldovich and Starobinsky (1976), Novikov et al (1979), Chechetkin et al (1982a), Khlopov, Polnarev (1982, 1985), Khlopov et al (1980), Khlopov, Chechetkin (1987), Khlopov (1999) and references therein.

Particle fluxes arising from the evaporation of primordial black holes may have the same effect on the physical processes in the Universe as products of decay of metastable particles. The difference lies in the fact that the evaporation of primordial black holes should produce, without exception, all types of particles with the mass $m < T_{PBH}$ predicted by the physics of elementary particles.

Regardless of the constants of their interaction, the rate of formation of particles of any type in PBH evaporation is determined by thermodynamic equilibrium at temperature T_{PBH}.

Due to the gravitational mechanism of particle creation, the evaporation of a black hole is a unique universal source for all of types of particles existing in nature whose mass does not exceed T_{PBH}.

2. The formation of black holes in the first order phase transitions

In the process of the first order phase transition bubble wall collisions can lead to the formation of primordial black holes (Hawking, Moss, Stewart 1982; Moss 1994), concentrating the kinetic energy of the wall within its gravitational radius. However, this concentration requires a very special conditions of the collision, say, the simultaneous collision of several walls, which strongly suppresses the probability of PBH formation.

The mechanism of PBH formation in the collision of only two walls (Konoplich et al 1998, 1999) removes this constraint thus making the spectrum of PBH a sensitive indicator for cosmological phase transitions.

The simplest example which leads to the cosmological first order phase transition with the birth of the bubbles is provided by the theory of a scalar field with two degenerate vacuum states. The state with lower energy is the true vacuum, and the state with higher energy corresponds to a false vacuum.

Being stable at the classical level, the state of false vacuum decays due to quantum effects that lead to the emergence of true vacuum bubbles and their subsequent growth in the false vacuum. The potential energy of the false vacuum is converted into kinetic energy of the walls, thus making within a short time their rate of expansion ultra-relativistic.

The bubble continues to expand until it collides with another bubble. As shown by Hawking et al (1982) and Moss (1994), the black hole can be produced directly by the collision of several walls. Research (Konoplich et al 1998, 1999) indicated a mechanism by which black holes can be formed with a probability of 1 in collision of the walls of only two bubbles. This leads to rapid formation of black holes, with significant cosmological implications.

2.1. Configuration of the field in the collision of bubble walls

Consider a field theory, in which the probability of decay of the false vacuum per unit volume is equal to Γ and the difference of energy of the false and true vacuum in volume is ρ_V.

Vacuum decay occurs through the formation of bubbles with a new phase, limited initially by stationary walls. Following Hawking et al (1982) and Watkins, Widrow (1992), we also assume for simplicity that the size of the horizon is much greater than the distance between the bubbles.

The resulting bubble begins to expand into the region of false vacuum. The speed of the walls quickly grows up to the speed of light

$$v = c = 1 \tag{4.52}$$

due to the transition of the energy of the false vacuum to their kinetic energy.

Consider the dynamics of the collision of two bubbles that were born in the points $(\vec{r}_1, t_1)$ and $(\vec{r}_2, t_2)$ and are expanding into the region of false vacuum.

Immediately after the collision, mutual penetration of the walls at distances comparable with the thickness of the wall, is accompanied by an additional increase in the potential energy (Konoplich 1980). Then the walls are reflected and travel in the opposite direction toward the true vacuum.

The space between them is filled with a field in a state of false vacuum, converting the kinetic energy of the walls back to the energy of the false vacuum and reducing the speed of the walls.

Meanwhile, the outer part of the wall, continuing to expand and accelerate, absorbs false vacuum in the outer region.

Inevitably, there is some point, depending on the parameters of the theory, when the central region of the false vacuum is separated and forms a separate 'bag' of false vacuum.

Successive stages of the formation of the false vacuum bag are shown in Fig. 4.1.

As shown by Hawking et al (1982) and Watkins, Widrow (1992), further evolution of the false vacuum bag comprises the following steps:

1) the bag grows to a certain size D_M, while the kinetic energy of the wall becomes zero;

2) the false vacuum bag starts to contract to size D^*, comparable to the thickness of the wall;

3) the volume expands again and then compresses again, so that the successive compression and expansion of the false vacuum bag continues.

This process of periodic change of the contraction and expansion (oscillation) results in a loss of energy of the false vacuum bag converting into oscillations of a classical scalar field. Watkins, Widrow (1992) and Belova, Kudryavtsev (1988) in their works have shown that only a few such oscillations can take place.

On the other hand, it is important to note that the secondary oscillations can occur only if the minimum size of the bag exceeds its gravitational radius,

$$D^* > r_g. \tag{4.53}$$

The reverse case

$$D^* < r_g \tag{4.54}$$

leads to the formation of a black hole with the mass of the false vacuum bag.

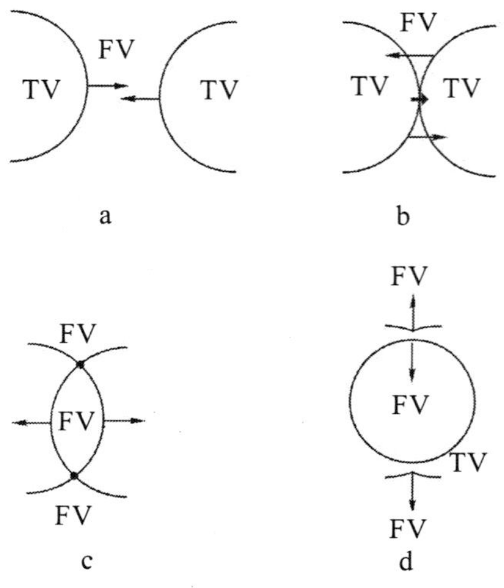

Fig. 4.1. Formation of a bag of the false vacuum in the collision of two bubble walls.

The basic idea of the indirect mechanism for the formation of black holes in the collision of two bubbles at the first order phase transition is to consider the contraction of the false vacuum bag within its gravitational radius. As we show below, the probability of a formation of a black hole in this process is of the order of 1 for a wide range of parameters of theories which predict first order phase transitions in the very early Universe.

It should be noted that the false vacuum bag is not spherically symmetric at the time of its separation from the outer wall, but the tension of the wall provides the spherical symmetry in the first compression.

2.2. Gravitational collapse of the false vacuum bag and the formation of PBH

We now consider in more detail the conditions for transforming a false vacuum bag into a black hole.

The mass M of the false vacuum bag can be expressed through the parameters of the theory and calculated in a coordinate system K' in which the colliding bubbles formed simultaneously.

In this system, the radius of each bubble in the first moment of the collision is equal to half the initial distance between the centers of bubbles. Obviously, the maximum size of the bag is of the same order as the size of the bubble, since this size is the only parameter with this scale:

$$D_M = 2b'C. \tag{4.55}$$

The parameter

$$C \cong 1 \tag{4.56}$$

can be calculated within each specific theory, but its numerical value does not have a strong influence on the result.

One can find the mass of the bag formed by the collision of two bubbles with radius b'

$$M = \frac{4\pi}{3}(Cb')^3 \rho_V. \tag{4.57}$$

This mass is contained in the compressed region of false vacuum. The minimum size of the bag D^* will be around or less than the thickness of the wall.

A black hole is born, if the minimum size of the false vacuum bag is less than its gravitational radius. This means that, at least, provided

$$\Delta < r_g = 2GM \tag{4.58}$$

it must be compressed into a black hole.

As an example, consider a simple model with the Lagrangian

$$L = \frac{1}{2}\left(\partial_\mu \Phi\right)^2 - \frac{\lambda}{8}\left(\Phi^2 - \Phi_0^2\right)^2 + \varepsilon\Phi_0^3\left(\Phi + \Phi_0\right). \tag{4.59}$$

In a thin wall approximation, the wall thickness can be expressed through the parameters of the Lagrangian as

$$\Delta = 2\left(\sqrt{\lambda}\Phi_0\right)^{-1}. \tag{4.60}$$

Using condition (4.58), it is easy to get that the false vacuum bag with the mass

$$M > \left(\sqrt{\lambda}G\Phi_0\right)^{-1} \tag{4.61}$$

should form a black hole with mass M.

Here we should note that this conclusion is true in the case where the false vacuum bag is completely within the cosmological horizon, namely when the following condition is valid:

$$M_H > \left(\sqrt{\lambda}G\Phi_0\right)^{-1}. \tag{4.62}$$

In the time of the transition, which may be related to the symmetry breaking scale Φ_0

$$t = \frac{m_{\mathrm{Pl}}}{\Phi_0^2}, \tag{4.63}$$

the mass M_H within the cosmological horizon is

$$M_H \cong \frac{m_{\mathrm{Pl}}^3}{\Phi_0^2}. \tag{4.64}$$

Thus, for the potential (4.59) provided

$$\lambda > \left(\frac{\Phi_0}{m_{\mathrm{Pl}}}\right)^2 \tag{4.65}$$

the formation of a black hole from a false vacuum bag is possible. It is easy to see that this condition is true for any realistic particle theory.

We now consider, following (Konoplich et al 1998), the distribution on the mass and velocity of black holes formed by the collision of bubbles assuming that their masses satisfy (4.58). These values depend on the location and time of birth of colliding bubbles, so that we have the mass M and velocity v of the black hole as a function of these parameters:

$$M = M\left(\left|\vec{r}_2 - \vec{r}_1\right|, t_2 - t_1\right) \tag{4.66}$$

and

$$v = v\left(\left|\vec{r}_2 - \vec{r}_1\right|, t_2 - t_1\right). \tag{4.67}$$

The probability dP of collision of two bubbles, which were born at a distance $\vec{r}_2 - \vec{r}_1$ from each other at times t_2 and t_1, has the form

$$dP = dP_1 dP_2 dP_-, \tag{4.68}$$

where

$$dP_1 = \Gamma dt_1 d\vec{r}_1 \tag{4.69}$$

is the probability of the birth of a bubble in a space-time point with coordinates $(t_1, \vec{r}_1)$, and

$$dP_2 = \Gamma dt_2 4\pi \left|\vec{r}_2 - \vec{r}_1\right|^2 d\left|\vec{r}_2 - \vec{r}_1\right| \tag{4.70}$$

is the probability of nucleation of the second bubble at a distance

$$\left|\vec{r}_2 - \vec{r}_1\right| \equiv 2b \tag{4.71}$$

from the first. In the integration over the angles we assume spatial isotropy.

The factor

$$P_- \equiv e^{-\Gamma\Theta} \tag{4.72}$$

determines the probability that the other bubbles are not produced in the 4-dimensional region Θ in which the two discussed bubbles collide. The probability density of decay of the false vacuum Γ is regarded as a free parameter. Calculation of the area Θ is carried out below.

After integration, assuming that the probability of decay of the false vacuum is independent of time, we get

$$\frac{dP}{V_c} = 32\pi\Gamma^2 e^{-\Gamma\Theta} b^2 dt_1 dt_2 db. \tag{4.73}$$

Here V_c is the volume within the cosmological horizon at the phase transition.

In the following we shall consider a system K' referred to above, which moves with the velocity (in this chapter we use a system of units where the speed of light is unity)

$$v = \frac{(t_1 - t_2)}{2b}. \tag{4.74}$$

Obviously, speed v is the speed of the false vacuum bag and of the formed black hole. The radius of the colliding bubbles is given by

$$b' = \frac{b}{\gamma}, \tag{4.75}$$

where

$$\gamma = \left(1 - v^2\right)^{-\frac{1}{2}}.$$

Using (4.57) and (4.73), we can easily obtain the distribution of the false vacuum bags, expressed in terms of new variables M, v, t, where M is the mass of the false vacuum bag (or BH), which is formed in the collision of bubbles, v is their velocity and (in this chapter we use a system of units where the speed of light is unity)

$$t = b + \frac{t_1 + t_2}{2} \tag{4.76}$$

is the first moment of contact of the bubbles. We find that this distribution is (Konoplich et al 1998)

$$\frac{1}{V_c}\frac{dP}{dv\,dM} = \int \frac{64\pi}{3}\Gamma^2 e^{-\Gamma\Theta}\gamma^4 \left(\frac{M}{C\rho_V}\right)^{\frac{1}{3}} \frac{dt}{C\rho_V}. \tag{4.77}$$

A very difficult problem of precise determination of the 4-dimensional region Θ can be considered in the following reasonable approximation (Konoplich et al 1998).

Namely, we assume that every bubble which reaches a sphere with a radius b' and centre at point O (Fig. 4.1), at the time of first contact of bubbles t' prevents the birth of bubbles. In this approximation, the region Θ can be estimated quite simply

$$\Theta = \int_0^{t'} d\tau' d^3 r' \theta(\tau' + r' - b' - t') = \frac{\pi}{3}\left[(b'+t')^4 - (b')^4\right]. \qquad (4.78)$$

Here, the parameter b' is related to the mass M according to (4.57), and the time t' in the system K' is

$$t' = \gamma t. \qquad (4.79)$$

Consequently, after integration over t the distribution on mass and velocity has the form (Konoplich et al 1998)

$$\frac{1}{V_c}\frac{dP}{dv\,dM} = \frac{64\pi}{3}\Gamma^2 \exp\left\{-\Gamma\frac{\pi}{3}\left(\frac{M}{C\rho_V}\right)^{\frac{4}{3}}\right\}\gamma^4\left(\frac{M}{C\rho_V}\right)^{\frac{1}{3}}\frac{I}{C\rho_V}, \qquad (4.80)$$

where

$$I = \int_{t_-}^{\infty} d\tau \exp\left\{-\frac{\pi}{3}\Gamma\left[\left(\frac{M}{C\rho_V}\right)^{\frac{1}{3}} + \tau\gamma\right]^4\right\} \qquad (4.81)$$

and

$$t_- = (1+v)\gamma\left(\frac{M}{C\rho_V}\right)^{\frac{1}{3}}. \qquad (4.82)$$

Let's compare the volume containing one bag of the false vacuum V_m with the volume of one bubble V_n at the end of the phase transition.

After numerical integration of expression (4.78)

$$V_m \approx V_{\mathrm{BH}} \approx 3.9\Gamma^{-\frac{3}{4}}. \qquad (4.83)$$

On the other hand, the average volume of one bubble

$$V_n = \frac{4\pi}{3}\left(\frac{3}{\pi}\right)^{\frac{3}{4}}\Gamma^{-\frac{3}{4}} \approx 4.0\cdot\Gamma^{-\frac{3}{4}}, \qquad (4.84)$$

where we have assumed that the bubbles are spherical in shape.

Performing the corresponding equality

$$V_m = V_n \tag{4.85}$$

fully confirms the above approximation.

In general, the distribution (4.80) can be expressed through the dimensionless variable μ

$$\mu \equiv \left(\frac{\pi}{3}\Gamma\right)^{\frac{1}{4}} \left(\frac{M}{C\rho_V}\right)^{\frac{1}{3}} \tag{4.86}$$

and has the form

$$\frac{1}{V_c}\frac{dP}{dv\,d\mu} = 64\pi\left(\frac{\pi}{3}\right)^{1/4} \mu^3 \exp\left(\mu^4\right)\gamma^3 J\left(v,\mu\right). \tag{4.87}$$

Here

$$J\left(\mu,v\right) = \int_{t_-}^{\infty} d\tau \exp\left(-\tau^4\right), \tag{4.88}$$

where

$$t_- = \mu\left(1+\gamma^2\left(1+v\right)\right). \tag{4.89}$$

The resulting distribution is very narrow: the number of black holes with masses up to 30 times more than the average should be suppressed by a factor of order 10^5.

The average value of the dimensionless 'mass' is equal to

$$\mu = 0.32. \tag{4.90}$$

Using this value (Konoplich et al 1998), we can relate the average mass of a black hole with the average volume containing it at the time of transition:

$$\langle M_{\mathrm{BH}} \rangle = 0.25 C\mu^3 \rho_V \langle V_{\mathrm{BH}} \rangle \approx 0.012 C\rho_V \langle V_{\mathrm{BH}} \rangle. \tag{4.91}$$

Recall that the constant C here is the model-dependent parameter of order 1 and its exact value is not essential for further evaluations.

2.3. First order phase transitions in the early Universe

Inflationary models in which inflation ends in a first order phase transition, occupy a prominent position in modern cosmology of the early Universe (see, for example, La, Steinhardt 1989; Holman, Kolb, Wang 1990; Adams, Freeze 1991; Copeland et al 1994; Occhionero, Amendola 1994).

In particular, within the framework of such models we can regard large-scale voids as remnants of the primary bubbles with a typical size of several tens of Mpc (Copeland et al 1994; Occhionero, Amendola 1994). A detailed analysis of cosmological phase transitions in the context of generalized inflation models can be found in (Turner et al 1992).

In what follows, we focus only on the last point of the inflationary stage when the phase transition has already been completed. Recall that the first order phase transition is regarded as completed immediately after the percolation regime is established in the true vacuum. The beginning of this regime roughly corresponds to the creation of at least one bubble in a single Hubble volume. Exact calculations (Turner et al 1992) showed that the phase transition is completed, if the condition

$$Q \equiv \frac{4\pi}{9}\left(\frac{\Gamma}{H^4}\right)_{t_{\text{end}}} = 1. \tag{4.92}$$

Here Γ is the bubble nucleation rate, t_{end} indicates the time of phase transition, and H is the Hubble constant during that period.

In the framework of inflationary models the true vacuum fills the space as a result of first order phase transition due to the collision of bubbles nucleated at the last moment of exponential expansion. The collisions between these bubbles occur when their size l is below or equal to the effective Hubble horizon

$$l \leqslant H_{\text{end}}^{-1} \tag{4.93}$$

in the phase transition era.

If we take

$$H_0 = 100h \frac{\text{km}}{\text{s} \cdot \text{Mpc}}$$

for the Universe with

$$\Omega = 1$$

the size of these bubbles is approximately equal to

$$l \sim 10^{-21} h^{-1} \text{ Mpc}. \tag{4.94}$$

So, at first glance, it seems that such bubbles rapidly thermalize, leaving no trace in the distribution of matter and radiation.

However, in the previous section it was shown that for realistic parameters of the field theory underlying the physics of the considered phase transition, binary collisions of bubbles lead to the formation of black holes with a probability of 1.

The mass of such black holes is

$$M_{\rm BH} = \gamma_1 M_n, \tag{4.95}$$

where

$$\gamma_1 < 10^{-2} \tag{4.96}$$

and M_n denotes the mass which may be in a bubble during the collision of bubbles, if the energy density of the bubble walls is completely thermalized.

The mechanism proposed in (Konoplich et al 1998) leads to a new possibility of PBH formation in the era of preheating in inflationary models involving the first order phase transition at the end of inflation.

Formation of primordial black holes in the evolution of small initial perturbations in the post-inflationary dust stage can occur with a fairly high probability, but is a very time-consuming process. The fast PBH formation from the tail of a Gaussian distribution of fluctuations in the case of small-amplitude fluctuations in the post-inflationary stage is exponentially suppressed.

A completely different situation arises at the end of inflation if there is a first order phase transition.

Namely, in the regime of percolation of true vacuum collision of the bubbles with the size of the order of the effective Hubble radius leads to the formation of black holes with masses

$$M_0 = \gamma_1 M_{\rm end}^{\rm hor}, \tag{4.97}$$

where $M_{\rm end}^{\rm hor}$ is the mass within the Hubble horizon at the end of inflation. According to (4.91), the initial mass fraction of such PBHs is given by

$$\beta_0 = \gamma_1 \exp(-1). \tag{4.98}$$

The factor of suppression

$$P_{\_} = \exp(-1) \tag{4.99}$$

giving rise to a slight difference with (4.91), is associated with the condition of lack of bubble nucleation inside the false vacuum bag, until it shrinks to PBH.

The expression for the mass of PBH (4.97) can be rewritten in terms of the energy scale of inflation H_{end} as

$$M_0 = \frac{\gamma_1}{2} \frac{m_{\text{Pl}}^2}{H_{\text{end}}}. \tag{4.100}$$

Take, for example, in (4.100) the estimate of the scale of inflation

$$H_{\text{end}} \approx 4 \cdot 10^6 m_{\text{Pl}}, \tag{4.101}$$

which follows from the observation of the anisotropy of the thermal electromagnetic background. Then we find that the fraction of the total density

$$\beta_0 = 6 \cdot 10^{-3} \tag{4.102}$$

is in the form of PBHs with mass

$$M_0 \approx 0.7 \text{ g} \tag{4.103}$$

immediately after the phase transition.

In the RD stage after heating of the Universe, the relative contribution of these primordial black holes to the total cosmological density increases as the scale factor, and this means that at the moment

$$t_1 \approx \left(\beta_0^2 H_{\text{end}}\right)^{-1} \tag{4.104}$$

over 50% of the total density is contained in the PBH.

Since PBHs behave like dust matter, the equation of state of the Universe takes the form

$$p = 0$$

starting at time t_1. The dominance of the PBH dust stage ends at the time of complete evaporation of primordial black holes:

$$t_2 = \frac{1}{g}\left(\frac{M_0}{m_{\text{Pl}}}\right)^3 t_{\text{Pl}}, \tag{4.105}$$

where g is the effective number of degrees of freedom of massless particles at this time.

In principle, black holes with a greater mass can form in the PBH-dominated stage. However, the probability of formation of more massive primordial black holes would be negligible for non-ultraviolet spectrum of density perturbations. If the initial spectrum is not growing to small scale, the amplitude of initial perturbations in the post-inflation stage does not exceed

$$\delta \cong 10^{-6}. \tag{4.106}$$

For such a small amplitude, the probability of direct formation of black holes in the dust stage is very small, and the evolutionary process of PBH formation requires much more time than the evaporation time of initial black holes with masses (4.103).

There are a set of observational constraints on the maximum allowable contribution of PBH in the total density corresponding to different ranges of masses of the PBHs (Zeldovich, Starobinsky 1976; Naselsky 1978; Miayama, Sato 1978; Lindley 1980; Rothman, Matzner 1981; MacGibbon, Carr 1991; Hawking 1974; Polnarev, Khlopov, 1985). Based on astrophysical observations, these restrictions have used different manifestations of the possible existence of primordial black holes and their results can be divided into two large groups.

The first group of constraints is based on an analysis of the effects of PBH evaporation due to the Hawking effect, while the second group – on analysis of only the gravitational effects of PBH.

PBH evaporation leads to potentially observable astrophysical consequences. The observations set the upper limit of the maximum number of PBHs allowed in the period of their evaporation. PBHs with mass

$$M_{ev} \leqslant 5 \cdot 10^{14}\,\mathrm{g} \tag{4.107}$$

evaporated prior to the modern era.

For more massive PBHs the evaporation effect is not significant, and they must be present in the modern Universe. Universal constraints on them follow from the condition that their density should not exceed the upper limit on the total density, which in the case of simple inflationary scenario is

$$\Omega_{\mathrm{PBH}} \leqslant 1. \tag{4.108}$$

It is widely believed that the evaporation process leads to a complete disappearance of the PBH (Hawking 1974), but there are also various arguments for the existence of stable remnants of evaporation (Markov 1984; Zeldovich 1981; Barrow et al 1992; Carr, Gilbert 1994; Alekseev, Pomazanov 1997; Dymnikova 1996).

If we assume that the black holes leave stable remnants with a mass of the order

$$m_\gamma = km_{\mathrm{Pl}}, \tag{4.109}$$

where

$$k \cong 1 \div 10^3, \tag{4.110}$$

then we can evaluate the current density of such remnants from all PBHs which evaporated by now.

In particular, we can estimate the density of the remnants of evaporation of low mass PBH, born during the first order phase transition after inflation.

Since the estimates provide a fairly high probability of formation of such primordial black holes, these PBH will soon begin to dominate in the Universe and implement an early dust stage of their dominance, which ends in the period of their evaporation. The fraction of the total density of the Universe, turning into the remnants of evaporation is thus

$$\alpha_\gamma = k\left(\frac{m_{\mathrm{Pl}}}{M_0}\right)\left(\frac{t_{eq}}{t_2}\right)^{\frac{1}{2}}, \tag{4.111}$$

where t_2 is the period of evaporation of PBHs with mass M_0 and the beginning of the modern MD stage

$$t_{eq} = 3.2\cdot 10^{10} h^{-4}\ \mathrm{s}. \tag{4.112}$$

From the condition that the current dominance of matter cannot begin with

$$t < t_{eq} \tag{4.113}$$

we find (Konoplich et al 1998) that the inequality

$$\alpha_\gamma < 1 \tag{4.114}$$

must be valid. Using (4.100). (4.109) and (4.110), we can express this condition as a restriction

$$\frac{H_{\rm end}}{m_{\rm Pl}} < \frac{0.6\gamma_1}{g^{1/5}k^{2/5}}\left(\frac{t_{\rm Pl}}{t_{eq}}\right)^{1/5} \leqslant 10^{-11}\frac{h^{4/5}\gamma_1}{g^{1/5}k^{2/5}}. \tag{4.115}$$

On the other hand, the restriction (Zeldovich and Starobinsky 1976), obtained from the condition that the evaporation of primordial black holes does not lead to overproduction of entropy in the Universe, is given by

$$\beta(M) < 10^{-8}\left(\frac{10^{11}\,\mathrm{g}}{M}\right) \tag{4.116}$$

and implies that evaporating PBHs can be the dominant part of the cosmological density during their formation only if their mass does not exceed 10^3 g. In principle, this means that all the observed entropy of the Universe may be caused by the evaporation of primordial black holes of small masses. So, PBHs with mass

$$M < 10^3\ \mathrm{g} \tag{4.117}$$

could be produced in the early Universe with a probability of the order of 1, without contradicting the observational constraints. Moreover, the entropy of the observable Universe could be attributed to evaporation of low mass primordial black holes.

For the PBHs born as a result of phase transition at the end of inflation, the condition that the observed entropy is a result of evaporation of primordial black holes implies the following lower limit on the energy scale of inflation

$$\frac{H_{\rm end}}{m_{\rm Pl}} \geqslant 10^{-9}\gamma_1. \tag{4.118}$$

The conditions (4.115) and (4.118) become incompatible if the hypothesis of the existence of stable remnants of evaporation of primordial black holes is confirmed.

We can conclude that the existence of stable remnants of evaporation of primordial black holes is difficult to reconcile with the presence of the first order phase transition at the end of inflation.

5

Relic particles in the period of Big Bang nucleosynthesis

1. Effects of new particles and nucleosynthesis

The simplest way to use the observed abundance of light elements to test the theory of elementary particles is to assess the effect of hypothetical new phenomena on the parameters of the theory of standard Big Bang nucleosynthesis (SBBN).

In this approach, the effects of new particles can be reduced to the parameters of the theory of Big Bang nucleosynthesis, so that the constraints on the possible properties of these particles follow from the constraints on these parameters.

To limit the parameters of the theory of Big Bang nucleosynthesis, the observed abundance of light elements, extrapolated to the initial value, is compared with the results of detailed calculation of this phenomenon (Wagoner et al 1967; see also Gamov 1946, 1948; review and references in Khlopov, Chechetkin 1987).

Let us consider the basic assumptions, parameters and results of these calculations.

Assumptions

1) Expansion of the Universe, considered on the basis of the metric theory of gravity;

2) Lack of strong inhomogeneity and anisotropy of the Universe during nucleosynthesis (about the effects of inhomogeneity of nucleosynthesis, see Zeldovich 1975; Scherrer, et al 1987; Alcock et al 1987; Fowler, Malaney 1988; Kurki-Suonio et al 1990. On the effects of anisotropy, see Rothman, Matzner 1984; Zeldovich, Novikov 1975, and references therein);

3) Existence of an early stage of the Universe with high temperature

$$T > 10^{10}\,\mathrm{K}. \qquad (5.1)$$

At this stage, there was an equilibrium between the nucleons, the neutrino–antineutrino, electron–positron pairs, and radiation. During the subsequent expansion, the β-processes come out of equilibrium, and there is 'freezing-out' of the ratio of neutrons and protons.

4) All antinucleon annihilation processes ended prior to the period of nucleosynthesis, and the effects of annihilation had practically no effect on nucleosynthesis (the effects of annihilation or non-equilibrium particles are discussed below).

Parameters

1) Baryon density. It is assumed that the cosmological expansion is adiabatic (see Chapter 3), so that the specific entropy (compare with (3.31)) is constant

$$s_b = \frac{T^3}{\rho_b} \propto \frac{n_\gamma}{n_b} = \text{const.} \tag{5.2}$$

If there was no additional energy generation increasing the density of photons (except for the annihilation of electron–positron pairs taken into account explicitly in the calculations), the value s_b is determined by the current ratio of baryons to photons.

2) The parameter ξ that determines the rate of expansion in the stage $p = 1/3\varepsilon$ as

$$\frac{dv}{v\,dt} = \frac{\ddot{a}}{a} = -\xi\sqrt{24\pi G\rho_n}. \tag{5.3}$$

Here a is the scale factor, and ρ_n is the cosmological density during nucleosynthesis, which takes into account all known types of particles (photons, the electron–positron pairs, neutrinos and antineutrinos). The relative contribution of baryons to the total cosmological density does not exceed

$$\frac{\rho_b}{\rho_{\text{tot}}} \leqslant 10^{-5}. \tag{5.4}$$

If the parameter $\xi > 1$, then the Universe must have contained some new types of relativistic particles.

Schvartsman (1969) first received from the results of cosmological nucleosynthesis the limit on the total possible number of types of light difficult to detect particles, i.e. the value ξ. In particular, assuming that all such particles are a new type of neutrino, one can obtain an upper limit on the number of neutrino types. According to (Shvartsman 1969), this upper limit on the number of types of neutrinos was

$$\Delta N_\nu < 8 \div 10 \tag{5.5}$$

and was of no practical interest for neutrino physics.

However, in subsequent studies (Steigman et al 1977; Yang et al 1979, Steigman et al 1986) the limit was reduced to

$$N_\nu < 1, \tag{5.6}$$

that has long been the strongest restriction on the number of neutrino types (Schramm, Copi 1990).

The strict limit, obtained in these studies, corresponds to the absence of new generations of quarks and leptons, except the three known generations,

$$\begin{pmatrix} \nu_e \\ e \\ u \\ d \end{pmatrix}; \begin{pmatrix} \nu_\mu \\ \mu \\ c \\ s \end{pmatrix}; \begin{pmatrix} \nu_\tau \\ \tau \\ t \\ b \end{pmatrix}.$$

related to the three known types of neutrinos ν_e, ν_μ, ν_τ.

The existence of additional families is consistent with a strict limit (5.6) if the corresponding neutrinos are massive, since the contribution to the total density of heavy neutrinos or hypothetical stable heavy leptons with the mass

$$m_\nu \gg 1 \text{ MeV} \tag{5.7}$$

is small

$$\frac{\rho_\nu}{\rho_{\text{tot}}} \propto \left(\frac{3 \text{ MeV}}{m_\nu} \right)^2 \tag{5.8}$$

due to the annihilation of most of these particles during their freezing-out in the early Universe at

$$T \leqslant m_\nu \tag{5.9}$$

(see reviews Dolgov, Zeldovich 1981, and Chapter 3).

In this case, complementing the cosmological limit (5.6) by the results of measuring the width of decay of Z^0-boson, we can exclude the existence of new heavy neutrinos with the mass less than

$$m_\nu \leqslant \frac{m_Z}{2}, \tag{5.10}$$

where m_Z is the mass of the Z^0-boson.

On the other hand, the upper limit on the value ξ imposes strong constraints on models of elementary particles predicting new types of light particles associated with the fundamental symmetries assumed in such models.

For example, in violation of *CP*-invariance the equivalence of left and right coordinate systems can be restored only if there are invisible twins of ordinary particles – their mirror partners (Lee, Yang 1956; Kobzarev et al 1966; Okun 1983).

Such particles hardly interact with ordinary particles, except gravity, and their existence can be verified only by astrophysical effects (Blinnikov, Khlopov 1982, 1983; Dubrovich, Khlopov 1989; Khlopov and others, 1991).

In this case there are predicted mirror electrons, positrons, photons, right-handed neutrinos and left-handed antineutrinos, with the same mass as ordinary particles, and they must be present in the Universe at the same concentration as their twins.

Therefore, the model predicts the mirror particles

$$\xi = \sqrt{2}, \tag{5.11}$$

which is equivalent to eight additional types of neutrinos.

This means that the restriction (Yang et al 1979, 1984) eliminates fully the possibility of strict symmetry between the ordinary and mirror matter in the Universe (Carlson, Glashow 1987). There remains the possibility of asymmetry either in the properties of the mirror particle or in their cosmological evolution. The first case corresponds to the shadow world with the sharp difference in the properties of ordinary and shadow particles, as in the case of superstring models. The second possibility is realized in a mirror-asymmetric inflation, which suppresses the contribution of mirror particles in the cosmological density.

It should be noted, however, that the severe restriction (5.6) on the value ξ follows from a comparison of the predictions of the theory of nucleosynthesis and the extrapolation of the observed abundance of light elements to their primordial pregalactic value. Critical analysis of observational data, given in Chapter 3, makes us more cautious than is usual in modern literature, to consider the possible ambiguities and to consider the model dependence of such extrapolations. We shall return to this subject in Chapter 10.

The frozen-out ratio of neutrons to protons n/p is determined by comparing the rate of expansion and the beta processes. The cross section of the beta reactions

$$\begin{aligned} &\nu_e n \to ep \\ &\bar{\nu}_e p \to e^+ n \\ &ep \to \nu_e n \\ &e^+ n \to \bar{\nu}_e p \end{aligned} \tag{5.12}$$

is determined, *inter alia*, by the value of the matrix element of the weak transition, determined from the data on the neutron lifetime.

In the calculations by Wagoner (1973), Wagoner et al (1976), Yang et al (1979,1984) the lifetime of the free neutron in the β-decay was chosen to be

$$\tau = \frac{926\,\mathrm{s}}{C}, \tag{5.13}$$

where the constant C takes into account the ambiguity in the experimental neutron lifetime.

Degeneracy of neutrinos during the nucleosynthesis (Yahil, Beaudet 1976; Linde 1979; Salam 1981), the effects of CP violation in non-equilibrium neutrino oscillations of the sterile neutrino states (Khlopov, Petkov 1981), the presence of non-equilibrium neutrino fluxes (Weiner, Naselsky 1977; Dolgov, Kirilova 1987) could affect the characteristic time of the β-process in the early Universe.

In the standard model of Big Bang nucleosynthesis (SBBN) it is assumed that that

$$C = \xi = 1 \tag{5.14}$$

and annihilation of the electron–positron pair is regarded as the sole source of energy release after nucleosynthesis.

The results

Analysis (see review Khlopov, Chechetkin 1987) of the dependence of the abundance of light elements predicted in SBBN on the parameters, given in (5.14), showed that the deuterium concentration is more sensitive to the value ρ_b. The higher ρ_b the greater is the suppression of the primordial abundance of deuterium because of its burnout. Thus, a large deficit is predicted for deuterium abundance X_{D}^{th} at

$$\rho_b > 10^{-30}\,\mathrm{g/cm^3} \tag{5.15}$$

compared with the average observed value of this abundance X_D^{th}

$$X_D^{th} < X_D^{obs} \sim 10^{-5} \div 10^{-4}. \qquad (5.16)$$

A qualitative physical explanation for this result belongs to Zeldovich.

In the chain of nucleosynthesis reactions after the formation of deuterium in the reaction (see Chapter 3)

$$n + p \to D + \gamma$$

its main part is converted to tritium in the reaction

$$D + D \to T + p.$$

The main channel through which deuterium burns is the reaction (see (3.130))

$$D + T \to {}^4He + n,$$

in which the relative concentration of deuterium is reduced in accordance with the kinetic equation

$$\frac{dX_D}{dt} = -X_D X_T (n_b \sigma v). \qquad (5.17)$$

Since at a sufficiently high baryon density

$$\rho_b > 10^{-31} g/cm^3 \qquad (5.18)$$

the relative concentration of tritium X_T is almost independent of baryon density, the frozen-out deuterium concentration depends on the baryon density exponentially

$$X_D^{th} \propto \exp\{-A\Omega_0\}. \qquad (5.19)$$

On the other hand, the abundance of ^4He depends greatly on ρ_b only at a very low baryon density. When

$$\rho_b < 10^{-31} g/cm^3 \qquad (5.20)$$

the frozen-out ratio n/p is higher than $\langle n/p \rangle$ in the nuclei formed as a result of nucleosynthesis because of the decay of free neutrons.

The dependence of the abundance ^4He on ξ (or the number of types of neutrinos N_ν) has been given in Chapter 3 for

$$\rho_b > 10^{-31}\,\mathrm{g/cm^3}.$$

The existence of the symmetrical mirror world corresponds to the mass concentration of ^{4}He

$$Y_{\mathrm{prim}} \sim 0.29 \tag{5.21}$$

for

$$\rho_b > 10^{-31}\,\mathrm{g/cm^3}.$$

If we assume that

$$Y_{\mathrm{prim}} \leqslant 0.25 \tag{5.22}$$

in accordance with the most severe restrictions, then the allowed value of the total number of light neutrino types can not exceed

$$N_\nu < 4. \tag{5.23}$$

1.1. Constraints *on the total cosmological density*

According to the standard Big Bang model, β-transitions of the weak interaction, namely, the reactions

$$ep \leftrightarrow \nu_e n, \quad e^+ n \leftrightarrow p\bar{\nu}_e \tag{5.24}$$

provide for

$$T > 1\ \mathrm{MeV} \tag{5.25}$$

the equilibrium concentration ratio of neutrons to protons

$$\frac{n}{p} = \exp\left(-\frac{\Delta m_{np}}{T}\right), \tag{5.26}$$

where Δm_{np} is the difference between the mass of the neutron and the proton.

When

$$T < 1\ \mathrm{MeV} \tag{5.27}$$

the rate of these reactions is less than the rate of cosmological expansion. The relative concentration of neutrons freezes out. Most of the neutrons are transferred to the ^{4}He nuclei as a result of nuclear reactions. Therefore, the abundance of primordial ^{4}He is essentially determined by the value of the frozen-out concentration ratio n/p and is very sensitive to the conditions of freezing-out of n/p in the early Universe.

In the absence of a significant excess of electron neutrinos (or electron antineutrinos) in the Universe during the freezing-out of n/p the rate of the beta reactions is determined solely by temperature T, so that the time scale of these reactions is

$$\tau(n \leftrightarrow p) = (n_\nu \sigma v)^{-1} = \left(AT^3 \frac{G_F^2 T^2}{\pi} \right)^{-1}, \tag{5.28}$$

where G_F is the Fermi constant of weak interactions and the numerical factor is

$$A = 0.1$$

At a given temperature, the rate of expansion depends on the equation of state. If it is relativistic

$$p = \frac{1}{3}\varepsilon$$

the expansion rate is determined by the total density of relativistic particles present in the Universe in the considered period. Therefore, the total density should take into account the contribution of both known and unknown types of particles,

$$\rho = \xi^2 \rho_n, \tag{5.29}$$

where ρ_n is the density of all known types of particles.

Increase of ξ increases the rate of expansion and the freezing-out condition is fulfilled earlier (at higher temperature T_f)

$$\tau(n \leftrightarrow p) = t, \tag{5.30}$$

with the defining relation

$$BG_F^2 T_f^5 = \xi \frac{T_f^2}{m_{\rm Pl}}, \tag{5.31}$$

where B is a constant.

The higher the frozen-out ratio (5.26), the greater is the amount of primordial ^{4}He produced.

When

$$\xi > 2 \tag{5.32}$$

the predicted abundance of primordial ^{4}He exceeds the observed abundance of ^{4}He in the Universe, which sets an upper limit on the allowed contribution of light weakly interacting particles to the cosmological density during freezing-out of n/p (Shvartsman 1969).

Limitation of the concentration of primordial ^{4}He (Steigman et al 1977; Olive et al 1979; Ellis et al 1986; Schramm, Copi 1996) by the value (cf. (5.22))

$$Y_{\text{prim}} \leqslant 25\% \tag{5.33}$$

prohibits the existence of even one additional type of the light two-component neutrino with the usual weak interaction, because (5.23) restricts the total number of types of neutrino

$$N_\nu < 4.$$

In this case, the right-handed neutrinos (and left-handed antineutrinos) can exist only under the condition that their cosmological abundance is considerably smaller than the number of conventional left-handed neutrinos (right-handed antineutrinos). Such suppression of the concentration of right-handed neutrinos could be the case if their interaction is much weaker than the usual weak interaction, so that the decoupling of right-handed neutrinos takes place significantly earlier than for ordinary neutrinos, at a higher temperature

$$T > 100 \text{ MeV}, \tag{5.34}$$

when a significantly greater number of species of relativistic particles is in equilibrium with the plasma and radiation.

Restriction on the primordial abundance ^{4}He, defined by (5.33), provides in this case the upper limit on the interaction section σ_R of the right-handed neutrino

$$\sigma_R \sim G_R^2 T^2, \tag{5.35}$$

where G_R is the dimensional constant of the four-fermion interaction of the right-handed currents.

If this interaction is realized by the appropriate gauge bosons, so that

$$G_R \propto M_R^{-2}, \tag{5.36}$$

the upper limit on the interaction cross section of right-handed currents following from the upper limit on the abundance of ^{4}He corresponds to the lower limit on the mass M_R of this boson.

If

$$G_R \propto G_F \left(\frac{M_W}{M_R} \right)^2, \tag{5.37}$$

we get (Olive et al 1979; Dolgov 1981)

$$M_R > 10 M_W. \tag{5.38}$$

A similar restriction

$$M_{Z'} > 10 M_Z \tag{5.39}$$

can be obtained for the neutral currents Z' of interaction between the right-handed neutrinos predicted by the models of superstrings $E_R \otimes E'_g$ (Ellis et al 1986).

In developing the original idea of Shvartsman (1969), such analysis can be applied to any form of matter predicted in the framework of particle physics. For example, Koikawa (1991) considered the contribution of gravitational waves, emitted by cosmic strings, to the cosmological density during nucleosynthesis and constraints on the parameters of cosmic strings. For an extensive review of these constraints see (Dolgov 2002a).

We now show that from the observational upper limit on the abundance of ^{4}He a limit can also be set for the allowable properties of massive non-relativistic particles, if they dominate in the Universe in the period of freezing-out of the ratio n/p.

Consider a particle with mass m and the relative frozen-out concentration

$$r = \frac{n_m}{n_r}, \tag{5.40}$$

where n_m and n_r are the densities of frozen-out and relativistic particles, respectively. The density of the considered particles ρ_m

$$\rho_m = m n_m = r m n_m, \tag{5.41}$$

begins to exceed the density of relativistic particles ρ_r which is approximately equal to

$$\rho_r \approx 3Tn_r, \tag{5.42}$$

at temperature

$$T_0 \approx \frac{1}{3} rm \tag{5.43}$$

that corresponds to the period (see Chapter 4)

$$t_0 \approx \frac{m_{\text{Pl}}}{T_0^2} \approx \left(\frac{3m_{\text{Pl}}}{rm}\right)^2 t_{\text{Pl}}. \tag{5.44}$$

When

$$t > t_0 \tag{5.45}$$

the Universe is dominated by frozen-out non-relativistic particles, which corresponds to the dust stage of expansion with the equation of state

$$p = 0.$$

At this stage, temperature decreases with time as

$$T = \left(\frac{t_0}{t}\right)^{2/3} T_0 \tag{5.46}$$

and time is related with the temperature as

$$t = \frac{m_{\text{Pl}}}{T_0^{1/2} T^{3/2}}. \tag{5.47}$$

If the ratio n/p is frozen-out at this stage at freezing-out temperature

$$T_{fr} < T_0, \tag{5.48}$$

then the inequality is valid

$$T_{fr} > T_n, \tag{5.49}$$

where the temperature T_n, defined as

$$T_n = \left(\beta G_F^2 m_{\text{Pl}}\right)^{-1/3} \approx 1 \text{ MeV} \tag{5.50}$$

with the numerical constant β, is the freezing-out temperature of the n/p ratio in SM with

$$\xi = 1.$$

Indeed, using (5.28), (5.30), (5.31) and (5.47), we obtain

$$T_{\text{fr}} = \left(\frac{T_0^{1/2}}{\beta G_F^2 m_{\text{Pl}}} \right)^{2/7} = \left(\frac{T_0}{T_n} \right)^{1/7} T_n. \tag{5.51}$$

From (5.51) we obtain, with (5.48) taken into account, that the concentration ratio n/p is frozen-out in the stage with $p = 0$ at temperature T_{fr} higher than in the stage $p = 1/3\varepsilon$ in the standard model without unknown particles. This corresponds to a higher value of this ratio.

Dicus et al (1978) studied the kinetics of nuclear reactions leading to the formation of ^{4}He in the stage $p = 0$ after freezing-out n/p. It was shown that the predicted abundance of ^{4}He is almost independent of the equation of state and is mainly determined by the value T_{fr} (or the frozen-out ratio n/p).

It is interesting to note that according to (Dicus et al 1978) and (Polnarev, Khlopov 1982) at a fixed current density of baryons in the presence of the stage $p = 0$ during the nuclear reactions that correspond to the time interval

$$1 < t < 100 \text{ s}, \tag{5.52}$$

leads to the prediction of a greater abundance of deuterium.

In the presence of inhomogeneities this effect is enhanced (Zeldovich 1975; Polnarev, Khlopov 1982) along with a slight increase in abundance of ^{4}He. Thus, the contradiction with the observed abundance of helium does not occur if

$$T_0 < T_n \tag{5.53}$$

and as a consequence, the n/p ratio is frozen-out at the RD stage, provided that

$$T_{\text{fr}} = T_n. \tag{5.54}$$

If massive hypothetical particles begin to dominate in the Universe before the epoch of cosmological nucleosynthesis at a temperature T_0 that satisfies the inequality

$$T_0 > 2T_n \tag{5.55}$$

the ratio n/p is frozen-out at the corresponding stage $p = 0$, leading to the predicted abundance of ^{4}He, exceeding its maximum observed value.

Thus, the existence of massive particles does not contradict the observational data on the abundance of ^{4}He, if

1) they begin to dominate the Universe at a temperature T_0, satisfying the condition

$$T_0 < 2T_n \tag{5.56}$$

which corresponds to the values given by

$$rm < 6 \text{ MeV} \tag{5.57}$$

(line 1 in Figure 5.1);

2) The stage $p = 0$ of dominance of massive particles is completed before freezing-out of n/p, i.e., the lifetime of massive particles with respect to decay or annihilation in the inhomogeneities formed in the stage $p = 0$, is less than 1 second. The corresponding constraint is given by line 2 in Fig. 5.1.

According to the strict limit condition (5.33)

$$Y_{\text{prim}} \leqslant 25\%$$

a much more severe constraint follows, also presented in Fig. 5.1 (dashed line).

Thus, the observed abundance of ^{4}He provides an opportunity to limit the contribution of any hypothetical particles and fields in the

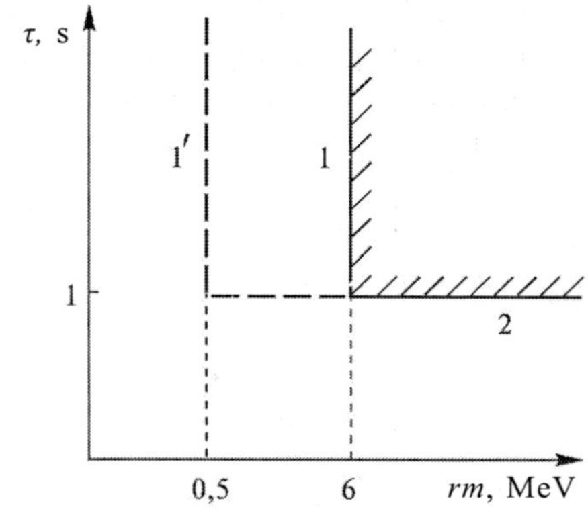

Fig. 5.1. Constraints on *rm* and τ of heavy particles from the observed (lines 1 and 2) and extrapolated to primordial abundance of helium (line 1').

first second of expansion. Generalization of constraints derived in (Shvartsman 1969) for the case of non-relativistic particles allows us to conclude that the contribution of any new particles to the cosmological density does not exceed the contribution of the known particles (or even its $1/16^{th}$ part) taken into account in the standard theory of nucleosynthesis.

This constraint is universal. It takes into account only the energy density of new particles. The corresponding contribution to the total density increases the expansion rate, which leads to higher freezing-out temperatures of n/p.

However, the impact of the hypothetical particles on the ratio of rates of β-processes and cosmological expansion, leading to a change in the frozen-out ratio n/p, can be caused not only by increasing the rate of expansion.

The existence of new particles, fields and phenomena can lead to a number of different effects during freezing-out of the n/p ratio thus influencing the rate of β-processes. Nevertheless, all these effects which differ in the physical nature exert a direct effect on the kinetics of β-reactions with the nucleons thus causing changes in the frozen-out n/p ratio.

1.2. Shift of the equilibrium rate of β-processes

Consider the effects that change the equilibrium rate of β-processes.

Analysis of the possible effects of the non-zero chemical potential of the electron neutrino on the rate of β-reactions in the theory of nucleosynthesis was carried out in (Steigman 1979; Linde 1979; review in Zeldovich, Khlopov 1981). The dependence of the excess of ^{4}He on the value of the chemical potential has been quantified in numerical calculations by Yahil, Beaudet (1976). A more recent review of the problem was published by Dolgov (2002b).

This dependence can be qualitatively simply illustrated (Fig. 5.2), as the non-zero chemical potential μ corresponds to the suppression of the density of antineutrinos or neutrinos, depending on the sign of the excess of the lepton charge.

In the first case neutrinos are dominant and this leads to a decrease of the frozen-out ratio n/p. In the second case, an excess of antineutrinos leads to an increase in this ratio.

In the presence of the chemical potential μ the frozen-out concentration instead of (5.26) is given by

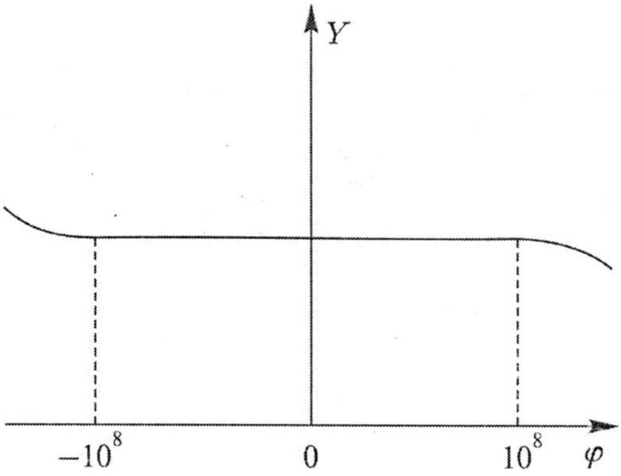

Fig. 5.2 Large chemical potential of electron neutrinos leads to a decrease in the abundance of primordial helium at $\varphi > 0$, and an increase at $\varphi < 0$.

$$\frac{n}{p} = \exp\left(-\frac{\Delta m_{np}}{T_f} - \varphi\right) \tag{5.58}$$

Here we introduce the dimensionless parameter φ defined as

$$\varphi = \frac{\mu}{T_f}. \tag{5.59}$$

In the framework of GUT models, the existence of a lepton asymmetry at the level 8–10 orders of magnitude larger than the baryon asymmetry seems to be impossible (Langacker, Pi 1980; Kolb, Turner 1983). As a rule, the value of lepton asymmetry which is formed in the early Universe with the baryon asymmetry in these models is of the same order of magnitude. Moreover, if there was a period in the early Universe when the processes with the lepton number violation were in equilibrium, then any initial lepton asymmetry vanishes.

At high temperatures of the order of or greater than the electroweak scale, i.e. at

$$T \geqslant m_W,$$

effects of the electroweak interactions beyond the framework of the perturbation theory (Chapter 3) enables fast processes in which the lepton number (L) and baryon number (B) are not conserved separately, but their linear combination B–L is conserved (Kuzmin et al 1985). These processes lead to vanishing of the $B+L$ combination, so that the absolute value of the baryon asymmetry ΔB must be equal in magnitude to the value of lepton asymmetry ΔL.

Thus, based on various theoretical arguments and the observed asymmetry ΔB, it seems possible to exclude large values of ΔL in the period of freezing-out of n/p. However, given the various possible modifications of the scenario of evolution of the very early Universe, the question of the dependence of the frozen-out ratio n/p on the value of the lepton number of the Universe still deserves attention (review by Dolgov 2002b).

Khlopov and Petkov (1981) revealed the possibility of significant differences between the concentrations of electron neutrinos and antineutrinos, which may occur during freezing-out of n/p even for the zero lepton number of the Universe.

This possibility is connected with the existence of 'sterile states' of the right-handed neutrinos (and left-handed antineutrinos) and the violation of CP-invariance in non-equilibrium oscillations of the electron neutrino to the left-handed sterile antineutrinos

$$\nu_e \leftrightarrow \bar{\nu}_L \tag{5.60}$$

and electron antineutrinos in the right-handed sterile neutrino

$$\bar{\nu}_e \leftrightarrow \nu_R. \tag{5.61}$$

Because of the superweak interaction of sterile neutrinos with ordinary particles, the concentration of sterile neutrinos in the Universe must be much less than the concentration of ordinary neutrinos (Chapter 3). The same is true if the existence of sterile neutrinos is associated with the existence of mirror particles interacting with sterile neutrino states in the same manner as ordinary particles interact with ordinary neutrinos, but there is not symmetry in the distribution of mirror and ordinary particles in the Universe.

Indeed, even if sterile neutrinos were in equilibrium with the ordinary particles, their decoupling should have taken place much earlier (at higher temperatures) than the decoupling of ordinary particles due to the extremely weak interaction of sterile neutrinos with plasma compared with the usual weak interaction. Therefore, the concentration of the decoupled sterile neutrinos must be less than the concentration of ordinary neutrinos. The concentration of sterile neutrinos should be even lower if there was no period in the early Universe when the sterile neutrinos were in equilibrium with other particles (Dolgov 1981, review in Dolgov 2002b).

If the neutrino has both Majorana and Dirac mass terms, apart from the oscillations between the normal states of left-handed neutrinos

$$\nu_e \leftrightarrow \nu_L \tag{5.62}$$

and right-handed antineutrinos

$$\bar{\nu}_e \leftrightarrow \bar{\nu}_R, \tag{5.63}$$

oscillations can take place between the states of the usual left-handed neutrino and sterile antineutrinos (5.60) and oscillations of ordinary electron antineutrinos in the sterile state of right-handed neutrinos (5.61).

If such transitions are 'switched on' prior to the decoupling of ordinary left-handed neutrinos (right-handed antineutrinos) from plasma and radiation, there is an effective mechanism for bringing sterile left-antineutrinos (right-handed neutrinos) to thermodynamic equilibrium.

This occurs when the oscillation length L_{osc} for the main fraction of neutrinos at a temperature T being about

$$L_{\text{osc}} \sim \frac{T}{\delta m^2}, \tag{5.64}$$

where δm^2 is the difference of the square of the masses of the different states of neutrinos, is less than the mean free path of the ordinary neutrinos in plasma, which is

$$l = \left(n_l \sigma_{\nu l}\right)^{-1} \sim \left(G_F^2 T^5\right)^{-1}. \tag{5.65}$$

In (5.65) we must take into account that at a temperature of 1–100 MeV, the main effect of opacity for neutrinos arises from the interaction of the neutrino with a relativistic gas of neutrino–antineutrino and electron–positron pairs, since the density of these gases n_l at temperature T will be about

$$n_l \sim T^3 \tag{5.66}$$

and $\sigma_{\nu l}$ is the effective cross-section corresponding to the weak interaction

$$\sigma_{\nu l} \sim G_F^2 T^2. \tag{5.67}$$

It can be seen that the condition

$$L_{\text{osc}} < l \tag{5.68}$$

takes place at $T > 3$ MeV and the difference of the squares of the neutrino mass states, satisfying the inequality (5.68)

$$\delta m^2 > 10^{-7}\,\mathrm{eV}^{-2}. \tag{5.69}$$

In this case, when the rate of the weak interaction of the ordinary left-handed neutrinos (right-handed antineutrinos) is less than the rate of expansion secondary decoupling of the sterile neutrino takes place. The concentration of the decoupled sterile neutrinos after second decoupling is equal to the concentration of ordinary neutrinos.

It should be noted that the oscillations of low-energy neutrinos with energy

$$E_\nu \ll T, \tag{5.70}$$

occur much earlier than the time when the condition

$$L_{\mathrm{osc}} \sim l \tag{5.71}$$

is fulfilled for neutrinos with

$$E_\nu \sim T.$$

However, the number of neutrinos with low energy is a small fraction of all the neutrinos, since the maximum of their energy distribution corresponds to

$$E_\nu \sim 3T. \tag{5.72}$$

That is why, to estimate the main effect of the oscillations, it is necessary to consider neutrinos with energies near the peak of energy distribution. In this approach, we find that because of equilibrium oscillations between ordinary and sterile neutrino states the number of types of neutrinos can double during freezing-out of the ratio n/p.

In principle, the condition (5.71) should take into account the amplitude of the oscillations. However, in (Enqvist, Kanulainen, Maalampi 1990) it is shown that in the event of a negative difference of the squares of the masses of the neutrino states, lying in the range

$$-10^{-5}\,\mathrm{eV} < \delta m^2 < 0, \tag{5.73}$$

there must be neutrino oscillations into sterile states due to the resonant transition of the MSW type (Mikheev, Smirnov 1985, 1986; Wolfenstein 1978, 1979). This should lead to doubling of neutrino types for a wide range of δm^2 and the neutrino mixing angles satisfying the condition

$$\sin 2\theta > 10^{-4}. \tag{5.74}$$

Prediction of doubling of the number of neutrino states due to neutrino oscillations, together with a strict constraint on the primordial abundance of ^{4}He (5.33) allows us to exclude a fairly wide range of the values of neutrino mixing angles (Dolgov 1980; Enqvist et al 1990).

Oscillations (5.60) and (5.61) of electron neutrinos and antineutrinos in sterile conditions with the difference of the squares of the masses lying in the range

$$10^{-10} < \delta m^2 < 10^{-8}\,\text{eV}^{-2}, \tag{5.75}$$

are 'switched on' in the period

$$T \sim 1 \text{ MeV} \tag{5.76}$$

that is directly during freezing-out of n/p.

The phenomenology of neutrino oscillations (Bilenky, Pontecorvo 1978; Bilenky et al 1980; Kobzarev, et al 1980) permits *CP*-violating effects in the oscillations in the case of more than two types of neutrinos. Such an effect is the cause of the asymmetric 'switching on' of oscillations (5.60) and (5.61). Thus, in the activation period of these oscillations there could be a significant excess of electron neutrinos as compared with electron antineutrinos (or vice versa), comparable to their full concentration.

Because of the existence of such an excess, the ratio between the rates of weak transitions

$$\nu_e p \to ep$$

and

$$\bar{\nu}_e p \to e^+ n$$

had to change, leading to a significant change in the frozen-out ratio n/p. In the case of oscillations causing a significant excess of neutrinos

$$n_{\nu_e} \gg n_{\bar{\nu}_e}, \tag{5.77}$$

ratio decreases, while the reverse case

$$n_{\nu_e} \ll n_{\bar{\nu}_e} \tag{5.78}$$

leads to its increase.

Thus, the observed concentrations can be used as a sensitive indicator of CP-violation in neutrino oscillations with

$$10^{-10} < \delta m^2 < 10^{-8} \mathrm{eV}^{-2}.$$

It should be noted that the existence of such oscillations can lead to seasonal changes in fluxes of monochromatic electron neutrinos from the Sun. However, one should take into account effects of decoherence of neutrino states in plasma, which can wash out CP-violating effects in neutrino oscillations to sterile states (Dolgov 2002b).

In the presence of even a small magnetic moment of neutrinos μ_ν, of the order

$$\mu_\nu \sim G_F m_\nu, \tag{5.79}$$

where G_F is the Fermi constant, for neutrinos with the Dirac mass m_ν, if there is a primordial magnetic field

$$B_{\text{prim}} > 4 \cdot 10^{-9} \text{Gauss} \, (1+z)^2 \left(\frac{1\,\text{eV}}{m_\nu} \right), \tag{5.80}$$

where z is the redshift, providing for the helicity flip of the neutrinos is possible before their decoupling from the plasma. The concentrations of the ordinary and sterile neutrinos become equal and this is excluded by the 'rigid' interpretation of observational data on the primordial abundance of ^{4}He (see review and references in Zeldovich, Khlopov 1981b; Dolgov 2002b; Enqvist et al 1995).

Time shift of freezing-out of the n/p ratio can be caused by time variations of the fundamental constants due to the effects of additional (compactified or large) dimensions.

In this case, the observed abundance of ^{4}He gives an upper limit for the time dependence of the compactification radius and the fundamental constants that determine the rate of β-processes (Kolb et al 1986).

1.3. Non-equilibrium particles and abundance of ^{4}He

The possible existence of non-equilibrium particles in the Universe during nucleosynthesis is a different class of effects that influence the ratio n/p.

For example, Vainer and Naselski (1977) estimated the variation of the n/p ratio due to interaction of nuclei with high-energy neutrinos from the evaporation of primordial black holes. It is easy to see that the interaction of nucleons with equal fluxes of electron neutrinos and antineutrinos from evaporating PBHs at $t < 10^2$ s (the evaporation of primordial black holes, see Chapters 4 and 6) increases the value of the ratio n/p. This gives a constraint on the possible neutrino flux and, consequently, the concentration of PBH evaporating at this time.

The same argument can be used to limit the concentration of any type of hypothetical massive particles whose decay products are predicted to contain neutrinos and antineutrinos at

$$t < 10^2 \text{ s}.$$

Depending on the mass of such particles, the ratio n/p may either decrease or increase.

For the energetic, non-equilibrium neutrinos the n/p ratio increases, because the interaction of energetic neutrinos with nucleons tends to equalize the concentration of protons and neutrons

$$\frac{n}{p} \to 1. \tag{5.81}$$

The existence of the hypothetical sources of non-equilibrium low-energy neutrinos

$$E_\nu < 5 \text{ MeV} \tag{5.82}$$

leads to a decrease in the n/p ratio because of threshold effects in weak reactions

$$\nu_e n \to ep, \quad \bar{\nu}_e p \to e^+ n, \tag{5.83}$$

which should take into account the mass difference between the neutron and the proton and the electron mass (Dolgov, Kirilova 1987; Sato and Kobayashi 1987).

It should be noted that in the period under review

$$t < 10^2 \text{ s}$$

the electron–positron pairs are in equilibrium with the plasma and radiation, and their concentration exceeds that of the nucleons by 8–10 orders of magnitude. Therefore, the Universe is opaque to neutrinos with energies

$$E_\nu > E_{th} \sim 100 \text{ MeV}\left(\frac{t}{1 \text{ s}}\right) \tag{5.84}$$

due to scattering of neutrinos on electrons and positrons, and these energetic neutrinos can not interact with nuclei. This causes the abundance ^{4}He to be sensitive to such non-equilibrium neutrino fluxes that are only 1–2 orders of magnitude less intense than the equilibrium neutrino fluxes.

The relatively weak sensitivity of n/p to the fluxes of non-equilibrium neutrinos is due to the weakness of neutrino–nucleon interactions. This sensitivity is increased for the sources of non-equilibrium particles with a stronger interaction with nucleons.

As was shown by Zeldovich et al (1977) on an example of evaporating primordial black holes, the existence of sources of anti-nucleons in the Universe in the period

$$1 < t < 100 \text{ s} \tag{5.85}$$

should lead to a more noticeable effect on the value of n/p.

In the case of evaporation of primordial black holes we can consider nucleon–antinucleon pairs among the evaporation products. Isotopic invariance of the evaporation of primordial black holes leads to an equal number of neutron–antineutron and proton–antiproton pairs produced with a concentration

$$n_n = n_p = n_{\bar{n}} = n_{\bar{p}} = n_1, \tag{5.86}$$

where the nucleon density n_1 is proportional to the relative contribution $\alpha(M)$ of the PBH to the total density during their evaporation

$$n_1 = f_N \frac{a(M)\rho_{\text{tot}}}{T_{\text{PBH}}}. \tag{5.87}$$

Here ρ_{tot} is the total density and T_{PBH} is evaporation temperature of PBH.

Production of an equal number of protons and neutrons in the given period of cosmological evolution leads to an increase in the ratio n/p. So after the annihilation of antinucleons n/p is equal to

$$\frac{n}{p} = \left(\frac{n}{p}\right)_f \tag{5.88}$$

rather than the value determined by freezing-out and designated as

$$\frac{n}{p} = \left(\frac{n}{p}\right)_i = \frac{n_0}{p_0}, \tag{5.89}$$

where

$$\left(\frac{n}{p}\right)_f = \frac{n_0 + n_1}{p_0 + n_1} > \frac{n_0}{p_0} = \left(\frac{n}{p}\right)_i . \tag{5.90}$$

From the observed abundance of ^{4}He it gives the strongest upper limit on the concentration of PBH evaporating in the Universe (Zeldovich et al,= 1977) in the period (5.85).

The same arguments give the strongest upper limit on the allowable value rn of hypothetical metastable particle with mass m, with the relative concentration

$$r = \frac{n_m}{n_r}, \tag{5.91}$$

decaying in the Universe during this period and containing the nucleon-antinucleon pairs among their decay products:

$$rm < 6.5\,\mathrm{eV}\frac{\Omega_b}{B_N}(1 - Y_{obs})^{-1}\left(1 + \frac{Y_{obs}}{2}\right)^{-1}(Y_{obs} - Y_0), \tag{5.92}$$

where Y_{obs} is the observed abundance of ^{4}He, the value Y_0 is defined as

$$Y_0 = 2\left(\frac{n}{p}\right)_i, \tag{5.93}$$

Ω_b is the relative cosmological density of baryons and B_N is the relative probability of the birth of the nucleon–antinucleon pairs in the decay of such particles. Constraint (5.92) is shown in Fig. 5.3, together with other constraints, from the analysis of effects of the hypothetical particles on the abundance of ^{4}He.

In addition to *CP*-violating effects in neutrino oscillations, discussed above, the abundance of ^{4}He is sensitive to other possible effects of certain mechanisms of *CP*-violation, namely, the existence of antimatter domains which are predicted in the baryon asymmetric Universe (Chechetkin et al 1982a) in models of inhomogeneous baryosynthesis based in particular on models with spatial dependence of the *CP*-violation, such as models of spontaneous *CP* violation (Chapter 3).

Because further annihilation after local annihilation of nucleon-antinucleon pairs is possible only at the domain boundaries, the domain structure of the distribution of matter and antimatter ensures that the large number of antinucleons is retained in the period when the antinucleons are absent in the case of homogeneous mixing of matter and antimatter. The larger domain the longer its annihilation time and the greater the period when the antinucleons exist in the Universe.

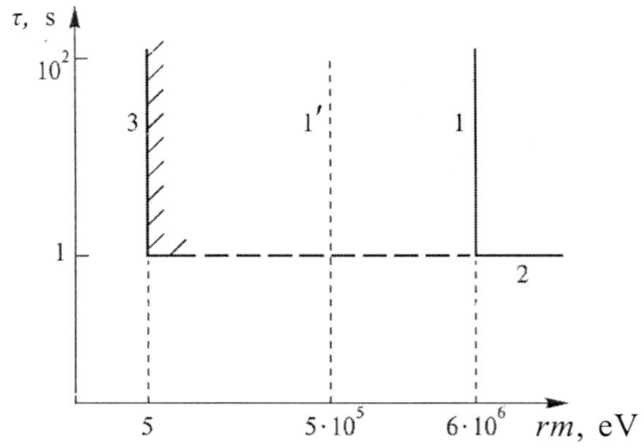

Fig. 5.3. Constraints on metastable particles from the analysis of the influence of nucleon–antinucleon pairs on the abundance of helium (line 3) compared with the constraints of the 'thermodynamic' effect on this abundance.

Annihilation of an antimatter domain in the period (5.85) reduces the concentration of neutrons diffusing more efficiently than protons to the boundary of the domain of matter (Steigman 1976). As a result, the ratio n/p decreases. The observed abundance of helium at the same time imposes an upper limit on the mean concentration of antinucleons annihilating in the Universe in the period under review. This limit follows from the lower estimate of abundance of ^{4}He.

Indeed, let N be the concentration of annihilating antinucleons. Due to the diffusion of neutrons to the domain boundaries and annihilation the ratio n/p decreases and is equal to

$$\left(\frac{n}{p}\right)_f = \frac{n_0 - N}{p_0} < \frac{n_0}{p_0} \tag{5.94}$$

after the annihilation. This estimate does not take into account the difference between the annihilation cross section of nucleon–antinucleon pairs in states with isotopic spin 0 and 1, but this does not lead to a noticeable change in the result.

Hence, the fraction of the annihilated antinucleons, defined as

$$f = \frac{N}{p}, \tag{5.95}$$

is restricted by the following inequality

$$f < \frac{n_0}{p_0} - \left(\frac{n}{p}\right)^{obs}_{\min}. \tag{5.96}$$

Evolution of domains of antimatter was discussed in the books by Zeldovich and Novikov (1975) and in reviews by Steigman (1976) and Chechetkin et al (1982). A new level of understanding of the possible existence of antimatter in the Universe as a manifestation of the mechanism of formation of matter is described in this book.

Thus, the observed abundance of ^{4}He allows us to obtain non-trivial constraints on the different effects of hypothetical particles during cosmological nucleosynthesis, corresponding to

$$1 < t < 100 \text{ s.}$$

In the next section we consider observational data on the spectrum of thermal electromagnetic background as a source of information on the possible cosmological effects of the hidden sector of the theory of elementary particles.

2. Metastable particles and the spectrum of thermal background radiation

2.1. Distortion of the spectrum of thermal electromagnetic background

As we mentioned earlier, the data on the spectrum of the thermal background give important observational information about the possible physical conditions at the RD stage, in addition to data obtained by observing the abundance of light elements. Data from the COBE satellite (Mather et al 1990) have shown that the spectrum of cosmic microwave background radiation corresponds to a Planck distribution with temperature

$$T = 2.7 \text{ K} \tag{5.97}$$

to within a few hundredths of one percent. This means that the possible distortion of the thermal spectrum of this radiation should be located within the experimental error range. Hence, according to Zeldovich and Sunyaev (1969, 1970) we can obtain constraints on the nature of non-equilibrium processes in the RD stage.

Some non-equilibrium processes are predicted by the standard Big Bang scenario. They are the annihilation of electron–positron pair and recombination of hydrogen. The predicted distortion of the thermal spectrum due to these processes is 1–2 orders of magnitude smaller than the sensitivity level available in present-day observations.

However, some hypothetical non-equilibrium processes are predicted at the RD stage by different models of particle physics. Decays of massive metastable particles, evaporation of primordial black holes, the creation of the particles in the decay of cosmological strings, and the annihilation of antimatter domains are examples of such non-equilibrium processes.

Interaction of non-equilibrium fluxes of particles from these sources with matter and radiation heat matter so that hot electrons appear. At a fairly late stage of expansion the interaction of these electrons with radiation can not provide the Planck form of the spectrum, so that the thermal spectrum is distorted (Zeldovich and Sunyaev 1969; Sunyaev Zeldovich 1970, Zeldovich and Novikov 1975).

The magnitude of distortion and, consequently, the possibility of detecting it is defined by the energy release and a period of cosmological evolution in which it occurred.

Observations impose an upper limit on the possible magnitude of the distortion and, consequently, on the characteristic time and energy released in the hypothetical sources (in particular, lifetimes, mass and concentration of hypothetical particles).

Let us briefly discuss the results of the theory of distortion of the spectrum of the thermal background. According to this theory (see Zeldovich and Novikov 1975; Zeldovich and Sunyaev 1969; Sunyaev and Zeldovich 1970; review and references in Khlopov and Chechotkin 1987) if the energy release occurs at a redshift

$$z > 10^8 \Omega_b^{1/2}, \tag{5.98}$$

where Ω_b is the relative baryon density in the Universe, that corresponds to the period

$$t > 10^3 \Omega_b^{-1} \text{ s}, \tag{5.99}$$

the reactions

$$ep \rightarrow ep\gamma, \quad \gamma e \rightarrow \gamma\gamma e \tag{5.100}$$

lead to the creation of additional soft photons, so that the Planck spectrum is restored after the energy release.

On the other hand, if the energy is released at a redshift

$$z < 10^8 \Omega_b^{1/2}$$

these reactions are ineffective for the creation of additional photons, and the Planck spectrum shape can not be established.

Depending on redshift z, corresponding to the energy release and satisfying

$$z < 10^8 \Omega_b^{1/2} \tag{5.101}$$

in theory (Zeldovich and Sunyaev 1969; Sunyaev and Zeldovich 1970; Zeldovich et al 1972) two cases are described:

1) early energy release, which corresponds to the redshift

$$4 \cdot 10^4 \Omega_b^{1/2} < z < 10^8 \Omega_b^{1/2} \tag{5.102}$$

and

2) late energy release at

$$z < 4 \cdot 10^4 \Omega_b^{1/2}. \tag{5.103}$$

In the first case, thermal equilibrium between the hot electrons and cold photons is established, subject to a fixed number of photons. Therefore, in this case, that is, during the period (5.102), the Planck distribution is not established, but due to elastic scattering of photons on the electrons, the equilibrium distribution of a fixed number of photons interacting with matter is established, that is, the Bose–Einstein distribution

$$F_{em}(\nu) = \frac{2h\nu^3}{c^2} \frac{1}{\exp\left\{\dfrac{h\nu + \mu k T_e}{k T_e}\right\} - 1}. \tag{5.104}$$

Here ν is frequency and T_e electron temperature. The dimensionless chemical potential μ determines the relative contribution of energy to the total energy density of the thermal electromagnetic background. It defines the difference of the spectrum (5.104) from the Planck spectrum shape, defining thus the distortion of the Planck spectral shape of the thermal electromagnetic background.

Observational data from the spectrum of the thermal background constrain an allowed energy release during the period (5.102) (Sunyaev and Zeldovich 1970; Khlopov and Chechetkin 1987; Mather et al 1990; Khlopov 1999)

$$\frac{\delta \varepsilon}{\varepsilon_\gamma} < \frac{1}{3}|\mu| < 1.1 \cdot 10^{-4} \tag{5.105}$$

for the distribution of the ideal Bose–Einstein gas, using a strict limit on the value of μ given in (Mather et al 1994) as

$$|\mu| < 3.3 \cdot 10^{-4}. \qquad (5.106)$$

Here $\delta\varepsilon$ is the specific energy release and ε_γ is the energy density of thermal radiation.

A detailed theory of spectrum distortions due to early energy release (Sunyaev and Zeldovich, 1970) takes into account the non-ideal shape of the distorted spectrum due to bremsstrahlung and Compton scattering, which introduces clarifications to the part of the spectrum which is most sensitive to distortion. From the COBE data we can derive more stringent restrictions on the energy release as a function of redshift (Wright et al 1994). However, such restrictions are strongly model-dependent and sensitive to the very special choice of the parameters of hypothetical sources.

If the energy release occurs at a rather late stage in the period (5.103), the equilibrium Bose–Einstein spectrum (5.104) can not be established. In this case, the distortion of the thermal spectrum is determined by the kinetics of heating of the photon gas by hot electrons.

The COBE observational constraints on parameter y, given in Mather et al (1994) as

$$|y| < 2.5 \cdot 10^{-5} \qquad (5.107)$$

show that late energy release should not exceed (Sunyaev and Zeldovich 1970; Khlopov and Chechetkin 1987; Mather et al 1990; Wright et al 1992)

$$\frac{\delta\varepsilon}{\varepsilon_\gamma} < 12|y| < 3 \cdot 10^{-4} \qquad (5.108)$$

for distortion of the spectrum due to Compton scattering.

Based on the constraints (5.105) and (5.106), we can, for example, put up a limit on the contribution f to the total density of heavy metastable particles with mass m and density n_m (relative to baryons)

$$f = \frac{m \cdot n_m}{m_p \cdot n_B}, \qquad (5.109)$$

decaying in the Universe during the period (5.101).

Indeed, the energy density of the thermal background ε_γ is given (Chapter 3)

$$\varepsilon_\gamma = 4 \cdot 10^{-13} \left(\frac{T_0}{2.7\ \mathrm{K}}\right)^4 (1+z)^4 \frac{\mathrm{erg}}{\mathrm{cm}^3}, \qquad (5.110)$$

where T_0 is the present-day temperature of the thermal background radiation. Assuming for simplicity that the particles decay at the moment

$$z = z_d \tag{5.111}$$

we obtain the density of energy released in these decays

$$\delta\varepsilon = a_{em} mc^2 n_m = a_{em} \frac{m \cdot n_m}{m_p \cdot n_b} \rho_b c^2 = 5 \cdot 10^{-9} a_{em} \Omega_\beta f (1+z)^3 \frac{\text{erg}}{\text{cm}^3}, \tag{5.112}$$

where

$$a_{em} \sim 0.5 \tag{5.113}$$

is the fraction of the rest energy of the metastable particles, which contribute to the heating of electrons.

From (5.110) and (5.112) we find that at redshift (5.111), the expression for the relative energy

$$\frac{\delta\varepsilon}{\varepsilon_\gamma} = 6 \cdot 10^3 \Omega_b f (1+z)^{-1} \left(\frac{2.7\ \text{K}}{T_0} \right)^4. \tag{5.114}$$

So that the upper limits (5.105) and (5.106) lead to the following constraints on the permissible value f:

$$f < 1.6 \cdot 10^{-4} \Omega_b^{-1} (1+z_d) \left(\frac{T_0}{2.7\ \text{K}} \right)^4 \cdot \begin{cases} 3 \cdot 10^{-4}, & z_d < 4 \cdot 10^4 \Omega_b^2 \\ 1.1 \cdot 10^{-4}, & 4 \cdot 10^4 \Omega_b^2 < z_d < 10^8 \Omega_b^{1/2} \end{cases}. \tag{5.115}$$

It follows from (5.115) that the constraints on f are rather weak at

$$z_d > 10^5 \tag{5.116}$$

that is, if the lifetime of the particle is

$$\tau < 10^9 \text{s}. \tag{5.117}$$

Constraint (5.115) for such particles is given by

$$f < 2 \cdot 10^{-7} (1+z_d) \tag{5.118}$$

at

$$\Omega_b = 1.$$

The relatively weak sensitivity of the distortion of the thermal background spectrum to the properties of hypothetical particles decaying in the RD stage, is explained by the dominance of the energy density of radiation in the total cosmological density at this period. Therefore, a significant effect on the spectrum of the thermal background takes place when the hypothetical sources of non-equilibrium particles release electromagnetic energy density exceeding a few tenths of a percent of the total density, which greatly exceeds the density of matter in the early periods of the RD stage.

However, the constraint on the energy release at redshifts (5.101) completely eliminates the possibility of a full 'recovery' of thermal background during this period.

In particular, at such redshifts the stage of massive metastable particles can not end because of decay or annihilation of such particles, if the products of their decay or annihilation interact with the plasma and radiation.

If the relativistic particles born in such decays (annihilation) do not interact with the plasma and radiation, they do not directly affect the spectrum of thermal radiation, but they may affect the development of gravitational instability and the formation of the large-scale structure of the Universe (see below).

The combination of constraints on the possible effect of non-equilibrium particles produced at the end of the dust stage, with constraints on the existence of such stages in the period of nucleosynthesis, excludes long stages with $p = 0$ after the first second of the expansion

On the other hand, the energy release at redshifts

$$z > 10^8 \Omega_b^{1/2},$$

corresponding to the cosmological time

$$t < 10^3 \Omega_b^{-1} \text{s},$$

leads to thermalization and restoration of the Planck radiation spectrum. Given the constraints on non-relativistic matter dominating at

$$t < 1 \text{ s},$$

derived in section 1.1, we can conclude that we can not exclude in principle the possibility (Doroshkevich, Khlopov 1983; Polnarev, Khlopov 1981) of existence of a relatively short stage of dominance of massive metastable particles in the period

$$1 < t < 10^4 \text{ s}. \tag{5.119}$$

2.2. Dominance of metastable particles during nucleosynthesis

The possibility of energy release in the period (5.117), corresponding to temperatures

$$30 \text{ keV} < T < 3 \text{ MeV} \tag{5.120}$$

can be associated with the decay of hypothetical metastable neutrinos with a lifetime in the range

$$1 \text{ s} < \tau < 10^4 \text{ s} \tag{5.121}$$

and mass m

$$30 \text{ keV} < m < 30 \text{ MeV} \tag{5.122}$$

For the considered parameters the frozen-out concentration of left-handed neutrinos and right-handed antineutrinos with respect to the density of relativistic particles, is (Dolgov, Zeldovich 1981; Lee, Weinberg 1977; Vysotsky et al 1977, Chapter 3):

$$r = \frac{(n_{\nu L} + n_{\bar{\nu} R})}{n_r} = f(m) = \frac{1}{k}\begin{cases} 1, \; m < 3 \text{ MeV} \\ \left(\dfrac{3\,\text{MeV}}{m}\right)^3, \; m \leqslant 3 \text{ MeV} \end{cases}, \tag{5.123}$$

where the numerical factor k that takes into account 'derelativisation' of the electron–positron pairs and their annihilation in the given period is in the interval

$$5 < k < 6. \tag{5.124}$$

Neutrinos with the mass in the range (5.122) begin to dominate the Universe at

$$t > t_0, \tag{5.125}$$

where

$$t_0 = 1 \text{ s} \left(\frac{rm}{1 \text{ MeV}}\right)^{-2}. \tag{5.126}$$

The dominance of such neutrinos ν_H in the period

$$t_0 < t < \tau \tag{5.127}$$

corresponds to a dust stage, with the equation of state

$$p = 0$$

The main decay mode of the neutrino ν_H may be a process

$$\nu_H \to e^+ e^- \nu_e, \tag{5.128}$$

if their mass is

$$m > 1 \text{ MeV}, \tag{5.129}$$

or the process (Kim, Uehara 1981)

$$\nu_H \to \nu_L \gamma. \tag{5.130}$$

The reasonable choice of parameters of such decays can lead to values comparable with the condition (5.119).

Indeed, consider the decay mode (5.128). This decay can occur due to mixing in the usual weak interactions of charged currents

$$j^e_\mu = \overline{\Psi}_e \gamma_\mu (1 + \gamma_5) \Psi_{\nu_e} \tag{5.131}$$

and

$$j^{H+}_\mu = \overline{\Psi}_{\nu_H} \gamma_\mu (1 + \gamma_5) \Psi_e. \tag{5.132}$$

The Fermi constant G_F in the interaction of these currents is suppressed by the mixing U_{He}, determined by the neutrino mass matrix, so that the Lagrangian of current–current interaction is given as

$$L = \frac{G_F}{\sqrt{2}} U_{He} j^{H+}_\mu j^e_\mu. \tag{5.133}$$

Under the assumption that the mass of ν_H

$$m = m_{\nu_H} \gg m_e, \tag{5.134}$$

where m_e is the electron mass, the probability of the considered decay is given by the expression similar to muon decay

$$W = \frac{G_F^2 |U_{He}|^2 m^5}{192\pi^3}. \tag{5.135}$$

So that the lifetime is

$$\tau = 10^4 \mathrm{s} \left(\frac{1 \text{ MeV}}{m} \right)^5 |U_{\mathrm{He}}|^{-2}. \tag{5.136}$$

General constraints on the mixing factor are given in the Review of Particle Physics (Nakamura et al 2010). The most stringent laboratory limit on this factor in this mass interval (5.122) follows from the data on the decay

$$\pi \to e\nu. \tag{5.137}$$

Differences from the conventional mode are searched for, for example, for the mode

$$\pi^+ \to e^+ \nu_e, \tag{5.138}$$

and the respective additional channel which is produced by mixing the neutrino with a massive neutrino state (Khlopov 1999)

$$\pi^+ \to e^+ \nu_H. \tag{5.139}$$

As a result of these experimental studies, the upper limit of the mixing factor is given as

$$|U_{\mathrm{He}}|^2 < 2.5 \cdot 10^{-3} \left(\frac{1 \text{ MeV}}{m} \right)^2 \tag{5.140}$$

and as follows from (5.136), the upper experimental limit of the mixing imposes a lower limit on the lifetime

$$\tau > 10^4 \mathrm{s} \left(\frac{10 \text{ MeV}}{m} \right)^3. \tag{5.141}$$

Therefore, the existing direct experimental constraints do not preclude the existence of heavy neutrinos with mass

$$10 \text{ MeV} < m < 30 \text{ MeV}, \tag{5.142}$$

that can dominate the Universe at

$$t < 10^4 \text{ s}. \tag{5.143}$$

The data on the direct measurement of neutrino species from the width of the Z^0-boson (Nakamura et al 2010), give a value

$$N_\nu = 2.984 \pm 0.008, \tag{5.144}$$

for a neutrino with the usual weak interaction and mass

$$m < \frac{m_Z}{2}, \tag{5.145}$$

where m_Z is the Z^0-boson mass.

Given the direct experimental upper limit on the mass $m(\nu_\tau)$ (Hagivara et al 2002),

$$m\ (\nu_\tau)\ < 18.2\ \text{MeV}$$

ν_τ could dominate in this period, if their mass was in the range

$$10\ \text{MeV} < m\ (\nu_\tau) < 18.2\ \text{MeV}. \tag{5.146}$$

However, the indirect data on neutrino oscillations (Nakamura et al 2010) exclude this possibility.

The electronic decay mode of ν_H at times (5.119) because of the successive thermalization and annihilation of positrons and electrons leads to an increase in the concentration of photons

$$n_\gamma = \left(\frac{\tau}{t_0}\right)^{1/2} \left(n_\gamma\right)_0. \tag{5.147}$$

The Planck form of the spectrum must be restored after thermalization. As shown in section 1.1, dominance of ν_H in the Universe at

$$t_0 > 1\ \text{s} \tag{5.148}$$

leads to almost no effect on the frozen-out ratio n/p. If the lifetime of these particles

$$\tau > 10^3\ \text{s} \tag{5.149}$$

their dominance in the period of SBBM thermonuclear reactions leads at a fixed baryon density to a small (<1%) increase in primordial ^{4}He abundance and a slight increase in primordial deuterium abundance. Both effects do not contradict the observational data. More serious problems for the schemes may arise due to their conflict with the data on supernovae explosions (Falk, Schramm 1978; Hut, White 1984).

Models of GUT, supersymmetric GUT, superstrings, and other particle models discussed above predict different types of massive metastable particles, and because their existence follows from various physical causes, they could all coexist and be present in the Universe.

Self-consistent interpretation of these features includes a consolidated analysis of multicomponent scenarios in which some of the predicted particles can dominate the Universe during the period (5.119).

Constraint (5.119) for the duration of the relevant stages $p = 0$ imposes constraints on every frozen-out concentration r, each mass m, each lifetime τ of such hypothetical particles predicted by each theory

$$10 \text{ keV} < rm < 3 \text{ MeV} \tag{5.150}$$

$$\max \{t_0, 1 \text{ s}\} < \tau < 10^4 \text{ s}, \tag{5.151}$$

where t_0 in (5.151) is given by equation (5.126).

The considered scenarios of short dust stages can lead to a change (decrease) in the predicted relative concentration of all other types of frozen-out particles, because the concentration of photons increases after the late stage of dominance of massive metastable particles, while the concentration of frozen-out stable primordial particles remains the same.

Suppose that in the period under review in the Universe, in the first second of expansion the frozen-out concentration of some stable particles with respect to the concentration of photons is

$$r = \frac{n_m}{n_\gamma}. \tag{5.152}$$

The maximum possible reduction in the relative concentration of such relic frozen-out particles after a short dust stage may be reached at

$$\tau \sim t_{\max} \sim 10^4 \text{s} \tag{5.153}$$

for

$$\Omega_b = 0.1$$

and

$$t_0 \sim 1 \text{ s}. \tag{5.154}$$

Recall that the upper limit on the duration of the dust stage follows from the condition that the Planck spectrum is restored after this stage is completed, and the lower limit follows from the constraints of the theory of nucleosynthesis on the possible contributions of massive particles to cosmological density.

Given these constraints, the maximum reduction in the relative concentration after the end of the dust phase is given as

$$r = \left(\frac{n_m}{n_\gamma}\right)_{t>\tau} = \left(\frac{\tau}{t_0}\right)^{-1/2}\left(\frac{n_m}{n_\gamma}\right)_{t<1\,\mathrm{s}}. \tag{5.155}$$

In the dust stage of dominance of massive neutrinos, the above experimental constraints (5.140) results in the minimal time at which the dust stage begins

$$t \sim 100 \text{ s}, \tag{5.156}$$

corresponding to the mass of unstable neutrinos

$$m_\nu \sim 10 \text{ MeV}. \tag{5.157}$$

With these parameters, the unstable massive neutrinos may dominate only in a limited time interval

$$t_0 \sim 100 \text{ s} < t < 10^4 \text{s}. \tag{5.158}$$

Therefore, the maximum possible reduction in the relative concentration of stable relic particles can not exceed a factor of 10.

Realistic scenarios of the very early Universe may contain possible non-trivial combinations of various cosmological effects of the theory of elementary particles. This makes the effect of short dust stages only one of the numerical factors which should be considered in the relic concentration of particles, creation, freezing-out or decoupling of which occurred in the very early Universe.

However, such effects may lead to a quantitative change in cosmology of massive neutrinos quantification of which is usually not in doubt.

The standard Big Bang model, it would seem, uniquely predicts the concentration of relic neutrinos on the basis of quantified relations with the observed number density of photons (Chapter 3). The numerical factor in this relation (cf. (3.118))

$$\frac{n_\nu + n_{\bar{\nu}}}{n_\gamma} = \frac{3}{4}\cdot\frac{4}{11} \tag{5.159}$$

is determined by the thermodynamic equilibrium of relativistic bosonic and fermionic gases, and by conservation of specific entropy in the equilibrium mixture of radiation and electron–positron pairs.

Accounting for the possible stages of the dominance of non-relativistic particles adds in this relationship a numerical factor, which is inversely proportional to the relative increase in the density of photons as a result of electromagnetic energy release at the end of a dust stage. This factor of 'dilution' of the relic radiation is given as

$$\frac{(n_\gamma)_{in}}{(n_\gamma)_f} = \left(\frac{t_0}{\tau}\right)^{1/2}. \quad (5.158)$$

Given such a dilution of the relic radiation due to electromagnetic energy release in the decays of massive metastable particles that dominate the Universe during the dust stage after 1 second of expansion, the generalized form of the concentration ratio of relic neutrinos and photons has the form

$$\frac{n_\nu + n_{\bar{\nu}}}{n_\gamma} = \frac{3}{4} \cdot \left(\frac{t_0}{\tau}\right)^{1/2}. \quad (5.161)$$

It should be noted that in several studies (Lindley 1980, 1984; Sarkar, Cooper 1984; Hut, White 1984) arguments were put forward against the dominance of the Universe by particles with the mass

$$5 \text{ MeV} < m < 10 \text{ MeV}, \quad (5.162)$$

decaying at

$$\tau > 10^2 \text{ s} \quad (5.161)$$

with the birth of photons and electron–positron pairs. These arguments are based on a rather detailed calculation of the photodisintegration of deuterium by products of these decays (Lindley 1984).

In particular, these results excluded the possibility of dominance of ν_τ in the early Universe and, together with constraints on the mixing and lifetime of ν_τ, they totally ruled out for ν_τ a mass greater than 1 MeV (Sarkar, Cooper 1984).

However, these calculations did not take into account that the neutron born as a result of the photodisintegration of deuterium in the reaction

$$\gamma D \to np$$

can then be captured by a proton in the inverse reaction

$$np \rightarrow \gamma D.$$

This possibility, first pointed out by Zeldovich et al (1977) when analyzing the evaporation of primordial black holes, and then considered in the general case of different sources of non-equilibrium particles (Chechetkin et al 1982; Khlopov, Chechetkin 1987) compensates for the effects of photodisintegration of deuterium. It is easy to evaluate (Zeldovich et al 1977) that at

$$t < 10^5 c \Omega_b^{2/3} \quad (5.162)$$

the characteristic time of neutron capture by a proton

$$t_{np} = \left(n_p \sigma v (np \rightarrow \gamma D)\right)^{-1} \quad (5.163)$$

will be less than the neutron lifetime

$\tau \approx 10^3$ s.

Thus, the neutrons have time to form deuterium before decay. Moreover, as we shall see, in most cases the decay products of metastable particles can disrupt ^{4}He, leading to an increase in the abundance of deuterium and ^{4}He.

6

Antiprotons in the Universe after Big Bang nucleosynthesis

In baryon-asymmetric Universe after the first millisecond of expansion the baryon–antibaryon pairs which are in equilibrium must be annihilated. An important consequence of the baryon asymmetry is the exponentially small number of antiprotons that have survived after such an annihilation.

The standard baryon-asymmetric cosmology predicts the exponential suppression of antibaryon abundance from the first microsecond after the beginning of the expansion until the time of galaxy formation when cosmic rays interacting with matter can lead to the formation of antibaryon components in cosmic rays.

We can therefore conclude that the occurrence of the non-exponentially small number of antiprotons and noticeable effects of the annihilation in the Universe from the first microseconds before the period of formation of galaxies is the clearest manifestation of new physics which was not taken into account in the standard scenario.

Based on the observational data, we can use this feature to limit the allowable parameters of the hypothetical sources of antiprotons in the RD stage after the first microsecond, and impose constraints on the parameters of particle physics which predicts such sources as its cosmological implications.

Particle theory leads to several non-trivial possibilities of formation of antiprotons in the Universe after the equilibrium annihilation of nucleon–antinucleon pairs, and even after a period of nucleosynthesis (Khlopov, Chechetkin 1987; Chechetkin et al 1982).

Primordial antinucleons could be preserved in the domains of antimatter predicted in some particle models, in particular in models of spontaneous *CP*-violation.

Another possibility is related to the creation of the nucleon–antinucleon pairs by some of the sources of non-equilibrium particles in the RD stage. These sources are:

a) evaporating primordial black holes formed in the very early Universe (Chapter 4),

b) decays of metastable particles with a mass exceeding 2 GeV,

c) residual annihilation of frozen-out massive stable particles.

Consider, first, the evaporation of primordial black holes (PBH) as the source of the nucleon–antinucleon pairs.

1. PBH evaporation as the source of the nucleon–antinucleon pairs

As was just pointed out, the possibility of evaporation of black holes described for the first time by Hawking (1975) is an important property in terms of possible effects of PBH in the RD stage.

Evaporation of a black hole is described as radiation from the surface of a black body with its temperature given by the equation

$$T_{\mathrm{PBH}} = \frac{1}{4\pi r_g} = \frac{m_{\mathrm{Pl}}^2}{8\pi M}.$$

Here M is the PBH mass, r_g its gravitational radius

$$r_g = 2GM = 2\frac{M}{m_{\mathrm{Pl}}^2}.$$

As shown in Chapter 4, the evaporation time of a black hole is equal to

$$t_{\mathrm{evp}} = 10^{-27}\,\mathrm{s}\left(\frac{M}{1\,\mathrm{g}}\right)^3.$$

Fluxes of particles from evaporating primordial black holes may have the same effect on the physical processes in the Universe, as the decay products of metastable particles.

Now consider (Chechetkin et al 1982; Khlopov, Chechetkin 1987), only one specific channel of PBH evaporation, namely, the creation of antinucleons during PBH evaporation.

Let us estimate the average number of antinucleons, F, evaporated from PBHs with mass M. It follows from (4.46) that the PBH with mass

$$M < M_{\bar{p}} = 10^{13}\,\mathrm{g} \tag{6.1}$$

has a surface temperature of more than 1 GeV, so that nucleons and antinucleons can form during evaporation of such a PBH.

The modern theory of strong interactions (Chapter 2) suggests that quarks and gluons form initially in hadronic processes at high energies. This is followed by their fragmentation into hadrons (hadronization).

We can expect that the same will happen when during evaporation the PBH emits gluons and quarks, and their distribution according to the thermodynamic equilibrium at

$$T = T_{\text{PBH}}$$

is implemented through a set of processes of production of individual particles. Each individual process is a process of creation in a strong gravitational field of a pair of particles, one of which goes to infinity, while the other disappears under the gravitational radius.

If a couple of coloured particles (quark–antiquark or gluon pair) is born, their hadronization can be considered like the birth of hadrons in electron–positron annihilation (or in hard hadronic processes).

Indeed, when the distance l between two colour charges reaches the confinement scale

$$l \sim \frac{1}{\Lambda_{QCD}} \sim 10^{-13}\,\text{cm} \tag{6.2}$$

hadronization occurs, and the emitted quark and gluon form a hadron jet.

Estimation of the formation of antiprotons in such jets can be based on arguments by Azimov et al (1983) who studied the creation of an antiproton in electron–positron annihilation. We obtain

$$F = \frac{\left(\dfrac{dN_{\bar{p}}}{dt}\right)}{\left|\dfrac{dM}{dt}\right|} = \frac{\sum\limits_{i=q,\bar{q},gl} \int dp_{\bar{p}} \int D_i^{\bar{p}}\left(p_{\bar{p}}, p_i\right) f\left(p_i\right) dp_i S}{\sum\limits_i \int f\left(p_i\right) \varepsilon_i dp_i S}, \tag{6.3}$$

where

$$f\left(p_i\right) = f\left(p_i, T_{\text{PBH}}\right) \tag{6.4}$$

is the distribution function of i-th type of particle emitted by PBH

$$S = 4\pi r_g^2 \tag{6.5}$$

–'surface' of the PBH

$$D_i^{\bar{p}}\left(p_{\bar{p}}, p_i\right)$$

– probability of fragmentation of the quark (q), antiquark ($\bar{q}$) or gluon (g) with momentum p_i in the antiproton, ε_i is the energy of the particle of i-th type.

Given the fragmentation of gluons to antinucleons which is more efficient than the fragmentation of quarks (antiquarks), (Azimov et al 1983), we have that (Khlopov 1999)

$$F = f_{\bar{p}} = \left\langle N_{\bar{p}} \right\rangle = \frac{1}{7}\left\langle N_g \right\rangle, \tag{6.6}$$

where the mean multiplicity of pions is associated with the average multiplicity of gluons by the following relation

$$\left\langle N_\pi \right\rangle = \left(\frac{T_{\text{PBH}}}{\mu}\right)^{1/2} \left\langle N_g \right\rangle \tag{6.7}$$

with

$$\mu \sim 1 \text{ GeV}. \tag{6.8}$$

Since the mass of PBHs decreases during the evaporation process, antiprotons can form only at a late stage of evaporation of primordial black holes, when their mass is reduced to 10^{13} g

$$M > M_{\bar{p}} = 10^{13} \text{ g}. \tag{6.9}$$

In this case, the multiplicity of antinucleons is

$$F = f_{\bar{p}}\left(\frac{M_{\bar{p}}}{M}\right). \tag{6.10}$$

For detailed analysis of hadron formation during evaporation of primordial black holes see (Golubkov et al 2000). Relativistic antiprotons emitted during PBH evaporation at

$$t < 10^2 \text{ s} \tag{6.11}$$

when the temperature of the plasma in the Universe is

$$T > 10^2 \text{ keV} \tag{6.12}$$

slowly slow down before annihilation, as presence in equilibrium with the radiation of electron–positron pairs makes the rate of Coulomb deceleration of antiprotons greater than the rate of their annihilation.

After the annihilation of positrons, i.e., when

$$t > 10^2 \text{ s} \tag{6.13}$$

the electron number density falls to the baryon density and the energy loss of antiprotons due to their Coulomb interaction with the plasma becomes small.

However, at high energies, the annihilation channel of antiprotons with protons is suppressed. The annihilation cross section σ_{ann} is of the order of the difference between the total cross sections of interactions of protons and antiprotons

$$\sigma_{\text{ann}} \sim \sigma^{\text{tot}}_{\bar{p}p} - \sigma^{\text{tot}}_{pp}, \tag{6.14}$$

which decreases with the total energy.

The cross section for inclusive inelastic reactions

$$\bar{p} + p \to \bar{p} + \ldots \tag{6.15}$$

contributes about half of the total cross section, and these are the reactions in which the antiprotons lose energy.

Therefore, even at

$$t > 10^2 \text{ s}$$

a significant proportion of antiprotons from evaporating primordial black holes slows down. In the RD stage, the rate of annihilation of relativistic antiprotons exceeds the rate of expansion. Thus, the antiprotons born in this period must be annihilated.

We now estimate the number density of antiprotons from primordial black holes with mass M (cf. (4.50)) evaporating at

$$t = t_e = 10^{-27} \text{s} \left(\frac{M}{1\text{g}} \right)^3. \tag{6.16}$$

The total number of antinucleons N_a from PBH evaporation is equal to

$$N_a = \int_0^{\min(M, M_{\bar{p}})} \left(\frac{dN_{\bar{p}}}{dM} \right) dM = \int_0^{\min(M, M_{\bar{p}})} f_a dM, \tag{6.17}$$

where

$$f_a \equiv f_{\bar{p}}$$

is given by equations (6.3) and (6.6). A rough estimate yields (Chechetkin et al 1982)

$$f_a = \frac{dN_{\bar{p}}}{dM} = \frac{g_g}{g_{\text{tot}}} \cdot \frac{\left(\frac{\langle N_{\bar{p}} \rangle}{\langle N_g \rangle}\right)}{3T_{\text{PBH}}} = \frac{g_g}{g_{\text{tot}}} \cdot \frac{\left(\frac{T_{\text{PBH}}}{\mu}\right)^{1/2}}{21T_{\text{PBH}}} = \tag{6.18}$$
$$= \frac{g_g}{21 g_{\text{tot}}} \cdot \left(\mu T_{\text{PBH}}\right)^{-1/2} = \frac{g_g}{21 g_{\text{tot}}} \cdot \left(\frac{8\pi M}{\mu}\right)^{1/2} \frac{1}{m_{\text{Pl}}}.$$

In equation (6.18) we assumed that the ratio of the multiplicity of gluons and the total multiplicity of particles emitted by the PBH is the ratio of their statistical weights, g,

$$\frac{\langle N_g \rangle}{\langle N_{\text{tot}} \rangle} = \frac{g_g}{g_{\text{tot}}} \tag{6.19}$$

and that the mean energy of particles that are emitted by evaporating PBHs is

$$E = 3T_{\text{PBH}}. \tag{6.20}$$

Averaged over the Universe, the number density of the antiprotons produced during PBH evaporation is equal to

$$n_{\bar{p}} = n_{\text{PBH}}(M) \cdot N_a, \tag{6.21}$$

where $n_{\text{PBH}}(M)$ is the concentration of PBHs with mass M in the Universe during their evaporation

$$n_{\text{PBH}}(M) = \frac{\rho_{\text{PBH}}(M)}{M} = \alpha(M)\frac{\rho_{\text{tot}}}{M}. \tag{6.22}$$

Here

$$\alpha(M) = \frac{\rho_{\text{PBH}}(M)}{\rho_{\text{tot}}} \tag{6.23}$$

is the relative contribution of PBHs with mass M to the total cosmological density in the period of their evaporation. From equations

(6.3)–(6.23) we obtain

$$n_{\bar{p}} = \frac{4}{63} \cdot \frac{g_g}{g_{\text{tot}}} \cdot \left(\frac{2\pi}{\mu}\right)^{1/2} \cdot \frac{1}{m_{\text{Pl}}} \cdot \min\left(M^{1/2}, \frac{M_{\bar{p}}^{3/2}}{M}\right) \cdot \alpha(M) \cdot \rho_{\text{tot}}. \quad (6.24)$$

Primordial black holes evaporating in the RD stage have masses

$$M < M_{\bar{p}} = 10^{13}\,\text{g}. \quad (6.25)$$

In this case,

$$n_{\bar{p}} = \frac{4}{63} \cdot \frac{g_g}{g_{\text{tot}}} \cdot \frac{\left(\frac{2\pi M}{\mu}\right)^{1/2}}{m_{\text{Pl}}} \cdot \alpha(M) \cdot \rho_{\text{tot}}. \quad (6.26)$$

Here, the total cosmological density is given as

$$\rho_{\text{tot}} = \frac{m_{\text{Pl}}^2}{t^2}. \quad (6.27)$$

In numerical estimates we will also use the expression

$$\rho = 5 \cdot 10^5 \frac{\text{g}}{\text{cm}^3} \left(\frac{1\text{s}}{t}\right)^2.$$

From equation (6.16) we can obtain the relationship between the mass of the PBH and the time of its evaporation

$$M = 10^9\,\text{g} \left(\frac{t_e}{1\text{s}}\right)^{1/3}. \quad (6.28)$$

Substituting the expression (6.27) for ρ_{tot} in equation (6.24) at

$$t = t_e, \quad (6.29)$$

we estimate the density of antiprotons from evaporating primordial black holes

$$n_{\bar{p}} = \frac{4}{63} \cdot \frac{g_g}{g_{\text{tot}}} \cdot \frac{\left(\frac{2\pi M}{\mu}\right)^{1/2}}{m_{\text{Pl}}} \cdot \alpha(M) \cdot \rho_{\text{tot}} = 6 \cdot 10^{21}\,\text{cm}^{-3} \frac{g_g}{g_{\text{tot}}} \left(\frac{M}{1\,\text{g}}\right)^{1/2} \alpha(M) \left(\frac{1\text{s}}{t_e}\right) =$$
$$= 2 \cdot 10^{26}\,\text{cm}^{-3} \frac{g_g}{g_{\text{tot}}} \alpha(M) \left(\frac{1\text{s}}{t_e}\right)^{11/6}. \tag{6.30}$$

Using the relation

$$n_\gamma = 2.5 \cdot 10^8 \Omega_b^{-1} n_b = 5 \cdot 10^{31}\,\text{cm}^{-3} \left(\frac{1\,\text{s}}{t}\right)^{3/2}, \tag{6.31}$$

we obtain

$$n_{\bar{p}} = 10^3 \cdot \frac{g_g}{g_{\text{tot}}} \cdot \alpha(M) \left(\frac{1\text{s}}{t_e}\right)^{1/3} \Omega_b^{-1} n_b. \tag{6.32}$$

Since nearly all antiprotons from primordial black holes evaporating at RD stage, annihilate, then from (6.32) we obtain the average number of annihilated antiprotons per baryon

$$r_a = \frac{n_{\bar{p}}}{n_b} = 10^3 \frac{g_g}{g_{\text{tot}}} \alpha(M) \left(\frac{1\text{s}}{t_e}\right)^{1/3} \Omega_b^{-1}. \tag{6.33}$$

Equation (6.32) holds for PBH evaporating at the RD stage. Primordial black holes with masses

$$10^{13}\text{ g} < M < 10^{15}\text{ g} \tag{6.34}$$

evaporate in the MD stage (the stage of dominance of matter) with

$$t > 10^{12}\text{ s}. \tag{6.35}$$

It follows from equation (6.24) that the average density of antiprotons formed by evaporation of such PBHs is equal

$$n_{\bar{p}} = \frac{g_g}{g_{\text{tot}}} \frac{4}{63} \left(\frac{2\pi}{\mu}\right)^{1/2} \frac{M_{\bar{p}}^{3/2}}{m_{\text{Pl}} M} \alpha(M) \left(\frac{1\,\text{s}}{t_e}\right)^{1/3} \rho_{\text{tot}} = \frac{g_g}{g_{\text{tot}}} \frac{4}{63} \left(\frac{2\pi}{\mu}\right)^{1/2} \frac{M_{\bar{p}}^{3/2} m_{\text{Pl}}}{M} \alpha(M) \left(\frac{1\,\text{s}}{t_e}\right)^{1/3} t_e^{-2} =$$
$$= \frac{g_g}{g_{\text{tot}}} \frac{4}{63} \left(\frac{2\pi}{\mu}\right)^{1/2} \frac{M_{\bar{p}}^{3/2} m_p}{m_{\text{Pl}} M} \alpha(M) \left(\frac{1\,\text{s}}{t_e}\right)^{1/3} \Omega_b^{-1} n_b. \tag{6.36}$$

where the total density is taken to be the critical density.

The relative concentration of antiprotons from evaporating primordial black holes is

$$r_a = \frac{n_{\bar{p}}}{n_b} = \frac{g_g}{g_{\text{tot}}} \frac{4}{63} \left(\frac{2\pi}{\mu} \right)^{1/2} \frac{M_{\bar{p}}^{3/2} m_p}{m_{\text{Pl}} M} \alpha(M) \left(\frac{1\,\text{s}}{t_e} \right)^{1/3} \Omega_b^{-1} = \tag{6.37}$$
$$= 4 \cdot 10^{-2} \frac{g_g}{g_{\text{tot}}} \left(\frac{M_{\bar{p}}}{M} \right) \alpha(M) \Omega_b^{-1}.$$

It should be noted that for a homogeneous distribution of baryons

$$n_b(x) = \langle n_b(x) \rangle \tag{6.38}$$

in the MD stage the value (6.37) does not coincide with the fraction f of antiprotons taking part in the annihilation, since the characteristic time for annihilation of relativistic antiprotons

$$t_{\text{ann}} = \left(n_b (\sigma v)_{\text{ann}} \right)^{-1} \tag{6.39}$$

exceeds the cosmological time at

$$t_e > t_{\bar{p}} = 2 \cdot 10^{14} c \Omega_b \tag{6.40}$$

so that the rate of their annihilation will be less than the rate of expansion, and most of the relativistic antiprotons do not participate in the annihilation. For the non-relativistic antiprotons Coulomb attraction strongly influences the rate of annihilation of the proton–antiproton pairs (Morgan, Hudges 1970; Cohen et al 1997)

$$(\sigma v)_{\text{ann}} = 6.5 \cdot 10^{-17} \frac{\text{cm}^3}{\text{s}} \left(\frac{c}{v} \right), \tag{6.41}$$

where v is the relative velocity. Thus, the rate of annihilation of the non-relativistic antiprotons may increase, and at a low velocity, corresponding to the condition

$$\frac{v}{c} < 10^{-4} \Omega_b \tag{6.42}$$

the rate of their annihilation exceeds the rate of expansion, even in the present-day Universe.

After the formation of galaxies at

$$t > t_s < 10^{16}\ \text{s} \tag{6.43}$$

neither matter nor the PBH were distributed uniformly. The PBH forming in this case a specific form of dark matter, should have been concentrated in a galactic halo (see below), and evaporation could have been a local (galactic) source of the antiproton component of cosmic rays (Chechetkin et al 1982).

The antiprotons produced in the period

$$t_{\bar{p}} < t < t_s \tag{6.44}$$

experience the redshift, but this redshift cannot slow down relativistic or semi-relativistic antiprotons born during PBH evaporation to these low velocities to start annihilation, as well as it cannot ensure their condensation into galaxies. The latter can only occur for very slow antiprotons, with the velocity during the evaporation period t_e being less than

$$\frac{v}{c} < 10^{-3}\left(\frac{t_s}{t_e}\right)^{2/3}, \tag{6.45}$$

but the rate of annihilation of slow antiprotons will be about the rate of expansion, so that a large proportion of them should annihilate. More energetic antiprotons should form an isotropic background of the antiprotons uniformly distributed over the Universe.

We now consider the possible effects of antinucleon annihilation in the RD stage. We first discuss the annihilation of the nucleons and antinucleons.

As we pointed out above, the antiprotons have to be slowed down by cosmic plasma until their annihilation starts. The annihilation of slow antiprotons with nucleons leads to the production of pions with the average multiplicity of 5 (Chechetkin et al 1982; Cohen et al 1997).

It is easy to estimate that at

$$t > 100\,\text{s}\ \Omega_b^{2/3} \tag{6.46}$$

the characteristic time of the pion–nucleon interaction is

$$t_{\pi N} \sim \left(n_b \sigma_{\pi N} c\right)^{-1} \tag{6.47}$$

and exceeds the lifetime of a charged pion, which is equal

$$\tau_{\pi^{\pm}} \approx 2 \cdot 10^{-8}\,\text{s}. \tag{6.48}$$

Therefore, both the neutral and charged pions from the nucleon-antinucleon annihilation should decay. Annihilation of slow antiprotons in the RD stage leads to cascades of pion and muon decays. As a result, the neutrinos are born, carrying away about 50% of the energy released during the annihilation, γ-quanta carrying away about 34% of this energy and the electron–positron pair carrying away the remaining 16% of energy.

The neutrinos born in the cascades of muon and pion decays of nucleon–antinucleon annihilation have energies E_0 not exceeding a few hundred MeV. These neutrinos barely interact with matter, and they must remain in the Universe, undergoing the redshift to energies

$$E = E_0\left(1 + z_{\text{ann}}\right)^{-1}, \tag{6.49}$$

where z_{ann} is the redshift, corresponding to the annihilation period. Fluxes of these neutrinos with an energy not exceeding for

$$z_{\text{ann}} > 10^5 \tag{6.50}$$

the value

$$E \sim 10^2\ \text{MeV} \tag{6.51}$$

cannot apparently be observed.

Gamma quanta, electrons and positrons from the annihilation of electrons interact with the cosmic plasma electrons, as well as positrons from annihilation annihilate with cosmic electrons. The energy released in these processes should lead to a distortion of the spectrum of the thermal background radiation.

However, the existing constraints on such a distortion (5.113) are not very sensitive to the annihilation of antinucleons in the early RD stage. The effect of this energy release should be a significant fraction of the total energy density of the thermal background. A significant effect means the number of antiprotons of the order of the total number of baryons in the Universe.

As we now show, the most severe constraints on the sources of antiprotons follow from the analysis of astrophysical effects of antiproton annihilation on ^{4}He nuclei. This circumstance is consistent with the general idea of cosmoarcheology that the most sensitive indicator of the new physics are the effects in the subdominant component of the Universe. At the RD stage this component is baryonic matter.

2. Effects of the annihilation of antiprotons with ^{4}He nuclei in the abundance of deuterium and ^{3}He

^{4}He in its abundance is an essential element of the Universe after hydrogen. Its mass concentration is

$$Y = \frac{4n\left(^4\mathrm{He}\right)}{\left(n(\mathrm{H}) + 4n\left(^4\mathrm{He}\right)\right)} \cong 0.25, \tag{6.52}$$

while the concentration of other light elements is much smaller. For example, the abundance of deuterium is estimated as

$$X_{\mathrm{D}} = 2.5 \cdot 10^{-5} \tag{6.53}$$

or the abundance of ^{4}He is given by

$$X_{^3\mathrm{He}} = 4.2 \cdot 10^{-5}. \tag{6.54}$$

The annihilation of antiprotons with ^{4}He nuclei could lead to the creation of deuterium and ^{3}He in the reactions

$$\overline{p} + ^4\mathrm{He} \to \mathrm{D} + p(n) + N_1\pi, \tag{6.55}$$

$$\overline{p} + ^4\mathrm{He} \to ^3\mathrm{He} + N_2\pi, \tag{6.56}$$

$$\overline{p} + ^4\mathrm{He} \to T + N_3\pi, \tag{6.57}$$

where N_1, N_2, N_3 are the numbers of pions.

If there were antiprotons in the Universe after nucleosynthesis, their annihilation would inevitably lead to the production of additional quantities of deuterium and ^{3}He. Comparison of the abundance of ^{4}He, deuterium and ^{3}He shows that at destruction of even a small fraction of ^{4}He ($\sim 10^{-4}$) interaction with antiprotons can lead to the production of all abundance of deuterium and/or ^{3}He observed at the present time.

We must also mention that in the early Universe deuterium could form not only directly in the reaction (6.55). The destruction of the nuclei of ^{4}He by antiprotons is also a source of free neutrons (Chechetkin et al 1982; Zeldovich et al 1977).

If the density of matter during annihilation in the Universe is sufficiently high, the neutrons will be involved in the reaction

$$n + p \to \mathrm{D} + \gamma, \tag{6.58}$$

thus forming an additional amount of deuterium.

Therefore, there are two mechanisms for the formation of deuterium in the annihilation of antiprotons with ^{4}He nuclei in the period after the cosmological nucleosynthesis:

1) in the direct reaction (6.55);

2) indirectly, due to the interaction of neutrons with protons after antiprotons destroy the ^{4}He nuclei.

The indirect mechanism is effective if neutrons can be captured by protons prior to neutron decay, that is, if the reaction time (6.58) is less than the neutron lifetime

$$\tau_n \sim 10^3 \text{ s}. \tag{6.59}$$

The following condition should be fulfilled

$$t_{np} = \frac{1}{n_p \sigma_{np} v} \leqslant \tau_n. \tag{6.60}$$

Here n_p is the concentration of protons and $\sigma_{np} v$ is the rate of neutron capture reactions. Evaluations in (Zeldovich et al 1977) show that the indirect mechanism is ineffective after

$$t_D = 10^5 \Omega_b^{2/3} \text{s}, \tag{6.61}$$

as

$$t_{np} > \tau_n \tag{6.62}$$

at

$$t > t_{\mathrm{D}}. \tag{6.63}$$

Thus, during the period (6.63) the direct mechanism is the only source of deuterium in the early Universe after primordial nucleosynthesis.

Let us consider the constraints on the value f following from analysis of the effect of interaction of antiprotons with ^{4}He on the abundance of light elements. The annihilation of antiprotons with ^{4}He nuclei must result in the formation of an additional amount of deuterium estimated from

$$\Delta n_{\mathrm{D}} = \begin{cases} n_{^4\mathrm{He}} (f_{\mathrm{D}} + f_n) f, \ 100\text{s} \leqslant t \leqslant t_{\mathrm{D}}, \\ n_{^4\mathrm{He}} f_{\mathrm{D}} f, \ t > t_{\mathrm{D}}, \end{cases} \tag{6.64}$$

$n_{^4\mathrm{He}}$ is the concentration of ^{4}He, and f_{D}, f is the mean multiplicity of the deuterium nuclei and neutrons, respectively. A similar equation is also obtained for ^{3}He

$$\Delta n_{^3\text{He}} = n_{^4\text{He}} f_{^3\text{He}} f. \tag{6.65}$$

Assuming that Δn_D does not exceed observed abundance of deuterium X_D, we obtain the following upper limit on the value f

$$f \leqslant \begin{cases} \dfrac{2X_\text{D}}{Y(f_\text{D} + f_n)}, & 100\text{ s} \leqslant t_\text{ann} \leq t_\text{D}, \\ \dfrac{2X_\text{D}}{Yf_\text{D}}, & t_\text{ann} \leqslant t_D, \end{cases} \tag{6.66}$$

where Y stands for the observed ^{4}He.

As for ^{3}He, it should be noted that the nuclei of ^{3}He can also be products of the decay of tritium produced in the reaction (6.57). So both reactions (6.56) and (6.57) contribute to the total amount of ^{3}He formed.

Using equation (6.65) we find, like (6.66), the limit for the observed abundance of ^{3}He

$$f \leqslant \frac{4}{3} \cdot \frac{X_{^3\text{He}}}{Yf^\text{eff}_{^3\text{He}}}, \tag{6.67}$$

where

$$f^\text{eff}_{^3\text{He}} = f_{^3\text{He}} + f_\text{T}. \tag{6.68}$$

It follows from equations (6.66) and (6.67) that the upper limit on the value f depends on the value f_A, where A = D, n, T and ^{3}He. To obtain more reliable information on possible sources of antiprotons in the RD stage (Chechetkin et al 1982) special experimental studies were carried out of the reaction of antiprotons with ^{4}He nuclei (Balestra et al 1980; Batusev 1984) in the experiment PS179 at LEAR. In this experiment, the following value was measured (Balestra et al 1980)

$$f_{^3\text{He}} = 0.102 \pm 0.008. \tag{6.69}$$

These experimental data were also used to evaluate the relative probability of tritium production

$$f_\text{T} = (1.32 \pm 0.05) f_{^3\text{He}}, \tag{6.70}$$

giving a value for the effective constant of helium-3

$$f^\text{eff}_{^3\text{He}} = 0.237 \pm 0.014. \tag{6.71}$$

Equation (6.67) shows that the maximum abundance of ^{3}He, allowed from observations

$$X_{^3\mathrm{He}} = 1.2\cdot 10^{-4}, \tag{6.72}$$

corresponds to the limit on the value f

$$f < 2.5\cdot 10^{-3}. \tag{6.73}$$

The data obtained by Batusov et al (1984) show that the value f_A for A = ^{3}He depends rather weakly on the momentum of the antiproton in the range 200–600 MeV/c. This confirms the theoretical arguments by (Shmatikova and Kondratyuk (1983)) in favour of the weak dependence of this value on the energy of the antiproton. Therefore, as a first approximation, we can assume that this quantity does not depend on the energy of the antiproton. In this approximation, the predicted concentration of ^{3}He, formed in the reactions of annihilation of antiprotons with ^{4}He nuclei, does not depend on the spectrum of antiprotons and their behaviour in cosmic plasma (in particular, the mechanism of their deceleration in the plasma in the case of energetic antiprotons).

The upper limit (6.73) in this approximation is true for all the hypothetical sources of antiprotons in the RD stage. This provides a reliable check of the possible physical nature of these sources and, as we shall see in the next chapter, gives some non-trivial information about the hidden sector of the particle theory, as well as some of the phenomena that reflect the physical mechanisms of inflation, baryosynthesis and dark matter.

It should be noted that the decays of metastable particles and evaporation of primordial black holes are not only a source of antiprotons but also of other particles, namely protons, mesons, hyperons, the electron–positron pairs, gamma rays and neutrinos. Besides, for all sources of antiprotons the proton–antiproton annihilation is also a source of γ-rays with energies around 100 MeV.

The interaction of ^{4}He nuclei with the protons and γ-rays with energies exceeding 20 MeV, can also lead to the destruction of ^{4}He, deuterium and ^{3}He (Ellis et al 1984; Dimopolous et al 1987; Levitan et al 1988). Analysis of these mechanisms examines in detail the evolution of the energetic nucleons and γ-rays in cosmic plasma, using the numerical calculation methods. Such analysis takes into account all energy loss mechanisms and interactions of non-equilibrium particles, as well as all the specific features of the considered non-equilibrium particles (in particular, the spectrum of particles emitted by the source).

As a first approximation, analysis of the effect of the antiproton-^{4}He interaction on the abundance of light elements avoids consideration of all these items, giving a reliable minimum estimation of the effects of antiproton sources. Additional production of deuterium and ^{3}He due to the destruction of ^{4}He by energetic particle enhances the effect and the corresponding constraints on the parameters of the sources of antiprotons in the RD stage.

This strengthening of constraints due to the additional production of ^{3}He as a result of destruction of ^{4}He by energetic particles of nucleonic cascades in the early Universe initiated by antinucleons with energy E_0 was calculated in the work of Levitan et al (1988) by the Monte Carlo method.

Detailed analysis of the predicted relationships of the abundance of light elements, corresponding to each specific source of non-equilibrium particles, reveals the effects of different sources and checks their existence and possible properties with the use of the observed chemical composition of the Universe.

Consider, finally, another aspect of the relationship between the abundance of light elements and parameters of the sources of antiprotons.

In the case of strict equality in equation (4.53) all observed deuterium can be attributed to the non-equilibrium processes of destruction of ^{4}He in the RD stage (Zeldovich et al 1977). Thus, there is a new mechanism for the production of deuterium, in which the abundance of deuterium does not depend on the density of baryons, as is the case in the standard theory of cosmological nucleosynthesis. Even in the case

$$\Omega_b = 1$$

when the standard theory of nucleosynthesis predicts (Chapter 3)

$$X_{\mathrm{D}} < 10^{-8}, \tag{6.74}$$

destruction of ^{4}He in the RD stage after nucleosynthesis could lead to the predicted amount of deuterium, corresponding to its observed abundance.

This possibility is of particular interest to the problem of dark matter in the Universe (see below), since the standard theory of nucleosynthesis indicates a low density of baryons

$$\Omega_b \ll 1,$$

giving thus a serious argument in favour of the non-baryonic dark matter/energy that dominates the Universe and increase the total density to the critical density, in accordance with the predictions of the simplest inflationary models.

Therefore, the possibility of non-equilibrium cosmological nucleosynthesis and, in particular, the antiproton–^{4}He mechanism of formation of deuterium deserve special consideration. Such a study, in general, requires a detailed consideration of the possible properties of the non-equilibrium source of hypothetical particles in the RD stage.

However, if the antiproton–^{4}He mechanism dominates in the non-equilibrium cosmological production of deuterium, we can verify its effectiveness regardless of the nature of sources of antiprotons (Chechetkin et al 1982).

For experimental verification of the effectiveness of this mechanism it is sufficient to compare f_A for A = D and A = ^{4}He.

If the experimental measurement of these variables will find that

$$f_{\mathrm{D}} < 0.1 f_{^3\mathrm{He}}, \tag{6.75}$$

the annihilation of antiprotons with helium should lead to the abundance of deuterium by an order of magnitude smaller than ^{3}He.

Attributing all the observed abundance of deuterium to the antiproton–^{4}He mechanism, we come to the overproduction of ^{3}He because, given all the uncertainties, the observed abundance of deuterium and ^{3}He cannot differ by an order of magnitude.

Thus, we can experimentally disprove the antiproton–^{4}He production mechanism of deuterium in the RD stage in measurements of the ratio f_A for A = D to A = ^{3}He.

3. Constraints on sources of antiprotons

We now consider the relationship between sources of antiprotons in the RD stage and the parameters of models of elementary particles. We will see how the constraints on the effects of antiproton annihilation in the RD stage limit such parameters.

3.1. Constraints on primordial black holes and the theoretical mechanism of their formation

In Chapter 4 it was shown that the spectrum of primordial black holes formed at the stage of dominance in the Universe of supermassive metastable particles (or scalar fields) can be related to the parameters of these particles (or fields). The upper bound on the number of antiprotons from primordial black holes with masses

$$10^{10} < M < 10^{13}\ \text{g}, \tag{6.76}$$

evaporated in the RD stage, constrains the possible contribution $\alpha(M)$ of such primordial black holes to the cosmological density during their evaporation.

Equation (6.33) shows that the fraction of the total density, corresponding to PBHs with mass

$$M < 10^{13}\ \text{g},$$

during their evaporation, may be related to the value f by the equation

$$\alpha(M) = 10^{-3} f\Omega_b \left(\frac{t_e}{1\text{s}}\right)^{1/3} k_{\bar{p}}^{-1}, \tag{6.77}$$

where

$$\alpha(M) = \frac{\rho_{\text{PBH}}(M)}{\rho_{\text{tot}}} \tag{6.78}$$

and

$$k_{\bar{p}} = \frac{g_g}{g_{\text{tot}}} \tag{6.79}$$

is the fraction of antiprotons in the spectrum of evaporating PBHs.

According to equation (6.77) the constraint (6.73) gives an upper limit on $\alpha(M = 10^{11}\ \text{g})$ for PBH evaporating at

$$t_e \sim 10^6\ \text{s}. \tag{6.80}$$

In equation (6.77) we can take into account

$$\Omega_b = 0.05$$

and

$$\frac{g_g}{g_{\text{tot}}} = \frac{8}{\frac{21}{4}N_q + \frac{7}{4}N_l + 9 + \frac{7}{8}N_\nu + N_{hs}}, \tag{6.81}$$

where $N_{q(l)}$ is the number of types of quark (lepton) flavours with masses

$$m < T_{\text{PBH}}, \tag{6.82}$$

N_ν is the number of types of neutrinos and N_{hs} is the effective number of two-component types of particles predicted in the hidden sector of the particle theory. Assuming

$$N_{hs} = 0 \tag{6.83}$$

that is, neglecting the effects of superweakly interacting particles, we obtain an upper limit

$$\alpha(M) < 10^{-5}. \tag{6.84}$$

In order to deduce from this upper limit the constraint on the probability W_{PBH} of PBH formation we must take into account the sequence of dust- and radiation-dominated stages, which follow the period of PBH formation.

Consider for definiteness the case when a PBH was formed in the dust stage of the dominance of superheavy particles in early Universe, so that this probability determines the fraction of particles that form a PBH. When the particles decay or annihilate inside an inhomogeneity at

$$t \sim t_f \tag{6.85}$$

the early dust stage comes to an end and the value W_{PBH} determines the contribution of PBHs to the cosmological density at

$$t \geqslant t_f. \tag{6.86}$$

At a later stage relativistic particles dominate in the Universe; these particles, born during the thermalization of decay products (or annihilation) of superheavy particles, provide the equation of state

$$p = \frac{1}{3}\varepsilon.$$

The relative contribution of PBHs to the cosmological density increases at this stage as

$$\frac{\rho_{\text{PBH}}(M)}{\rho_{\text{tot}}} \propto \frac{M}{T} \propto a(t) \propto \left(\frac{t}{t_f}\right)^{1/2}. \tag{6.87}$$

Consequently, during the evaporation of primordial black holes in the RD stage, i.e. when

$$t_e = \left(\frac{M}{m_{\text{Pl}}}\right)^3 t_{\text{Pl}} \tag{6.88}$$

the relative contribution of PBHs to the cosmological density is

$$\alpha(M) = W_{\text{PBH}}\left(\frac{t_e}{t_f}\right)^{1/2} = W_{\text{PBH}}\left(\frac{M}{m_{\text{Pl}}}\right)^{3/2}\left(\frac{t_{\text{Pl}}}{t_f}\right)^{1/2}. \tag{6.89}$$

If the model of elementary particles predicts one or more additional types of superheavy metastable particles, so that at

$$t > t_f$$

we have one or more stages of the dominance of metastable non-relativistic particles, the right-hand side of equation (6.89) should contain an additional factor

$$A = \prod_i \left(\frac{t_{fi}}{t_{0i}}\right)^{1/2}, \tag{6.90}$$

where t_{0i} and t_{fi} correspond to the beginning and end of the i-th stage. Factor A takes into account that the relative contribution of PBHs to the total density does not increase during the dust stages. But the possibility of such a sequence of dust stages is limited by the observed baryon asymmetry of the Universe, because after each stage the specific entropy is increased by a factor

$$S \propto \left(\frac{t_{fi}}{t_{0i}}\right)^{1/2}, \tag{6.91}$$

so that the initial baryon asymmetry is suppressed by a factor

$$\frac{n_b}{n_\gamma} \propto A^{-1}. \tag{6.92}$$

The looser constraints on the probability of PBH formation, the stronger the constraints on the baryosynthesis mechanism.

The upper limit (6.84) on the value $\alpha(M)$ imposes an upper limit on the probability of W_{PBH} which depends on the amplitude of density fluctuations δ. This provides a check for the existence of superheavy particles and fields, predicted by the particle theory.

Indeed, the mass spectrum of primordial black holes formed during the stage of dominance of superheavy particles is limited within the mass interval from

$$M_0 = m_{\text{Pl}}\left(\frac{m_{\text{Pl}}}{rm}\right)^2 \tag{6.93}$$

up to the maximum mass

$$M_{\max} = m_{\mathrm{Pl}} \frac{\tau}{t_{\mathrm{Pl}}} \delta^{3/2} \tag{6.94}$$

and the spectrum in this interval is determined by the spectrum of initial fluctuations $\delta(M)$. In the framework of inflationary models, the shape and amplitude of this spectrum are associated with the parameters of the inflaton potential.

The parameters of the heavy particles and fields define the lower limit of the spectrum of the PBH given in Chapter 4. The condition that the lower limit of the spectrum does not contradict the PBH upper limit on this spectrum, given above, provides non-trivial constraints on the masses, concentrations, and lifetimes of superheavy metastable particles predicted by the particle theory at extremely high energies. Such constraints on the values of *rm* and τ, depending on the

$$\delta(M) = \delta = \text{const} \tag{6.95}$$

are represented in the review (Polnarev, Khlopov 1985). Some examples of these constraints were discussed in Chapter 4.

It should be noted that, according to Carr (1975) and Zabotin Naselsky (1982), the PBH can form at the stage

$$p = \frac{1}{3}\varepsilon$$

through the tail of high-amplitude outliers in the Gaussian distribution of the amplitude of fluctuations. In this case, the PBHs with mass M form at

$$t \sim \left(\frac{M}{m_{\mathrm{Pl}}}\right) t_{\mathrm{Pl}} \tag{6.96}$$

and the probability of PBH formation is determined only by the value of δ (Zabotin, Naselsky 1982; Khlopov et al 1984) and is given as

$$W_{\mathrm{PBH}} \sim \exp\left\{-\frac{1}{18\delta^2}\right\}. \tag{6.97}$$

The upper limit on $\alpha(M)$, obtained previously from the effects of PBH evaporation on the primordial chemical composition, set strong constraints on the inhomogeneity of the early Universe when

$$t < 10^{-28} \text{ s.} \tag{6.98}$$

Since the relative contribution of PBHs to the cosmological density increases as

$$\frac{\rho_{\rm PBH}}{\rho_{\rm tot}} \propto \frac{M}{T} \tag{6.99}$$

during the RD stage, the upper limit to $\alpha(M)$ imposes the following constraint on the probability of PBH formation at times

$$t = \frac{M}{m_{\rm Pl}} t_{\rm Pl} \tag{6.100}$$

given by

$$W_{\rm PBH} \sim \exp\left\{-\frac{1}{18\delta^2}\right\} = \beta(M) = \frac{m_{\rm Pl}}{M}\alpha(M) < 2\cdot 10^{-20}\left(\frac{10^{10}\,{\rm g}}{M}\right), \tag{6.101}$$

thus providing an upper limit on the possible amplitude of fluctuations on a small scale:

$$\delta < 3.5\cdot 10^{-2}\left(1+\frac{1}{46}\ln\left(\frac{M}{10^{10}\,{\rm g}}\right)\right)^{-1/2}. \tag{6.102}$$

This fact limits the possible parameters of the scalar field which is introduced in inflationary models (Khlopov et al 1984).

On the other hand, in Chapter 4 it was shown that relatively long stages of coherent oscillations of the scalar field can be predicted within the framework of inflationary models (Linde 1984; Starobinsky 1980, see review in Khlopov 1999). PBHs can form in these phases (Khlopov et al 1985a, b). Upon completion of these stages as a result of the of the decay of a scalar field (Starobinsky 1980), its decay products thermalize, and the Universe is heated to temperature T_R determined by the duration τ of the stage of coherent oscillations of the field (Chapter 4)

$$T_R \sim \left(\frac{m_{\rm Pl}}{\tau}\right)^{1/2}. \tag{6.103}$$

As the duration of the stage and the probability of PBH formation in this stage are determined by the parameters of the same inflationary model, inflaton scalar field, in the simplest case, the upper limit on the probability of PBH formation imposes constraints on these parameters (Khlopov, Chechetkin 1987).

Note that the constraint on the value of τ imposes a lower limit on the temperature of heating after inflation T_R, if

$$\delta > 6.4 \cdot 10^{-4}. \tag{6.104}$$

These lower limits are closely connected with constraints on the parameters of the local supersymmetric models (Balestra et al 1984; Khlopov, Linda 1984; Ellis et al 1984; Chapter 2), which predict the existence of gravitinos with mass (Ellis et al, 1983)

$$m_G \sim 100 \text{ GeV}. \tag{6.105}$$

3.2. Problem of relic gravitinos

The gravitino has a special place in supersymmetric models due to the extremely weak interaction with ordinary particles. The cross sections of absorption or emission of gravitinos have the order of magnitude

$$\sigma \sim \frac{\alpha}{m_{\rm Pl}^2} \sim 10^{-66}\,\text{cm}^{-2} \tag{6.106}$$

where

$$\alpha = \frac{g^2}{4\pi} \sim 10^{-2} \tag{6.107}$$

and g is the gauge constant. For this reason, if there are other lighter supersymmetric particles the unstable gravitino is a very long-lived particle with lifetime (Weinberg 1982; Nanopoulos et al 1983)

$$\tau_G = \left(\frac{m_{\rm Pl}}{m_G}\right)^3 t_{\rm Pl}. \tag{6.108}$$

Therefore, for a gravitino with mass (6.105), predicted in many minimal supersymmetric models (Nanopoulos et al 1983), the lifetime exceeds

$$t \sim \tau_G \sim 10^8\,\text{s}. \tag{6.109}$$

Thus, supersymmetric models predict (Chapter 2) a massive particle, which will be absolutely stable (if it is the lightest supersymmetric particle and the R-parity is conserved), or metastable with a lifetime of more than 1 second at

$$m_G < 10^5 \text{ GeV}. \tag{6.110}$$

This prediction leads to some difficulties in cosmology (Weinberg 1982).

Indeed, in the case of decoupling of the gravitino from the plasma at a temperature of about

$$T \sim m_{\text{Pl}} \tag{6.111}$$

standard arguments about the decoupling of weakly interacting particles in the early Universe (Chapter 3) lead to the primordial concentration of the the gravitino of the order of

$$r_G = \frac{n_G}{n_\gamma} \sim \frac{1}{N}, \tag{6.112}$$

where N is the effective number of relativistic types that are in equilibrium in the Universe at $T \sim m_{\text{Pl}}$. Such an abundance of stable gravitinos with mass (6.105) corresponds to their density in the Universe equal to

$$\rho_G = m_G n_G = 10^2 m_p \cdot \frac{1}{N} \cdot n_\gamma \sim 3 \cdot 10^7 \rho_{\text{cr}}, \tag{6.113}$$

which exceeds the critical density by seven orders of magnitude (Weinberg 1982).

Contradiction with the observations does not eliminate the case of the metastable gravitino which decays in the RD stage. Decay products of gravitinos interact with the plasma and radiation, which leads to a distortion of the Planck spectral shape of the thermal electromagnetic background.

Indeed, the energy density ε_G released in the decay of gravitinos during

$$t \sim \tau_G \sim 10^8 \text{ s},$$

is

$$\varepsilon_G = m_G n_G \sim 100 \text{ MeV} \cdot n_\gamma \cdot \left(\frac{m_G}{100 \text{ GeV}} \right), \tag{6.114}$$

while the energy density of the thermal background in this period is only

$$\varepsilon_\gamma \sim 100 \text{eV} \cdot n_\gamma. \tag{6.115}$$

We obtain the energy release six orders of magnitude greater than the energy density of the thermal background in the period in which the energy release cannot exceed the total energy density of radiation.

Attempts were made to resolve the contradiction on the basis of inflationary models (Ellis et al 1983).

Since the Planck temperature is not reached after the end of inflation, the equilibrium between the gravitino and other particles predicted by these models is also not established. The abundance of primary gravitinos is not defined in this case by the equilibrium concentration during decoupling.

However, according (Nanopoulos et al (1983)), the non-vanishing gravitino abundance is predicted even in this case. There is a small but non-zero probability of creation of gravitinos during collisions of particles and their supersymmetric partners.

The main source of gravitinos after inflation (Nanopoulos et al 1983) may be associated with the process

$$X + V \rightarrow X + G, \qquad (6.116)$$

where X are scalar particles interacting with the gauginos V and the gravitino G. The cross section of this process is of the order

$$\sigma \sim \frac{\alpha}{m_{\mathrm{Pl}}^2}, \qquad (6.117)$$

where

$$\alpha = \frac{g^2}{4\pi} \sim 10^{-2}$$

and g is the gauge constant. Through this process, the relative concentration of gravitino increases after the end of inflation at a rate (Khlopov, Linde 1984)

$$\frac{dr_G}{dt} \sim \alpha \cdot N(T) \cdot \frac{T^3}{m_{\mathrm{Pl}}}. \qquad (6.118)$$

Here

$$N(T) \sim 100 \qquad (6.119)$$

is the effective number of degrees of freedom, defined as the number of types of particles with mass

$$m < T \qquad (6.120)$$

at temperature T according to their statistical weights.

From (6.118) we obtain that at the heating temperature of the Universe after inflation

$$T = T_R \tag{6.121}$$

the gravitino abundance increases from almost zero value to

$$r_G \sim \frac{\alpha\sqrt{N(T)}T_R}{m_{\mathrm{Pl}}} \sim 10^{-1}\frac{T_R}{m_{\mathrm{Pl}}}. \tag{6.122}$$

Consider, for definiteness, minimum ($N = 1$) supergravity (Khlopov, Linde 1984) coupled with the matter in which the photino and gluino are lighter than the gravitino so that the decays of the gravitino become possible

$$G \to \gamma\tilde{\gamma} \tag{6.123}$$

and

$$G \to g\tilde{g}. \tag{6.124}$$

The lifetime of the gravitino in this case is

$$\tau_G = \left(\frac{m_{\mathrm{Pl}}}{m_G}\right)^3 t_{\mathrm{Pl}}. \tag{6.125}$$

If

$$m_{\tilde{\gamma},\tilde{g}} \ll m_G \tag{6.126}$$

the relation between the relative probabilities of the considered decay modes of the gravitino is

$$\frac{Br(G \to \gamma\tilde{\gamma})}{Br(G \to g\tilde{g})} = \frac{1}{8}. \tag{6.127}$$

The concentration of antinucleons, which appeared in the decay of the gravitino, is given by

$$n_a = n_G \cdot Br(G \to g\tilde{g}) \cdot \int dp_a \int dp D_g^a(p_a, p) \cdot \delta\left(p - \frac{m_G}{2}\right), \tag{6.128}$$

where $D_g^a(p_a, p)$ is the probability of fragmentation of the gluon with momentum p in the antinucleon with momentum p_a and δ is the Dirac delta function.

We can estimate the fraction of antinucleons in the decay of gravitinos by the method similar to our approach to the evaluation of the flux of antiprotons from evaporating primordial black holes (see above). Evaluation of this quantity, defined as

$$f_a = \frac{n_a}{n_G} \tag{6.129}$$

in this case gives (Khlopov, Linde 1984; Balestra et al 1984) the value

$$f_a \sim 1. \tag{6.130}$$

According to (Khlopov, Linde 1984; Balestra et al 1984), the experimental data obtained by Batusev et al (1984), leading to inequality (6.73), correspond to the constraint on the relative concentration of the gravitino

$$r_G = \frac{n_G}{n_\gamma} < 10^{-12}. \tag{6.131}$$

Comparison of (6.131) with the theoretical estimates of the primordial concentration of gravitino gives a fairly tight upper limit on the heating temperature of the Universe after inflation:

$$T_R < 10^8 \text{ GeV}. \tag{6.132}$$

This constraint leads to serious problems in explaining the observed baryon asymmetry of the Universe. The bosons, predicted in the GUT theory (Chapter 3), cannot be represented in sufficient quantity in the Universe at such temperatures. Taking into account (6.132), a similar problem occurs for particles with masses of 10^{10}–10^{11} GeV, predicted by supersymmetric models and considered as a source of the cosmological baryon excess of their decays (Linde 1984).

Thus, a serious problem of relic gravitinos appears in the local supersymmetric GUT models. Analysis of the effect of decays of the gravitino on the primordial chemical composition leads to a low temperature T_R so that there is a problem with implementing baryosynthesis.

In addition, low temperature T_R can correspond to a long stage of coherent oscillations of the scalar field in which PBHs evaporating in the RD stage should be formed. The upper limit of the concentration of PBHs of this type imposes additional constraints on the parameters of the theory which determines the spectrum of small-scale density perturbations.

In conclusion we consider constraints on another source of antinucleons, annihilating in the RD stage, namely, the domains of antimatter predicted by the particle models, models of spontaneous CP-violation, in particular.

3.3. Constraints on the structure of domains of antimatter

Analysis of dissipation of antimatter domains (see Chapter 3 and Chechetkin et al 1982, Zeldovich and Novikov 1975; Steigman 1976) shows that the domains containing the average baryon number N_A, are completely annihilated during the period

$$10^3 < t < 10^{12} \text{ s}, \tag{6.133}$$

if N_A is in the range

$$10^{37} f\Omega_b < N_A < 10^{59} f\Omega_b. \tag{6.134}$$

Here f is the fraction of antibaryons contained in such domains.

The data on the annihilation of antiprotons with ^{4}He can be used to constrain f on the basis of analysis of the effect of annihilation on the abundance of light elements. We obtain

$$f < 2.5 \cdot 10^{-3}. \tag{6.135}$$

Analysis of possible distortions of the thermal electromagnetic spectrum gives weaker limits on such domains, because (5.118) gives

$$f < 2 \cdot 10^{-2} \left(\frac{1+z}{10^5} \right) \tag{6.136}$$

at a redshift

$$z > 10^5 \tag{6.137}$$

that corresponds to the period

$$t < 10^9 \text{ s}. \tag{6.138}$$

Based on the GUT models, the scale of domains will be determined by the nature of the phase transition in the early Universe (Stecker 1978; Brown, Stecker 1979; Senjanocic, Stecker 1980, Kuzmin et al 1981; Chechetkin et al 1982) and in models with spontaneous CP-

violation the value f can be expressed through the phase of the soft φ_s and hard φ_h CP-violation

$$f = \frac{\varphi_s - \varphi_h}{\varphi_s + \varphi_h}. \tag{6.139}$$

The condition (6.135) then imposes a very severe constraint on the CP-violation mechanism.

If

$$\varphi_s > \varphi_h \tag{6.140}$$

and the formation of domains of antimatter is possible, then φ_s must be very close to φ_h, so that the following condition should be satified

$$\frac{\varphi_s - \varphi_h}{\varphi_s + \varphi_h} < 2.5 \cdot 10^{-3}. \tag{6.141}$$

Such a constraint cannot be obtained in the analysis of possible distortions of the spectrum of thermal electromagnetic background.

So, tracing only one particular effect of the hypothetical sources of non-equilibrium particles in the RD stage, namely the influence of the interaction of antiprotons with ^{4}He nuclei on the abundance of light elements, we can verify quite effectively the models of elementary particles, predicting sources of antiprotons annihilating at $t > 10^3$ s.

The abundance of light elements is a much more sensitive detector of the possible effects associated with non-equilibrium particles in the RD stage than the spectrum of thermal electromagnetic background.

It is clear that a more complete analysis of the effects associated with non-equilibrium particles in the RD stage is a more sensitive indicator of the presence of the hypothetical sources of such particles in the early Universe.

7

Non-equilibrium effects as a test of new physics

Decays of metastable particles, evaporation of primordial black holes (PBHs), and the fragmentation of cosmic strings into small loops (Smith, Vilenkin 1987) with subsequent decay of such loops to high-energy particles are sources of hadronic, electromagnetic and neutrino cascades of particles with energies much higher than the average thermal energy of the equilibrium particle in the RD stage.

These processes are also sources of significant quantities of particles that are not in equilibrium or their equilibrium concentration is negligible.

Examples of such sources for the latter case are antiprotons in the baryon-asymmetric Universe at $t > 10^{-4}$ s (see Chapter 6) or positron sources after the first 100 seconds of expansion.

Obviously, the physical conditions at the RD stage should change in the presence of hypothetical sources of non-equilibrium particles.

Moreover, the example of the antiproton–helium annihilation has shown that the astrophysical data are much more sensitive to the non-equilibrium effects in the RD stage than the effects of shifting the equilibrium parameters in this stage.

Therefore, we turn to the general analysis of the effects associated with non-equilibrium particles in the RD stage. Based on this analysis, we can explore the sensitivity of the astrophysical data to the presence of hypothetical sources of non-equilibrium particles in the early Universe, thereby enhancing verification of the cosmological and physical phenomena determining the properties of these sources.

1. Non-equilibrium cosmological nucleosynthesis

Consider the sequence of processes leading to the sources of non-equilibrium particles in the RD stage.

Elementary particles (quarks, leptons, photons, gluons,...) are the primary products of the source, such as the decay of massive particles and evaporation of the PBHs. Sources of ultrahigh energy particles can contain massive particles, for example, $W^{\pm}$- and Z^0-bosons, as their products.

The subsequent decay of unstable particles and the fragmentation of quarks, antiquarks and gluons into hadrons lead ultimately to the fluxes of non-equilibrium photons, neutrino–antineutrino, electron–positron and nucleon–antinucleon pairs. Under the conditions of the Universe in the RD stage, after the first second of expansion all the other particles decay prior to their interaction with the plasma and radiation. The interaction of energetic non-equilibrium particles with the equilibrium plasma in the RD stage slows these particles down.

If the source can be regarded as stationary, then a stationary non-equilibrium spectrum of particles forms as a result of emission of particles and their interactions with the environment.

Interaction of non-equilibrium fluxes of particles with nuclei – products of nucleosynthesis – leads to a secondary flux of non-equilibrium nuclei whose slowdown in the plasma is accompanied by reactions with the nuclei of equilibrium plasma.

Analysis of the chain of nuclear reactions induced by non-equilibrium particles, depending on the source parameters, has been the subject of studies in the theory of non-equilibrium cosmological nucleosynthesis.

This theory has not yet been developed fully. In the literature there are only some elements of this theory and analysis of some special cases. But the general outlines of a future theory can be fairly certain.

1.1. Kinetic equation for non-equilibrium particles

Of interest here are two physically different types of sources of non-equilibrium particles in the Universe.

Data on the spectrum of the thermal background (see Chapter 5) imposes an upper limit on the possible energy release at

$$t > 10^4 \text{ s} \tag{7.1}$$

which is

$$\frac{\delta\varepsilon}{\varepsilon_\gamma} < 1.1 \cdot 10^{-4}, \tag{7.2}$$

where ε_γ is the energy density of blackbody background radiation.

During this period, the sources of non-equilibrium particles cannot dominate the Universe and the existence these sources does not affect the dynamics of cosmological expansion, except in the case of pure sources of weakly interacting particles, e.g., pure sources of neutrinos.

Sources that give a negligible contribution to the cosmological density are called the *weak sources*.

At

$$t < 10^4 \text{ s} \tag{7.3}$$

plasma density is high enough for efficient processing of non-equilibrium energy of the particles and the formation of the Planck spectrum of background radiation.

The value $\delta\varepsilon$ may exceed the value ε_γ in this period, implying a change in the law of expansion in the period when there were non-equilibrium particle in the Universe. Thus, during the period (7.3) *strong sources* of non-equilibrium particles could in principle exist in the Universe.

The existence of such sources strongly modifies the standard picture of nucleosynthesis. It also affects the thermodynamic conditions of nuclear reactions between the equilibrium particles and the kinetics of the processes due to the non-equilibrium particles.

Unstable relic gravitinos, discussed in Chapter 6, are examples of weak sources of non-equilibrium particles.

The unstable massive τ-neutrinos, discussed in Chapter 5, are an example of a strong source of non-equilibrium particles.

We now turn to a quantitative description of non-equilibrium cosmological nucleosynthesis. For simplicity (Sedelnikov et al 1995) consider the case of uniformly distributed sources of non-equilibrium particles in the Universe.

This approach is valid when the inhomogeneity of distribution of sources is significant in scales much smaller than the scale of the present-day inhomogeneities, as in this case strongly inhomogeneous distribution of sources at small scales, when averaged over a large scale, would lead to a redefinition of the parameters of homogeneously distributed sources.

A more subtle redefinition of the parameters allows the use of this approach, even if the Universe is inhomogeneous in the RD stage, or in case of an inhomogeneous distribution of baryons at this stage (Zeldovich 1975; Polnarev, Khlopov 1985).

In the case of a homogeneous medium at the time t the momentum distribution p_i of particles of i-th type that are in thermal equilibrium at temperature $T(t)$, has the distribution function $f_i(p_i,T(t),t)$. Such a distribution for non-equilibrium particles of the i-th type is described by distribution function $\varphi_i(p_i, t)$. The number density of particles of i-th type is defined by both the equilibrium and non-equilibrium particles, i.e. by the quantity

$$F_i = f_i + \varphi_i \tag{7.4}$$

and equals

$$n_i(t) = \int F_i(p_i)\,dp_i. \tag{7.5}$$

The source of non-equilibrium particles is characterized by its intensity $Q_i(p_i, t, \tau)$, defined as the distribution of the density of particles of i-th type emitted with momentum p_i per unit time in period t. Here τ is the time scale of the source. If the source of non-equilibrium particles are decays of metastable particles, then τ has the meaning of their lifetime. In the case of evaporation of primordial black holes τ is the time of evaporation.

By analogy with the case of photons in an expanding Universe (cf. Zeldovich and Novikov 1975), we can derive a system of kinetic equations for distribution functions $F_i \equiv F_i(p_i, t)$ (Filippov et al 1993; Sedelnikov, Filippov, Khlopov 1995)

$$\frac{\partial F_i}{\partial t} + 2HF_i - Hp_i\frac{\partial F_i}{\partial p_i} = I^+ - I^- + Q_i. \tag{7.6}$$

Here $H = H(t)$ is the Hubble constant in the period t, and the collision terms define the inflow and outflow of particles of i-th type in a spherical shell in the momentum space of radius p_i with width Δp_i due to their interaction with other particles.

The structure of the collision term has the form

$$I^+(p_i) = \begin{pmatrix} \int\sum_j F_j \dfrac{d\langle\Gamma_j\rangle}{dp_i}dp_j + \\ + \iint\sum_{j,k} F_jF_k \dfrac{d\langle\sigma^{jk}\rangle}{dp_i}dp_jdp_k + \ldots \end{pmatrix} A_i \tag{7.7}$$

and

$$I^-(p_i) = F_i\left(\langle\Gamma_i\rangle + \int\sum_j F_j\langle\sigma^{ij}\rangle dp_j + \ldots\right), \tag{7.8}$$

where the dots denote the terms corresponding to simultaneous collisions of three or more particles ($n \geqslant 3$),

$$A_i = 1 \pm \frac{F_i(p_i)}{4\pi p_i^2} \tag{7.9}$$

takes into account the effects of quantum statistics ('+' sign corresponds to the absorption of bosons and the sign '–' – fermions) and the brackets $\langle ... \rangle$ denote averaging over the final states according to their quantum statistics

$$\langle S_i \rangle = \prod_j \int d^3 p_j \frac{dS_i}{dp_j} A_j. \tag{7.10}$$

The system of equations (7.6) together with a system of cosmological Friedmann–Robertson–Walker equations can be considered as the most general framework for the analysis of non-equilibrium processes in a homogeneous and isotropic Universe.

To analyze the impact of sources of non-equilibrium particles at an abundance of light elements, it is appropriate to move from the equations (7.6)–(7.8) to a simpler description. We can take into account the difference in the effects of equilibrium and non-equilibrium components and very fast (compared to the rate of expansion) formation of the spectrum of non-equilibrium particles.

If

$$H\tau_i \ll 1, \tag{7.11}$$

where τ_i is the characteristic time of interaction (decay) of a particle of i-th type, we can neglect in the left side of the equation (7.6) the effects of expansion and redshift which are proportional to H.

In the opposite case

$$H\tau_i \gg 1 \tag{7.12}$$

effects of particle interactions are not significant and the contribution of the collision terms on the right side of equation (7.6) can be neglected.

In order to analyze the impact of non-equilibrium fluxes on an abundance of light elements, the presence of low energy non-equilibrium charged particles, photons and neutrino can be ignored for

$$E \ll 10 \text{ MeV}. \tag{7.13}$$

We also will not take into account the quantum statistics effects setting

$$A_i = 1. \tag{7.14}$$

1.2. Kinetic equation for weak sources in the RD stage

The source of non-equilibrium particles is weak if the energy density of their products is small compared with the energy density of the thermal background. This condition can be formulated as

$$\sum_i \int \varepsilon_i Q_i (p_i, t, \tau) dp_i \Delta t \ll \varepsilon_\gamma, \tag{7.15}$$

with

$$\Delta t \sim t \tag{7.16}$$

for all t.

In the presence of weak sources the equilibrium concentration after thermalization of non-equilibrium particles did not significantly change. Then, we can assume that the expansion law has the form

$$a(t) \propto t^{1/2}$$

and examine the impact of non-equilibrium particles on the background of the processes with the equilibrium particles corresponding to the standard scenario of the Big Bang evolution in the RD stage.

If condition (7.11) holds, then the system (7.6)–(7.8) leads to a system of kinetic equations for processes with non-equilibrium particles in a stationary equilibrium. The following equation is then derived for the distribution function φ_i of non-equilibrium particles of i-th type

$$\begin{aligned} \frac{\partial \varphi_i}{\partial t} = & \sum_{j,k} \varphi_j F_k \frac{d(\sigma v)^i_{jk}}{dp_i} dp_j dp_k + \sum_j \varphi_j \frac{d\Gamma_j}{dp_i} - \\ & -\varphi_i \left(\Gamma_i + \sum_j \left(n_j (\sigma v)^{ij} (p_i) + \int \varphi_j (\sigma v)^{ij} dp_j \right) \right) + Q_i (p_i, t, \tau). \end{aligned} \tag{7.17}$$

In the case of equilibrium particles, we can use the number density of particles $n_i(T)$ at a temperature T instead of the distribution function f_i. We have

$$n_i(T) = \int f_i(p_i, T, t)\,dp_i, \tag{7.18}$$

where $T(t)$ is defined by the law of expansion. Thus, we obtain the following equation for the equilibrium particles

$$\begin{aligned}\frac{\partial n_i(T)}{\partial t} = -n_i\left(\Gamma_j + \sum_j n_j \langle\sigma v\rangle_{ij} + \sum_j \varphi_j(p_j)\langle\sigma v\rangle_{ij}\,dp_j\right) + \\ + \sum_{j,k} n_j n_k \langle\sigma v\rangle^i_{jk} + \delta n_i(T) + \sum_j n_j \Gamma^i_j.\end{aligned} \tag{7.19}$$

Here Γ_i is the total probability of decay of unstable particles of i-th type, Γ^i_j is the relative probability of the creation of a particle of i-th type in the decay of a particle of type j, and the term

$$\delta n_i(T) = \int_0^{p_{\max}} \varphi_i(p_i)\,dp_i \tag{7.20}$$

takes into account the high rate of deceleration of non-relativistic charged particles in space plasma and threshold values of the reaction energy.

We now consider in more detail the sequence of processes occurring in the Universe under the influence of weak sources of non-equilibrium particles.

In general, the primary products of these sources are lepton-antilepton and quark–antiquark pairs, photons and gluons. The sources of ultrahigh-energy particles can produced also more massive fundamental particles, such as W- and Z^0-bosons.

Decays of unstable particles and the fragmentation of quarks, antiquarks and gluons into hadrons lead to the formation of fluxes of non-equilibrium pairs of neutrino–antineutrinos, electrons, positrons, photons, nucleons, antinucleons and possibly charged pions and muons. These fluxes interact with the equilibrium particles.

Such interaction with the nuclei formed during primordial nucleosynthesis leads to the formation of non-equilibrium nuclei. The slowdown of the latter in cosmological plasma is accompanied by their reactions with equilibrium nuclei.

Thus, the non-equilibrium particles may be both a direct product of the source and the result of interaction of other non-equilibrium particles with plasma and radiation.

Detailed analysis involves identifying the most significant types of particles and processes. They include obviously stable particles (photons, neutrinos and antineutrinos, electrons and positrons, nucleons and antinucleons) and processes caused by them.

With regard to metastable particles with lifetimes $t_i(p_i)$, they should be considered in the reactions with particles of type j only if the condition is valid

$$\left(n_j \langle \sigma v \rangle_{ij}\right) t_i(p_i) > 1, \tag{7.21}$$

where

$$t_i(p_i) = t_i(0) \frac{\left(p_i^2 + m_i^2\right)^{1/2}}{m_i} \tag{7.22}$$

and $t_i(0)$ is the lifetime of particles of type i at rest. The above condition means that the rate of interaction of particles is greater than the rate of their decay.

Thus, the charged pions, whose lifetime is

$$t_\pi(0) \cong 10^{-8} \text{ s}$$

and the cross section of interaction with the nucleons is of the order of magnitude

$$\sigma_{\pi N} \sim 10^{-26} \text{ cm}^2,$$

should be considered in the reactions with nucleons and nuclei at

$$2\Omega_b \left(\frac{H_0}{70 \dfrac{\text{km}}{\text{s} \cdot \text{Mpc}}} \right)^2 \left(\frac{2.7 \text{ K}}{T_0} \right)^3 \left(\frac{E_\pi}{m_\pi} \right) > 1, \tag{7.23}$$

where

$$\Omega_b = \frac{\rho_b}{\rho_{\text{cr}}},$$

ρ_b, H_0, T_0 are the present-day baryon density, Hubble constant and the temperature of the electromagnetic background, respectively, and t is the cosmological time. The above estimates show that the charged pions with energy

$$E_\pi < 100 \text{ GeV} \tag{7.24}$$

decay before their interaction with nucleons and nuclei at

$$t > 10^{-3} \text{ s}. \tag{7.25}$$

Definition (7.15) shows that a weak source of all types of non-equilibrium particle must satisfy

$$\int \varepsilon_i \varphi_i dp_i \ll \varepsilon_\gamma. \tag{7.26}$$

In particular, limiting our considerations to the case

$$\omega_\gamma \gg kT, \tag{7.27}$$

where ω_γ is the energy of the non-equilibrium photon, k is the Boltzmann constant, and T is the temperature of background radiation, we get from (7.26) that the number density of non-equilibrium photons n_γ^{non} satisfies the inequality

$$n_\gamma^{\text{non}} = \int \varphi_\gamma d\omega_\gamma \ll \left(\frac{T}{1\ \text{MeV}}\right) n_\gamma \sim \left(\frac{1\ \text{s}}{t}\right)^{1/2} n_\gamma, \tag{7.28}$$

where t is the cosmological time and n_γ is the number density of equilibrium photons. This means that when

$$t \gg 1\ \text{s} \tag{7.29}$$

the number density of energetic non-equilibrium photons is much smaller than that of the equilibrium photons. Therefore, in each case when the process can go through the thermal photons, the contribution of non-equilibrium photons in the process can be neglected.

On the other hand, the condition (7.26) allows for the number densities of non-equilibrium electrons, positrons, nucleons and antinucleons the following cases:

$$n^{\text{non}}_{e^-(e^+)} = \int \varphi_{e^-(e^+)} dp_{e^-(e^+)} \gg n_e \tag{7.30}$$

for electrons (positrons) at

$$t > 100\ \text{s}$$

and for nucleons (antinucleons)

$$n^{\text{non}}_{N(\bar{N})} = \int \varphi_{N(\bar{N})} dp_{N(\bar{N})} \gg n_N \tag{7.31}$$

at

$$t > 10^{-3}\ \text{s},$$

where n_e and n_N are respectively the number densities of electrons and nucleons in equilibrium plasma.

In these cases, the non-equilibrium electrons (positrons) and nucleons (antinucleons) play a dominant role in the development of respectively electromagnetic and nucleon cascades.

To formulate the general problem of weak sources of non-equilibrium particles, we present the structure of the system of equations (7.17) explicitly

$$\begin{aligned}
\frac{\partial \varphi_\gamma}{\partial t} &= \left\langle \varphi_\gamma n_{e^\pm} \to \gamma \right\rangle + \left\langle \varphi_\gamma \varphi_{e^\pm} \to \gamma \right\rangle + \left\langle n_\gamma \varphi_{e^\pm} \to \gamma \right\rangle - \\
&\quad - \varphi_\gamma \left(\left\langle n_{e^\pm} \to \right\rangle + \left\langle n_\gamma \to \right\rangle + \left\langle \varphi_{e^\pm} \to \right\rangle + \left\langle \varphi_\gamma \to \right\rangle \right) - \\
&\quad - \varphi_\gamma \left(\left\langle n_p \to \right\rangle + \left\langle n_4 \to \right\rangle \right) + Q_\gamma \left(p_\gamma, \tau, t \right) \\
\frac{\partial \varphi_{e^\pm}}{\partial t} &= \left\langle \varphi_{e^\pm} n_\gamma \to e^\pm \right\rangle + \left\langle \varphi_\gamma \varphi_{e^\pm} \to e^\pm \right\rangle + \left\langle n_{e^\pm} \varphi_\gamma \to \gamma \right\rangle + \\
&\quad + \left\langle n_p \varphi_\gamma \to e^\pm \right\rangle + \left\langle n_4 \varphi_\gamma \to e^\pm \right\rangle - \varphi_{e^\pm} \left(\left\langle n_p \to \right\rangle + \left\langle n_4 \to \right\rangle \right) - \\
&\quad - \varphi_{e^\pm} \left(\left\langle n_{e^\pm} \to \right\rangle + \left\langle n_\gamma \to \right\rangle + \left\langle \varphi_{e^\pm} \to \right\rangle + \left\langle \varphi_\gamma \to \right\rangle \right) + Q_{e^\pm} \left(p_{e^\pm}, \tau, t \right) \\
\frac{\partial \varphi_{N,\bar N}}{\partial t} &= \left\langle \varphi_{N,\bar N} n_p \to N, \bar N \right\rangle + \left\langle \varphi_{N,\bar N} n_4 \to N, \bar N \right\rangle + \\
&\quad + \left\langle \varphi_{N,\bar N} \varphi_{N,\bar N} \to N, \bar N \right\rangle - \varphi_{N,\bar N} \left(\left\langle n_p \to \right\rangle + \left\langle n_4 \to \right\rangle + \left\langle \varphi_{N,\bar N} \right\rangle \right) - \\
&\quad - \varphi_{N,\bar N} \left\langle \ \right\rangle_{Coul} + Q_{N,\bar N} \left(p_{N,\bar N}, \tau, t \right) \qquad (7.32) \\
\frac{\partial \varphi_d}{\partial t} &= \left\langle \varphi_{N,\bar N} n_4 \to d \right\rangle + \left\langle \varphi_\gamma n_4 \to d \right\rangle + \left\langle \varphi_n n_p \to d \right\rangle + \\
&\quad + \left\langle \varphi_n \varphi_p \to d \right\rangle - \varphi_d \left(\left\langle n_p \to \right\rangle + \left\langle n_4 \to \right\rangle + \left\langle n_\gamma \to \right\rangle + \left\langle \varphi_\gamma \to \right\rangle \right) - \\
&\quad - \varphi_d \left\langle \ \right\rangle_{Coul} \\
\frac{\partial \varphi_{T,{}^3\mathrm{He}}}{\partial t} &= \left\langle \varphi_{N,\bar N} n_4 \to T, {}^3\mathrm{He} \right\rangle + \left\langle \varphi_\gamma n_4 \to T, {}^3\mathrm{He} \right\rangle - \\
&\quad - \varphi_{T,{}^3\mathrm{He}} \left(\left\langle n_p \to \right\rangle + \left\langle n_4 \to \right\rangle \right) - \varphi_{T,{}^3\mathrm{He}} \left\langle \ \right\rangle_{Coul} \\
\frac{\partial \varphi_4}{\partial t} &= \left\langle \varphi_{N,\bar N} n_4 \to {}^4\mathrm{He} \right\rangle - \varphi_4 \left(\left\langle n_p \to \right\rangle + \left\langle \varphi_{N,\bar N} \right\rangle \right) - \varphi_4 \left\langle \ \right\rangle_{Coul}
\end{aligned}$$

Here the brackets ⟨⟩ denote the corresponding collision terms, and φ_i and n_j represent the distribution functions of non-equilibrium particles of type i and the number density of equilibrium particles of type j, respectively, $\langle\rangle_{\text{Coul}}$ corresponds to the Coulomb deceleration in the plasma, $\varphi_i(n_d)$ is the distribution function (number density) of non-equilibrium (equilibrium) ^{4}He nuclei.

Due to the strong Coulomb deceleration in plasma, in a first approximation we can neglect the non-equilibrium flux of nuclei with charge $Z > 2$ and consider changes in their concentration under the condition that these nuclei are in thermal equilibrium with the plasma.

The reactions of neutrinos (antineutrinos) and the evolution of neutrino cascades are not included in the equations (7.32) and will be discussed below (see §3).

1.3. Non-equilibrium nucleosynthesis. Some special cases

The complete system of equations (7.32) is quite complicated and so far has not been studied in detail, even within the above mentioned simplifying assumptions. To date, there are only the results of simplified studies of individual processes of non-equilibrium cosmological nucleosynthesis in the electromagnetic and nuclear cascades from some hypothetical sources of non-equilibrium particles.

According to (Burns, Lovelace 1982; Aharonian, Vardanian 1985; Khlopov 1999) the following reaction plays an important role in the development of electromagnetic cascades from high-energy photons in astrophysical conditions and especially in the RD phase

$$\gamma + \gamma \to e^+ + e^-. \tag{7.33}$$

The production of pairs dramatically cuts the non-equilibrium photon spectrum

$$E_{\max} \cong \frac{m_e^2}{25T} \ln\left(15\Omega_b\right), \tag{7.34}$$

where

$$\Omega_b \sim 0.05$$

is the density of baryons and m_e is the electron mass.

In (Dimopoulos et al 1987) the spectrum of non-equilibrium photons has been expressed in terms of energy distribution $\xi_\gamma(E)$, defined as the number density of the photons per unit energy range originating from the two-particle decay of a hypothetical supermassive particle X with mass M, in which high energy photons are emitted. This distribution function has the form

$$\xi_\gamma(E) = \begin{cases} \dfrac{M_X}{2E_{\max}^{1/2}} \cdot \dfrac{1}{E^{3/2}}, \; E < E_{\max} \\ 0, \; E > E_{\max}, \end{cases} \tag{7.35}$$

where E_{max} is given by (7.34).

In Dimopoulos et al (1987,1988), the system of equations (7.32) was reduced to a system of equations for reaction rates, taking into account non-equilibrium fluxes (Dimopoulos et al 1987a), and there was found a simple analytical solution which is well consistent with numerical solutions of these equations and gave an interesting possibility for the formation of light elements.

The analytical approach (Dimopoulos et al 1988) was based on numerical calculations (Dimopoulos et al 1987) of values ξ_i which are defined as the number of nuclei of i-th type per baryon from X decay and are determined by the chain of nuclear reactions induced by nucleons and antinucleons from X decay.

It was assumed that the abundance of ^{4}He is reduced due to reactions with baryons, while the contribution of photodisintegration is negligible. The change of the relative concentration of the ^{4}He nuclei with time, defined as

$$f_4 = \frac{n_4}{n_\gamma}, \tag{7.36}$$

where n_4 and n_γ are the number densities of ^{4}He nuclei and photons, respectively, and is given by

$$\frac{\partial f_4}{\partial t} = f_X^0 \Gamma_X \exp\{-\Gamma_X t\} r_b^* \xi_4, \tag{7.37}$$

where

$$f_X^0 = \frac{n_X}{n_\gamma} \tag{7.38}$$

is the frozen-out relative concentration of superheavy particles X,

$$\tau_X^{-1} = \Gamma_X \tag{7.39}$$

is the probability of decay X, r_b^* is the effective relative probability of decay channels with baryons in the final state, and ξ_4 is negative.

The equations for the abundance of deuterium, ^{3}He, ^{6}Li and ^{7}Li are derived under the assumption that their formation occurs mainly in the hadronic cascades, and use appropriate ξ_i, and their destruction is dominated by photodisintegration. These equations have the form:

$$\frac{\partial f_i}{\partial t} = f_X^0 \Gamma_X \exp\{-\Gamma_X t\} r_b^* \xi_i - n_i \int_{Q_i}^{E_{\max}(t)} f_\gamma \sigma_{\gamma i}(E) dE, \quad (7.40)$$

where Q_i is the photodisintegration threshold, n_i is the number density and $\sigma_{\gamma i}$ is the photodisintegration cross section of nuclei of type i, and $E_{\max}(t)$ is given by (7.34) taking into account the fact that the standard temperature dependence on the time in the RD stage is defined as

$$T \propto t^{-1/2}.$$

The spectrum of non-thermal photons $f_\gamma(E)$ is determined by its value at a fixed point, i.e. by solving the equation

$$\frac{\partial f_\gamma(E)}{\partial t} = f_X^0 \Gamma_X \exp\{-\Gamma_X t\} \xi_\gamma(E) - n_e \sigma_c(E) f_\gamma(E) \quad (7.41)$$

at

$$\frac{\partial f_\gamma(E)}{\partial t} = 0. \quad (7.42)$$

We obtain

$$f_\gamma(E) = \frac{\xi_\gamma(E) f_X^0 \Gamma_X \exp\{-\Gamma_X t\}}{n_e \sigma_c(E)}. \quad (7.43)$$

Assuming that equation (7.40) reaches its fixed point, Dimopoulos et al (1988) obtained values for the abundance of deuterium, ^{3}He, ^{6}Li and ^{7}Li

$$f_i = 30 \xi_i Q_i^{1/2} E_{\max}^{1/2}(t_{fi}) \frac{r_b^*}{M_X} f_b, \quad (7.44)$$

where f_b is the concentration ratio of baryons to photons and is

$$f_b = \frac{n_b}{n_\gamma} = 1.5 \cdot 10^{-8} \Omega_b \left(\frac{H_0}{70 \frac{\text{km}}{\text{s} \cdot \text{Mpc}}} \right)^2 \left(\frac{2.7\,\text{K}}{T_0} \right)^3, \quad (7.45)$$

cross-section ratio is equal to

$$\frac{\sigma_{\gamma i}(E)}{\sigma_c(E)} = \frac{1}{30} \tag{7.46}$$

and t_{fi} is the freezing-out time when photodisintegration and formation of nuclei by the flux of baryons is terminated. The condition

$$E_{\max}(t_{fi}) > Q_i \tag{7.47}$$

and the constraint on the photodisintegration of ^{4}He leads to a rather narrow range of lifetimes

$$4 \cdot 10^5 < \tau_X < 8 \cdot 10^5 \text{ s}, \tag{7.48}$$

for which the values (7.44) are valid at a fixed point. This yields for the following ratio of abundances

$$\text{D}:^3\text{He}:^6\text{Li}:^7\text{Li} = 1:1:10^{-5}:10^{-6}, \tag{7.49}$$

which is mainly determined by the ratio ξ_i. The absolute values of these abundances correspond to those observed when

$$2 \cdot 10^4 < \frac{M_X}{r_b^*} < 10^5 \text{ GeV}. \tag{7.50}$$

Assuming a significant decrease in the abundance of ^{6}Li (Dimopolous et al 1988) there was shown the principal possibility of interpreting the observed abundance of light elements on the basis of nucleosynthesis in the era of keV temperatures.

This approach removes the strong dependence of the predicted abundance on the number of neutrino species or the ratio of baryons to photons, but also involves a hypothetical particle X with very specific properties.

The allowed range for the lifetimes τ and for the ratio of the mass to the relative probability of baryon decay modes M_X/f^*_b is very narrow and may disappear altogether when the considered nuclear processes are considered in greater detail. In any case, it illustrates the possibility of nuclear cosmoarcheology which relates the abundance of light elements to the relic 'fingerprint' of hypothetical physics in the early Universe.

Non-equilibrium cosmological nucleosynthesis sets a new type of relations between cosmological data and the nuclear experiments.

The result of standard cosmological nucleosynthesis is mainly determined by the thermodynamic conditions, so that a slight change of thermodynamic parameters compensates for any error in the estimates of nuclear reaction rates.

The results of non-equilibrium cosmological nucleosynthesis are directly determined by the properties of nuclear reactions themselves, their total and differential cross sections.

Sometimes it turns out that the data on nuclear reactions are not available.

Experimental nuclear cosmoarcheology formulates the experimental task to fill these gaps and by appropriate experimental measurements fills the missing link in the cosmoarcheological chain of reasoning.

To prove that the source of non-equilibrium particles is responsible for the observed abundance of light elements it is necessary to carry out a detailed self-consistent analysis of all channels for formation of light elements in all non-equilibrium cascades caused by the sources in non-equilibrium cosmological nucleosynthesis. This approach shows what **should be** the parameters of hypothetical sources.

However, we can only use one particular trace of the source to find out what parameters **should not be**, that is, to exclude those values that correspond to the overproduction of some elements.

The mere existence of antinucleons in the Universe after nucleosynthesis is quite a profound signature of the sources of non-equilibrium particles.

We saw above that for any source of antinucleons their annihilation in the Universe in the RD stage is strictly limited by the observed abundance of deuterium and/or ^3He. The only fact that antiprotons are present in the Universe and annihilate with ^4He was sufficient to constrain the sources of antiprotons discussed in Chapter 6.

However, the sources of the nucleon–antinucleon pairs in the RD stage, such as the decay of the gravitino or the evaporation of primordial black holes, produce high-energy nucleons and antinucleons, leading to a nuclear cascade. One can expect that the inclusion of the cascade of non-equilibrium particles should increase the sensitivity of cosmoarcheological analysis.

Levitan et al (1988) analyzed by the Monte Carlo method the production of deuterium and ^3He in nuclear cascades induced by high-energy nucleons and antinucleons from hypothetical sources of the nucleon–antinucleon pairs.

There are several numerical factors leading to increased production of deuterium and ^3He in these stages in comparison with the case of annihilation of slow antiprotons with ^{4}He nuclei (Chapter 6).

Due to the suppression of the annihilation of relativistic antinucleons, the main mechanism of production of deuterium and ^{3}He by high-energy antiprotons is inelastic inclusive processes

$$\bar{p} + {}^4\mathrm{He} \rightarrow \bar{p} + \mathrm{D} + ... \tag{7.51}$$

$$\bar{p} + {}^4\mathrm{He} \rightarrow \bar{p} + {}^3\mathrm{He} + ... \tag{7.52}$$

$$\bar{p} + {}^4\mathrm{He} \rightarrow \bar{p} + \mathrm{T} + ... \tag{7.53}$$

and the dominant mechanism of energy loss of relativistic antiprotons is associated with inelastic scattering on the nuclei (mainly on the hydrogen nuclei, the most common element, i.e. the protons) in the process

$$\bar{p} + p \rightarrow \bar{p} + p + ... \tag{7.54}$$

Therefore, relativistic antiprotons, before they slow down to non-relativistic energies, may participate in more than one collision, leading to the formation of deuterium, ^{3}He and tritium.

The following circumstance applies to recoil protons, which are obtained after the interaction of antiprotons with protons or destruction of helium by the antiproton with the energy exceeding the threshold of ^{4}He destruction by the proton. The destruction of ^{4}He by the secondary protons is an additional source of deuterium, ^{3}He and tritium in the case of primary high-energy antiprotons.

The last but not least factor is related to the interaction of primary protons produced in pair with the antiproton. These protons, as well as secondary protons, produce nuclear cascades, also forming deuterium, ^{3}He and tritium.

Quantitative estimation of the relative importance of these factors as a result of numerical calculations are presented in a study by Levitan et al (1988).

It was found that in the case of the primary antiproton in the formation of abundance of deuterium (^{3}He) the antiprotons with maximum energy are dominant. The second peak at low energies arises from the annihilation of antiprotons at rest. The minimum in the formation of deuterium (^{3}He) at low kinetic energies

$$E \leqslant 0.1 \text{ GeV} \tag{7.55}$$

of slowing down antiprotons with an initial energy

$$E_0 > E \tag{7.56}$$

comes from the strong Coulomb deceleration of non-relativistic charged particles in cosmic plasma. The effective cross section of Coulomb scattering is proportional

$$\sigma_{\text{Coul}} \sim v^{-4}, \tag{7.57}$$

where v is the velocity of the charged particle.

Therefore, the probability that a charged particle will destroy ^{4}He before it slows down to thermal energies as a result of Coulomb interaction with plasma is strongly suppressed.

Levitan et al (1988) found a simple analytical formula giving good approximation of the numerical results.

Consider, for example, the probability S_{D}^{A} of formation of deuterium in the antiproton reactions. The rate of slowing down of antiprotons is much higher than the rate of change of the parameters of the medium or the rate of evolution of the sources. The probability distribution $\varphi_a(E, E_0)$ of the antiprotons with energy E for their initial energy E_0 is stationary. This probability is determined by the rate of deceleration of antiprotons with energies

$$\varepsilon > E \tag{7.58}$$

to the energy E and the rate of further slowing down of antiprotons with energies E up to lower energy

$$\int_E^{E_0} \varphi_a(\varepsilon, E_0) \frac{d(\sigma v)^{\text{in}}(\varepsilon)}{dE} n_p d\varepsilon = n_p (\sigma v)^{\text{tot}} \varphi_a(E, E_0). \tag{7.59}$$

At high energies, we can use the approximation

$$\sigma^{\text{tot}}(\varepsilon) = \text{const}, \tag{7.60}$$

$$\sigma^{\text{ann}} \ll \sigma^{\text{tot}}, \tag{7.61}$$

$$\frac{d(\sigma v)^{\text{in}}(\varepsilon)}{dE} = \frac{\sigma^{\text{in}}}{\varepsilon} = \frac{2}{3} \cdot \frac{\sigma^{\text{tot}}}{\varepsilon}, \tag{7.62}$$

which gives the solution of equation (7.59)

$$\varphi_a(E, E_0) \propto E^{-2/3}. \tag{7.63}$$

The probability of formation of deuterium in the interaction of antiprotons with helium is

$$S_{\mathrm{D}}^{A} = f_{\mathrm{D}}^{a} \int \varphi_{a}(E, E_{0}) n_{\mathrm{He}} \sigma_{\bar{p}\mathrm{He}}^{\mathrm{tot}}(E) v \tau \, dE, \tag{7.64}$$

where f_{D}^{a} is the relative fraction of deuterium amongst the products of the antiproton–helium interaction

$$f_{\mathrm{D}}^{a} = \mathrm{const} \tag{7.65}$$

and

$$\tau = \frac{1}{n_{p} \sigma_{\bar{p}p}^{\mathrm{tot}} v} \tag{7.66}$$

is the characteristic time scale of cascade development. We find from the above mentioned equations that the energy dependence is defined as

$$S_{\mathrm{D}}^{A} \propto E_{0}^{2/3}. \tag{7.67}$$

The examples given by (Dimopoulos et al 1987, Filippov et al 1987, Levitan et al 1988, see also Sedelnikov 1999, 2000) reveal a direct correlation between the data on the interactions of particles with nuclei, as measured in experimental nuclear physics, and an abundance of light elements, determined astronomically.

Merging of the experimental and astronomical data suggests fairly definite conclusions about the possible existence of sources of non-equilibrium particles in the early Universe, thus providing a detailed indirect verification of ultrahigh-energy physics, unattainable in the laboratory, which determines the parameters of these sources.

A new type of relations between the astronomical, the experimental nuclear and theoretical investigations arising in the context of nuclear cosmoarcheology may be illustrated by the ASTROBELIX project, discussed in the next section.

2. ASTROBELIX project

2.1.1. Astro-nuclear experiment ASTROBELIX

The underlying idea of the ASTROBELIX project follows from the fact that antiprotons cannot exist in the baryon asymmetric Universe from the end of the era of local nucleon–antinucleon annihilation at

$$t < 10^{-3} \text{ s},$$

and before the period of formation of galaxies at

$$t > 10^{16} \text{ s}$$

when the interaction of cosmic rays with the matter is capable of producing antiprotons. Thus, the emergence in the early Universe of annihilation of antiprotons is a clear indication of new cosmological phenomena associated with physics outside the framework of the standard model (SM). Indeed, the sources of the annihilation of antiprotons can be related to the cosmological consequences of the models of elementary particles (Chechetkin et al 1982a, b), so that the analysis of the possible effects of antiproton annihilation allows us to define the permissible parameters of these models.

Since different particle models can provide inflation, predict the observed baryon asymmetry of the Universe and offer candidates for dark matter, it is necessary to identify additional effects related to the cosmological implications of such models, in order to distinguish them and make the selection among them. Sources of annihilating antiprotons after nucleosynthesis refer to these additional effects associated with broad classes of physical mechanisms for inflation, baryosynthesis and dark matter.

From a phenomenological point of view there are two possible types of existence of antiprotons in the baryon-asymmetric Universe, after ending the local annihilation of equilibrium nucleon–antinucleon pairs:

a) antiprotons may survive in the domains of antimatter,

b) they can be produced in a pair with the nucleons in the sources of energetic particles.

Considering the possibility a), we can find that almost every implementation of baryosynthesis (Chapter 3) can naturally lead under certain conditions to the formation of local antibaryon excess with the dominant global excess of baryons.

For instance, in baryosynthesis mechanisms related to the non-equilibrium baryon non-conserving process of generation of baryon excess special selection of the *CP*-violating phase takes place. The possibility of spatial variation of the magnitude and sign of this phase gives for the same mechanism an antibaryon excess in some areas (Chapters 3 and 6).

In mechanisms described in (Affleck, Dine 1985; Linde 1985) (Chapter 3) the existence of the baryon asymmetry is the result of the independence of the supersymmetric potential of the baryon charge of the condensate of scalar quarks, which leads to the natural features of the spatial variation of the density of the baryonic charge in such a condensate.

Possibility b) refers to the existence of new (approximately) conserved charges in almost all extensions of the SM. The lightest particles with such a charge are metastable and should be present in the Universe long after their freezing-out in the early Universe.

The existence of relic metastable particles directly leads to the sources of energetic particles in the RD stage due to their decay.

Thus, in the case of the gravitino (Chapter 6) with a mass

$$m_G \sim 100 \text{ GeV}$$

decays of relic gravitinos at

$$t \sim 10^8 \text{ s}$$

via the gluino–gluon channel

$$G \to \tilde{g} + g$$

lead to hadronic jets containing a nucleon–antinucleon pair due to the fragmentation of the gluon

$$g \to N\bar{N} + \text{hadrons.}$$

Supermassive metastable particles may not be sufficiently long-lived to survive until the period of cosmological nucleosynthesis. But they can be a source of non-equilibrium particles in the RD stage indirectly, forming PBHs at the stage of their dominance in the early Universe (Chapter 4).

Evaporation of primordial black holes in the RD stage leads to the production of the nucleon–antinucleon pairs among the products of evaporation (Chapter 6).

The influence of the annihilation of antiprotons with ^{4}He nuclei on the abundance of light elements is the most sensitive test for the existence of hypothetical sources of antiprotons in the RD stage (Chapter 6).

At an early stage of development of nuclear cosmoarcheology the permissible parameters of hypothetical sources of antiprotons annihilated in the RD stage, were investigated by analysis of the production of deuterium and ^{3}He in the annihilation of antiprotons with ^{4}He nuclei.

The missing link in the logical chain linking the evaporating PBH, annihilating antimatter domains and decaying gravitino as a source of antiprotons, which are predicted based on GUT, supersymmetry and other particle models with the observed abundance of deuterium and ^{3}He, was completed using the data of experiment PS179 at CERN-LEAR. This experiment was conducted to measure the yield of D and ^{3}He in antiproton annihilation on a ^{4}He nucleus.

As a result of this experiment it was shown that the predicted overproduction of ^{3}He in antiproton annihilation on ^{4}He imposes the most stringent constraints on the possible number of antiprotons annihilating in the RD stage and, in particular, causes serious difficulties in the simplest form of local supersymmetric models (Chapter 6).

Since the abundance of lithium, beryllium, and boron is six orders of magnitude smaller than ^{3}He, the sequence of nuclear processes

$$\bar{p} + {}^4\mathrm{He} \to {}^3\mathrm{He} + \ldots \Rightarrow {}^4\mathrm{He} + {}^3\mathrm{He} \to {}^7\mathrm{Be} + \gamma, \qquad (7.68)$$

$$\bar{p} + {}^4\mathrm{He} \to \mathrm{D} + \ldots \Rightarrow \mathrm{He} + {}^4\mathrm{D} \to {}^6\mathrm{Li} + \gamma, \qquad (7.69)$$

$$\bar{p} + {}^4\mathrm{He} \to \mathrm{T} + \ldots \Rightarrow T + {}^4\mathrm{He} \to {}^7\mathrm{Li} + \gamma, \qquad (7.70)$$

$$\bar{p} + {}^4\mathrm{He} \to \mathrm{T} + \ldots \Rightarrow \mathrm{T} + {}^4\mathrm{He} \to {}^6\mathrm{Li} + n \qquad (7.71)$$

offers a more sensitive indicator to verify the existence of antiprotons in the RD stage by comparing the abundance of lithium, beryllium and boron predicted for this sequence of reactions, with the observed abundance.

However, the logical chain that connects the cosmological consequences of particle theory with the astronomical observations of the abundance of light elements is incomplete because there is no information about the distribution of ^{3}He, D, T, produced in the annihilation of antiprotons with ^{4}He nuclei at large momenta, as well as the pregalactic abundance of Li, Be and B is almost unknown.

To complete this logical chain, in the ASTROBELIX project (Khlopov 1990, 1992) it was assumed to:

a) measure the distribution of ^{3}He, D, T in the study in experiment OBELIX in LEAR antiproton annihilation on ^{4}He nuclei;

b) improve methods to measure the primordial chemical composition using narrow-band distortions in the Rayleigh–Jeans part of relic radiation.

Although each source of antiprotons is also a source of gamma rays, the interaction γ + ^{4}He cannot lead to a significant excess of lithium, beryllium or boron because

1) reaction cross sections of photodisintegration of ^{4}He

$$\gamma + {}^4\mathrm{He} \to \mathrm{T} + p, \tag{7.72}$$

$$\gamma + {}^4\mathrm{He} \to {}^3\mathrm{He} + n \tag{7.73}$$

have a peak near the threshold and

2) the γ-spectrum of the electromagnetic cascade generated in the plasma in the RD stage is also reduced with the energy E of the γ-quantum

$$f_\gamma(E) \propto E^{-3/2}. \tag{7.74}$$

Therefore, in contrast to the formation of ^{3}He, the resultant photodisintegration sequence of nuclear reactions

$$\gamma + {}^4\mathrm{He} \to \mathrm{T} + p \Rightarrow \mathrm{T} + {}^4\mathrm{He} \to {}^6\mathrm{Li} + n, \tag{7.75}$$

$$\gamma + {}^4\mathrm{He} \to \mathrm{T} + p \Rightarrow \mathrm{T} + {}^4\mathrm{He} \to {}^7\mathrm{Li} + \gamma, \tag{7.76}$$

$$\gamma + {}^4\mathrm{He} \to {}^3\mathrm{He} + n \Rightarrow {}^3\mathrm{He} + {}^4\mathrm{He} \to {}^7\mathrm{Li} + \gamma \tag{7.77}$$

cannot compete with the formation of lithium, beryllium and boron, due to the interaction of antiprotons with helium nuclei.

The full scheme of the ASTROBELIX project included:

1) theoretical studies of the relations between the parameters of the particle models and sources of antiproton annihilation in the early Universe;

2) numerical calculation of the annihilation of antiprotons in the plasma in the RD stage and the kinetics of the reactions of the non-equilibrium cosmological nucleosynthesis which should identify gaps in data on the interaction of antiprotons with helium and astronomical observations;

3) experimental measurement of the cosmologically significant parameters of the antiproton–helium interaction in the experiment OBELIX in LEAR;

4) improvement of radioastronomy methods to measure the primordial chemical composition on the basis of the search for narrow-band distortions in the spectrum of thermal electromagnetic background radiation.

The project revealed an interdisciplinary relationship of the main major research areas in cosmoparticle physics: theoretical, numerical, experimental and observational.

In particular, in the context of the approach of cosmoparticle physics research has reached a new, astronuclear level of the relationship of nuclear physics and astrophysics.

While nuclear astrophysics uses reliable experimentally proven laws of nuclear physics, astronuclear experiments combine nuclear and astrophysical studies to investigate the unknown physical laws hidden in the scale of ultrahigh energies.

Full implementation of the ASTROBELIX project could not be achieved. The specificity of the OBELIX experiment was inadequate for solving problems of the measurement of secondary nuclei in the antiproton–helium interaction. But cosmoarcheological studies, mainly in the framework of the ASTROBELIX project, revealed the importance of further development in this direction and gave birth to a project entitled 'Astroparticle studies of Dark, Decaying and Annihilating matter in the Universe' (ASTRODAMUS), which we will discuss later.

2.2. Constraints on antiprotons annihilating in the RD stage, from the observed abundance of lithium and beryllium

Among the results of cosmoarcheological searches, initiated by the ASTROBELIX project, let us briefly discuss the theoretical proof of the higher sensitivity of lithium, beryllium and boron abundances to the parameters of the sources of antiprotons compared to the case of deuterium and ^{3}He.

As shown by Khlopov et al (1994), even with the lack of direct experimental data we can make quite certain cosmoarcheological conclusions on the basis of observational data on the abundance of lithium. Based on calculations by the Monte-Carlo method, it was shown that the existing uncertainties in the exact form of the distribution of secondary nuclei from the interaction of antiprotons with ^{4}He lead to rather small uncertainties in the ratio between the number density of antiprotons and the concentration of lithium and beryllium formed in the cascades initiated by antiprotons.

The relative stability of this relation follows from the fact that virtually all existing nuclear models of the distribution of secondary nuclei from the destruction of ^{4}He by antiprotons are predicting the maximum distribution at energies exceeding the threshold for subsequent nuclear processes. Thus, in a first approximation, the amount of lithium is determined by the integral number of secondary nuclei which are much less uncertain.

Based on the observed abundance of ^{6}Li, Khlopov et al (1994) found a limit on the relative amount of antimatter

$$f < 1.1 \cdot 10^{-4} \left(\frac{1\ \text{GeV}}{E_0} \right)^{1/2}, \tag{7.78}$$

where E_0 is the initial energy of the antiproton. Constraint (7.78) imposes an upper bound on the concentration of gravitinos (Chapter 6)

$$\frac{n_G}{n_\gamma} < 3.8 \cdot 10^5 \frac{n_{^6\mathrm{Li}}}{n_\gamma}. \tag{7.79}$$

This upper limit was obtained for the gravitino with mass

$$m_G = 100 \text{ GeV}$$

taking into account the ratio between the average energy of the antiprotons and the gluon E_g

$$E_{\bar{p}} \sim \left(1 \text{ GeV} \cdot E_g\right)^{1/2}. \tag{7.80}$$

If the mass of the gluino in decay into gravitinos

$$G \to \tilde{g} + g$$

is negligible compared to the mass of the gravitino

$$m_{\tilde{g}} \ll m_G \tag{7.81}$$

then the initial energy of the gluon is

$$E_g = \frac{m_G}{2} = 50 \text{ GeV}, \tag{7.82}$$

and, therefore, the initial energy of the antiproton is

$$E_0 \approx 7 \text{ GeV}. \tag{7.83}$$

Since the concentration of relic gravitinos is related to the heating temperature T_R at

$$\frac{n_G}{n_\gamma} = 0.1 \frac{T_R}{m_{\mathrm{Pl}}} \tag{7.84}$$

we finally obtain the upper limit on the temperature of heating on the basis of the observed abundance of ^{6}Li

$$T_R < 3.8 \cdot 10^6 \text{ GeV}. \tag{7.85}$$

Comparison with the corresponding limit obtained in (6.132) from the observed abundance of ^{3}He shows that, taking into account the sequence of nucleosynthesis processes initiated by antiprotons, the resultant constraint is more than one order of magnitude stronger.

This example reveals the importance of the theoretical framework of non-equilibrium nucleosynthesis as a sensitive tool in cosmoarcheological studies.

3. Cosmological backgrounds of non-equilibrium particles

3.1. High-energy neutrinos on the RD stage

The weakness of neutrino interactions makes neutrino cascades very inefficient in non-equilibrium cosmological nucleosynthesis, since it is virtually impossible to use an abundance of light elements to verify the presence of non-thermal sources of neutrinos in the RD stage.

The same reason makes it impossible to use the spectrum of the thermal background radiation as a calorimeter for neutrino sources.

But because of the weakness of neutrino interaction we can verify directly the existence of high-energy neutrino sources. Since the non-thermal neutrinos could have survive until the present time, it is possible to search for them in neutrino observatories and large volume neutrino experiments.

The higher the neutrino energy, the higher the probability of finding them and check the presence of their hypothetical sources in the early Universe.

There are several possibilities for the occurrence of energetic neutrinos in the reactions of energetic particles.

Neutrinos can be produced directly from the evaporating primordial black holes or in the decays of supermassive metastable particles

$$M \to \nu + ... \tag{7.86}$$

Sources can produce high-energy pions and kaons, which have the neutrinos amongst the decay products.

Development of nucleon cascades is accompanied by the inelastic nucleon interaction, generating pions and kaons, with their subsequent decay with the production of the neutrino.

The existence of thermal background neutrinos imposes an upper limit on the energy of non-thermal neutrinos.

Because of the neutrino–neutrino interaction at energies

$$E > E_{\max} \tag{7.87}$$

their energy must be lost in the scattering on the thermal background of neutrinos. We can estimate the value of energy E_{max} (Khlopov, Chechetkin 1987)

$$E_{max} \sim 1\,\mathrm{GeV}\left(\frac{t}{1\,\mathrm{s}}\right), \tag{7.88}$$

which defines the opacity of the thermal background of neutrinos for energetic neutrinos.

If the source gives rise to energetic neutrinos with energy (7.87), the neutrino-neutrino interaction leads to the neutrino cascade, since the scattering of energetic neutrinos with

$$E \gg E_{max} \tag{7.89}$$

on thermal neutrinos with energy

$$E_T \sim 3T, \tag{7.90}$$

where T is the temperature of thermal neutrinos, results in the average recoil energy of the neutrino

$$E' \sim \frac{E}{2}. \tag{7.91}$$

So the number of energetic neutrinos can increase.

The totality of all of these processes forms the spectrum of energetic neutrinos caused by ultrahigh-energy sources.

In general, the shape of this spectrum can be determined only by detailed numerical calculations. This requires detailed information about all the particle cascades generated by the sources. However, some fundamental features of the spectrum and the possibility of detection of the spectrum can be estimated on the basis of general properties of the source as its lifetime τ, the relative contribution of its high-energy products to the total cosmological density and maximum particle energy.

Consider, for definiteness, the case of superheavy particles with mass m, the relative concentration

$$r = \frac{n_m}{n_\gamma}$$

and lifetime τ. Let the quantity $\xi_\nu(E)$ be the number of neutrinos with energy E (in the energy range with width dE close to E) per one decay of such a particle.

The spectrum must be cut off at high energies

$$E > \min\{E_{\max}, m\}. \tag{7.92}$$

To evaluate the sensitivity of existing or proposed detectors of cosmic neutrinos, we must first compare the maximum energy of neutrinos from the considered sources in the present-day Universe with the threshold energy of neutrino detectors.

Consider, for definiteness, the case

$$E_{\max} < m. \tag{7.93}$$

The maximum energy of non-thermal neutrinos undergoes a redshift in the Universe to

$$E_{\max}^{\mathrm{mod}} = E_{\max}\left(1+z_d\right)^{-1}. \tag{7.94}$$

Here z_d is the redshift corresponding to the period of decay of the particles

$$t \sim \tau$$

and related with the lifetime of the particles by

$$\frac{t_U}{\tau} = \left(1+z_d\right)^{3/2}\begin{cases}1, & \tau > t_{\mathrm{MD}}\\ \left(\dfrac{1+z_d}{1+z_{\mathrm{MD}}}\right)^{1/2}, & \tau < t_{\mathrm{MD}}\end{cases}. \tag{7.95}$$

where

$$z_{\mathrm{MD}} = 10^4\left(\frac{H_0}{50\dfrac{\mathrm{km}}{\mathrm{s}\cdot\mathrm{Mpc}}}\right)^2 \tag{7.96}$$

for $\Omega = 1$, H_0 is the modern Hubble constant,

$$t_U = 4 \cdot 10^{17}\ \mathrm{s} \tag{7.97}$$

is the age of the Universe,

$$t_{MD} = 4 \cdot 10^{11} \text{ s} \tag{7.98}$$

is the beginning of the present-day non-relativistic stage of dominance of matter in the Universe.

Equation (7.95) should be changed taking into account the accelerated mode of the present-day expansion as well as possible changes in the scenarios of formation of the large-scale structure by the unstable dark matter (see below). However, this modification is negligible for further discussion.

From equation (7.95) it follows that the current value E_{max} in the case of supermassive particles decaying in the RD stage, i.e., when

$$t \sim \tau < t_{MD} \tag{7.99}$$

given by

$$E_{max}^{mod} = E_{max} \left(1 + z_{MD}\right)^{-1/4} \left(\frac{\tau}{t_U}\right)^{1/2}. \tag{7.100}$$

Using the estimate given by (7.100), we find that the present-day maximum energy of non-thermal neutrinos arising from the decays of supermassive unstable particles with mass, satisfying the condition

$$m > E_{max}$$

and the lifetime determined by the equation (7.99) is given by (Khlopov, Chechetkin 1985)

$$E_{max}^{mod} = 2 \cdot 10^{-10} \text{ GeV} \left(\frac{\tau}{1\text{s}}\right)^{3/2}. \tag{7.101}$$

From (7.101) we can easily find that for

$$\tau > 10^5 \text{ s} \tag{7.102}$$

the maximum energy of non-thermal neutrinos exceeds the threshold energy for such neutrinos in the neutrino laboratory.

The sensitivity of neutrino detectors to the sources of ultrahigh-energy particles in the RD stage is also determined by the intensity of the high-energy neutrino flux.

In the case of superheavy particles the flux $F_\nu(E)$ of non-thermal neutrinos is determined by the relative concentration r and quantity $\xi_\nu(E)$.

The number density of the non-thermal neutrinos in the phase space is given by

$$\frac{dJ_\nu}{E^2 d\Omega} = \frac{d^3 F_\nu}{d^3 E} = \frac{d^3 n_\nu}{d^3 E} r n_\nu c. \qquad (7.103)$$

Here c is the speed of light, and the spectrum is normalized in respect of the present-day density of relic neutrinos n_ν. Density

$$\frac{d^3 n_\nu}{d^3 E}$$

is associated with the spectrum $\xi_\nu(E)$ of neutrino cascades caused by the decay of superheavy particles as follows

$$\frac{d^3 n_\nu}{d^3 E} = \iint \xi_\nu \left(\frac{E}{(1+z(t))} \right) \cdot \exp\left(-\frac{t}{\tau} \right) \cdot \frac{d^3 n_m}{d^3 p} \cdot dt \cdot d^3 p, \qquad (7.104)$$

where the dependence of ξ_ν on $\left(\frac{E}{1+z}\right)$ takes into account the redshift of the neutrino energy and the exponential factor in the integrand corresponds to the exponential decay law. Expression (7.104) also takes into account the momentum distribution of the decaying particles.

3.1.2. Ultrahigh-energy neutrino background as a test of new physics

Comparison of differential J_ν and integral F_ν spectra of non-thermal background neutrinos with the relevant characteristics of atmospheric neutrinos J^a_ν and F^a_ν allows us to estimate the sensitivity of neutrino telescopes to the existence of sources of high-energy particles in the early Universe.

The high-energy neutrinos are most interesting for such a comparison, since the calculations showed a rapid fall in the spectrum of atmospheric neutrinos with increasing energy (see review in Khlopov 1999).

To estimate the spectrum of neutrinos from the decay of superheavy metastable particles we assume an instantaneous decay, that is, in equation (7.104) the Dirac delta function $\delta(t-\tau)$ replaces $\exp\{-t/\tau\}$. We can also ignore the momentum distribution of superheavy particles

$$\frac{d^3 n_\nu}{d^3 E} = \delta(\vec{p}). \qquad (7.105)$$

In this case, the distribution of neutrinos coincides with the spectrum

$$\frac{d^3 n_\nu}{d^3 E} = \xi_\nu \left(\frac{E}{(1+z_d)} \right). \tag{7.106}$$

To find function $\xi_\nu(E)$, it is important to consider the relation between the mass of the metastable particle m and magnitude E_{max}.

In the case

$$m < E_{max} \tag{7.107}$$

interactions of the neutrinos with the thermal background is negligibly weak, and the function $\xi_\nu(E)$ is determined by the detailed properties of the metastable particles and their decay products. In the case

$$m \gg E_{max} \tag{7.108}$$

$(\nu\nu)$ interaction is important.

Consider the last case, assuming that the effects of neutrino interaction dominate in the formation of the non-thermal spectrum.

The $(\nu\nu)$-scattering matrix element due to neutral currents in the standard theory of electroweak interactions has the form

$$M = \frac{G_F}{4\sqrt{2}} \bar{\nu}_2 \gamma_\mu (1+\gamma_5) \nu_1 \cdot \bar{\nu}_4 \gamma_\mu (1+\gamma_5) \nu_3, \tag{7.109}$$

where G_F is the Fermi constant.

The differential $(\nu\nu)$-scattering rate is given by

$$\frac{d(\sigma v)}{2\pi p_\nu^2 dp_\nu} = \frac{G_F^2 (1-\cos\theta)^2 E_1 \omega_1}{8\pi^2 p_\nu^2 p}, \tag{7.110}$$

where E_1 and ω_1 are the initial neutrino energies, θ is the angle between their momenta,

$$E_\nu = p_\nu \tag{7.111}$$

is the final energy of the neutrinos, and

$$p^2 = E_1^2 + \omega_1^2 + 2E_1\omega_1 \cos\theta. \tag{7.112}$$

Provided

$$\omega_1 \gg E_1 \tag{7.113}$$

the differential $(\nu\nu)$-scattering rate becomes

$$\frac{d(\sigma v)}{dp_\nu} = \frac{G_F^2(1-\cos\theta)^2 E_1}{4\pi}. \tag{7.114}$$

In the case of scattering of identical neutrinos the right-hand side of (7.110) and (7.114) must be multiplied by a factor of ½.

Consider (Suslin et al 1982) the kinetic equation for the distribution of neutrinos in a homogeneous environment.

The distribution function of the neutrino $f_\nu(E,t)$ in this case is the sum of the equilibrium distribution $f_\nu^{(0)}$ of thermal neutrino background and the distortion of this distribution $f_\nu^{(1)}$, produced by supermassive metastable particles.

The concentration of thermal neutrinos is unavoidably much greater than the concentration of energetic neutrinos, i.e. the following inequality is fulfilled

$$n_\nu^{(0)} = \int f_\nu^{(0)} d^3p \gg \int f_\nu^{(1)} d^3p. \tag{7.115}$$

We also assume the dominance of the energy density of thermal neutrinos above the energy density of non-thermal particles, that is, consider the case of weak sources of non-equilibrium neutrinos

$$\varepsilon_\nu^{(0)} = \int E f_\nu^{(0)} d^3p \gg \int E f_\nu^{(1)} d^3p. \tag{7.116}$$

Then, we obtain a system of linear equations for the functions $f_i (i = 1,\dots,N)$ of neutrino species (N=3)

$$\begin{aligned}\frac{df_{\nu i}^{(1)}}{dt} = \sum_{j,k} \int d^3p_j \int d^3p_k f_{\nu j}^{(1)} f_{\nu k}^{(0)} \frac{d^3(\sigma v)^{jk}}{dp_i} - \\ -f_{\nu i}^{(1)} \sum_{j,k} \int d^3p_k f_{\nu k}^{(0)} (\sigma v)^{ik} + S_i(p_i).\end{aligned} \tag{7.117}$$

Here

$$\frac{d^3(\sigma v)^{jk}}{dp_i}$$

is the differential ($\nu\nu$)-scattering rate, and $S_i(p_i)$ is the source function.

If the neutrinos of all types have the same unperturbed distribution function, the value

$$f = \sum_i f_{\nu i}^{(1)}$$

satisfies the same equation for $f_\nu^{(1)}$ as in the case of only one type of neutrino

$$\frac{df_{\nu i}^{(1)}}{dt}=\sum_{j,k}\int d^3p_j f_\nu^{(0)}(\omega)\cdot\omega\int_E^\infty dx f_\nu^{(1)}(x)\frac{G_F^2(1-\cos\theta)^2}{8\pi}-$$
$$-f_{\nu i}^{(1)}E\int d^3p f_\nu^{(0)}(\omega)\frac{G_F^2(1-\cos\theta)^2}{8\pi}+S_\nu(E,t). \quad (7.118)$$

So we can consider the case of only one type of neutrino.

Functions $f_\nu^{(1)}$ and S_ν can be conveniently normalized as

$$f_\nu^{(1)}=\frac{1}{m}f\left(\frac{E}{m}\right)n_m \quad (7.119)$$

and

$$S_\nu(E,t)=n_m\frac{d\Gamma}{\Gamma dE}\delta(t-\tau)=S(E)\delta(t-\tau), \quad (7.120)$$

where

$$S(E)=\frac{1}{m}S\left(\frac{E}{m}\right)n_m \quad (7.121)$$

and the normalization conditions are used

$$\int_0^\infty E\cdot f_\nu^{(1)}(E)dE=\frac{mn_m}{a_\nu} \quad (7.122)$$

and

$$\int_0^m E\cdot S(E)dE=mn_m\int_0^1 x\cdot S(x)dx=\frac{mn_m}{a_\nu}. \quad (7.123)$$

Here a_ν^{-1} is the fraction of the rest energy of the decaying particles corresponding to the total energy of the initial flux of non-equilibrium neutrinos.

A dimensionless variable

$$x=\frac{E^{\text{mod}}(1+z_d)}{m}, \quad (7.124)$$

can also be introduced, where z_d is the redshift corresponding the decay period and E^{mod} is the present-day energy of non-equilibrium neutrinos.

Then the equation for

$$f(x) \equiv f_\nu^{(1)}(x) \tag{7.125}$$

has the form

$$\frac{df(x,t)}{dt} = 2B\int_0^\infty dy\; f(y,t) - Bf(x,t) + S_\nu(x,t), \tag{7.126}$$

where

$$B = \frac{G_F^2 \varepsilon_\nu m}{6\pi} \tag{7.127}$$

and ε_ν is the energy density of thermal background neutrinos.

Equation (7.126) can be solved using the method of Laplace transformation. It has the form

$$f(x,t) = \exp(-B\tau x)\left\{ S(x) + 2B\tau \int_x^1 S(y)dy + (B\tau)^2 \int_x^1 yS(y)dy \right\}. \tag{7.128}$$

The quantity

$$B\tau \sim n_\nu \sigma_{\nu\nu} v t \tag{7.129}$$

is the optical depth of the thermal background of neutrinos for energetic neutrinos with energy

$$E \sim m. \tag{7.130}$$

From (7.128) we find that the more opaque is the source of for the initial energetic neutrinos, the better is the approximation of the distribution function which is dominated by a term proportional to

$$\int_x^1 yS(y)dy.$$

This term corresponds to the 'multiplication' of high-energy neutrinos due to their scattering on the thermal background of neutrinos.

Putting

$$a_\nu \sim 1 \tag{7.131}$$

we obtain

$$f(E) = \exp\left\{-\frac{E}{E_0}\right\}\frac{mn_m}{E_0^2}, \tag{7.132}$$

where

$$E_0 = \left(\frac{G_F^2 \varepsilon_\nu \tau}{6\pi}\right)^{-1} \approx E_{\max}. \tag{7.133}$$

Finally, the modern spectrum of high-energy neutrinos predicted in this case has the form

$$J_\nu(E) = \frac{rm\, n_\gamma}{(1+z_d)\left(E_{\max}^{\mathrm{mod}}\right)^2} \cdot \exp\left\{-\frac{E}{E_{\max}^{\mathrm{mod}}}\right\} \cdot \frac{c}{8\pi}. \tag{7.134}$$

Since the neutrino telescopes do not register a significant excess of atmospheric neutrinos over the calculated neutrino background, we can set an upper limit on the flux of non-thermal high-energy cosmological neutrinos

$$F_\nu(E > 1\,\mathrm{GeV}) > F_\nu^a(E > 1\,\mathrm{GeV}). \tag{7.135}$$

Using equations (7.134) and (7.135), we find a constraint on the quantity rm and τ for supermassive metastable particles with mass

$$m \gg 10^6\ \mathrm{GeV}, \tag{7.136}$$

decaying at

$$t > 10^6\ \mathrm{s}. \tag{7.137}$$

This constraint is shown in Fig. 7.1.

Despite the ambiguity of the spectrum shape in the case of (7.107), we can assume the absence of a strong suppression of the high-energy tail of the spectrum up to energies

$$E \leqslant \frac{m}{2}. \tag{7.138}$$

Then the constraints outlined in Fig. 7.1 also applies in this case.

3.3. Large volume experiments as a tool for cosmoparticle physics

Large volume experiments like ANTARES or IceCube open new

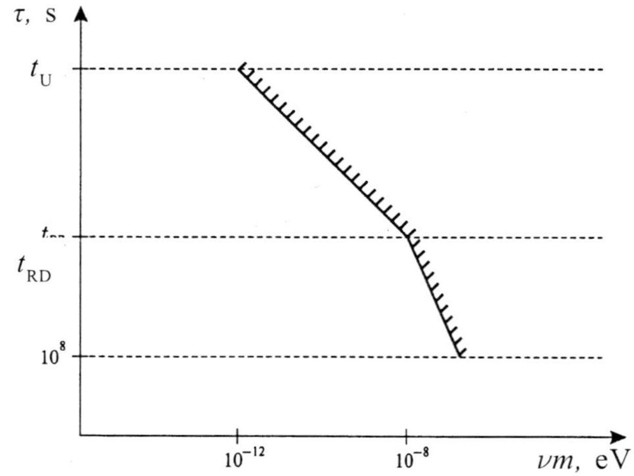

Fig. 7.1. Constraints on the value νm and lifetime τ of supermassive metastable particles from measurements of fluxes of high-energy neutrinos in DUMAND experiment.

opportunities to test the existence of sources of high-energy particles in the early Universe. To illustrate these opportunities we follow early work by (Khlopov, Chechetkin 1987) in which such possibilities were studied for deep under-water muon and neutrino detectors (DUMAND).

The minimum sensitivity of these detectors at

$$E \sim 1 \text{ TeV} \tag{7.139}$$

was taken in (Khlopov, Chechetkin 1987) equal to

$$F_\nu \,(E > 1 \text{ TeV}) = 10^{-11} \,(\text{cm}^{-2}\cdot\text{s}\cdot\text{ster})^{-1} \tag{7.140}$$

In this case, from equation (7.134) it follows that DUMAND is able to detect the effects of the decays of relic particles with

$$rm > 10^{-10}\,\text{eV}\cdot(1+z_d). \tag{7.141}$$

The neutrino detection threshold was taken equal to 200 GeV. From equations (7.100) and (7.101) we find that the condition

$$E_{\max}^{\text{mod}} > 200 \text{ GeV} \tag{7.142}$$

is valid for

$$\tau > 10^8 \text{ s}. \tag{7.143}$$

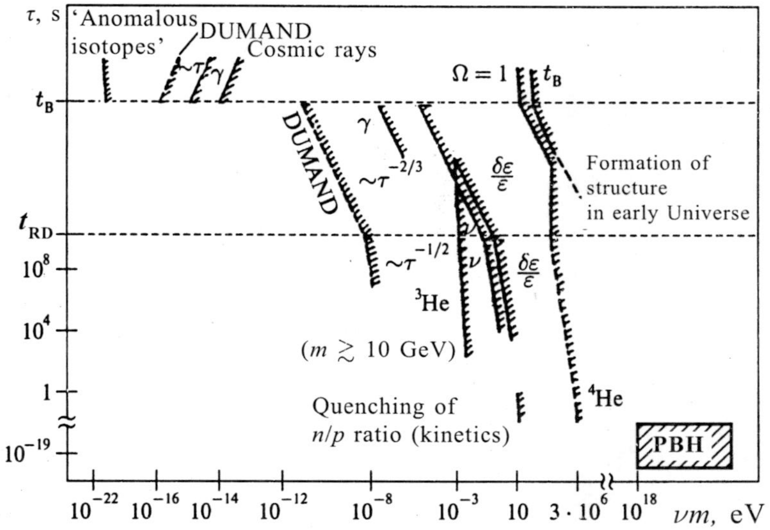

Fig. 7.2. Constraints on the value νm of metastable particles in relation to their lifetime τ.

Values rm and τ available on large volume experiments are presented in Fig. 7.2 together with constraints on these values, which follow (Khlopov and Chechetkin 1987) from the data of neutrino telescopes, data on the abundance of light elements (Chapter 6) and the observed spectrum of cosmic relic radiation (Chapter 5). Figure 7.2 shows that DUMAND can play the role of a unique detector that can detect the decays of particles with masses exceeding

$$m > 200 \text{ GeV} \cdot (1 + z_d), \tag{7.144}$$

during the period (7.143).

This can be useful for the study of 'white spots' on the map (rm, τ), making it possible to test such cosmological effects of ultrahigh-energy physics which may not affect either the abundance of light elements, or the thermal background spectrum.

The existence of metastable particles with Planck mass

$$m = m_{\text{Pl}} \tag{7.145}$$

electrically charged maximons may be a non-trivial consequence of quantum gravity (Markov 1978 1993). The decay of these particles may be accompanied by superhigh-energy neutrino cascades. This case corresponds to the condition (7.108)

$$m \gg E_{\max}$$

so that the spectrum of neutrinos from the decay of maximons is given by (7.134).

It should be noted that the evolution of cosmic strings may be an interesting example of a source of neutrinos of ultrahigh energies. Large volume experiments can check in this case the relative contribution of cosmic strings to the total density, which exceeds (Khlopov, Chechetkin 1985, 1987)

$$\frac{\rho_s}{\rho_{\text{tot}}} \geqslant 10^{-11}. \tag{7.146}$$

Elementary particle theory relates this contribution to the linear energy density of cosmic strings, determined by the energy scale Λ at which symmetry breaking takes place, which leads to the cosmic strings as topological defects so that

$$\frac{\rho_s}{\rho_{\text{tot}}} = \Omega_s \sim \frac{\Lambda^2}{m_{\text{Pl}}^2}. \tag{7.147}$$

The existence of cosmic strings is associated with any broken $U(1)$ symmetry, or with symmetry violation, leaving unbroken any strict discrete subgroup of the symmetry (Chapter 2).

Thus, large volume detectors can be sensitive to the structure of symmetry breaking at scales greater than

$$\Lambda > 10^{14}\ \text{GeV}. \tag{7.148}$$

Hypothetical particles, PBHs and other objects, which are predicted in the framework of modern particle theory and discussed in previous chapters, can be regarded as different forms of dark matter in the Universe. Analysis of the possible impact of these forms of dark matter on various processes in the Universe allows us to use the observational data for definitive conclusions about the permissible extent of such influence thus imposing constraints on the possible properties of hypothetical objects and, as a consequence, the parameters of the models that predict them.

Such an analysis leads to a constraint on the admissible properties of dark matter in the Universe. On the other hand, there are areas of modern cosmology in which we have to assume the existence of dark matter for self-consistent explanation of the entire set of observational data. In the next chapter we will see that such a case occurs when comparing the theory of formation of the large-scale structure of the Universe with observations.

8

New physics in formation of the large-scale structure

1. The problems of large-scale structure

1.1. The problem of initial fluctuations

The structure of the inhomogeneities in the expanding quasi-uniform Friedmann–Robertson–Walker Universe is the result of a long stage of growth of small initial fluctuations. Such fluctuations in cosmological models should be set initially. They cannot be attributed to statistical density fluctuations of particles.

Indeed, in the observed supercluster of galaxies with mass

$$M > 10^{13} M_{\odot}, \tag{8.1}$$

where $M_{\odot} = 2 \cdot 10^{33}\,\mathrm{g}$ is the mass of the Sun, there are

$$N > 10^{70} \tag{8.2}$$

baryons.

The relative amplitude of statistical density fluctuations for these scales is

$$\left.\frac{\delta\rho}{\rho}\right|_{\mathrm{st}} \sim \frac{\delta N}{N} \sim \frac{1}{\sqrt{N}} < 10^{-35}. \tag{8.3}$$

On the other hand, the density fluctuations grow in an expanding Universe in accordance with law

$$\frac{\delta\rho}{\rho} \propto (1+z)^{-1}. \tag{8.4}$$

To form the inhomogeneity of the order

$$\frac{\delta\rho}{\rho} \sim 1 \tag{8.5}$$

at the present moment, we need the density fluctuations with an amplitude

$$\delta = \frac{\delta\rho}{\rho} \sim 10^{-3} \tag{8.6}$$

during the recombination period corresponding to the redshift

$$z = z_{\text{rec}} \sim 10^3. \tag{8.7}$$

There were two different possibilities to explain the initial density fluctuations in the old cosmological models, which assume that the only components of the Universe were the baryonic matter and radiation. (Here we do not consider the so-called vortex theory (Ozernoi, Chernin 1966)).

According to the first possibility, the density fluctuations were generated by metric perturbations in matter and radiation, so that the specific entropy did not change. Such fluctuations were called adiabatic, and the theory treating them as the basis of galaxy formation was called the adiabatic theory of galaxy formation (A-theory).

Another possibility was the fluctuation of the ratio of baryons to photons at a uniform total density (baryons and photons) in the RD stage. In this case, the specific entropy of matter was perturbed and such fluctuations were referred to as entropy fluctuations, and the theory of their evolution leading to the formation of galaxies was called the entropy theory (E-theory) of galaxy formation.

Both the adiabatic and entropy theory suggested that effective formation of inhomogeneities occurred under the dominance of the baryonic matter after recombination, when the radiation pressure did not affect the development of gravitational instability in the baryonic matter.

But the E- and A-theories differed from the standpoint of the formation of some objects.

They also differ in the predictions for large-scale distribution of matter. Starting from the late seventies or early eighties of the twentieth century, a large amount of observational evidences for the existence of a developed cell structure in the Universe (De Vaucouleuers 1959; Einasto et al 1980; Tully, 1982) was found.

The existence of this structure was predicted in the A-theory of galaxy formation (see review Shandarin et al 1983), and was a specific feature of this theory.

In contrast to the entropy theory, according to A-theory the large-scale inhomogeneities (the so-called pancakes) (Zeldovich, 1970) should be formed first. The subsequent fragmentation of 'pancakes' led to the formation of galaxies and clusters of galaxies inside pancakes.

In most cases, the entropy theory leads naturally to the hierarchical clustering of inhomogeneities (Jones, 1976; Shandarin et al 1983), so that small-scale inhomogeneities (on the scale of globular clusters) were formed first. Their subsequent clustering led to the formation of larger inhomogeneities (galaxies and galaxy clusters).

The adiabatic theory of galaxy formation was in fact the theory of formation of the large-scale structure of the Universe. The existence of 'pancakes' (superclusters of galaxies) and giant (on the order of tens of Mpc) 'void' (De Vaucouleuers 1959; Einasto et al 1980; Tully, 1982; Shandarin et al 1983), in which galaxies are virtually absent, it seemed, gave a final confirmation of the A-theory.

However, the theory of formation of the large-scale structure was faced with numerous difficulties in explaining the observational facts. One of these difficulties has been ambiguous data on the average density of the Universe, estimated by different methods. These issues were addressed in Chapter 3.

Data on the density of luminous matter of galaxies and the data on the abundance of light elements testified in favour of low-density matter in the Universe

$$\Omega < 0.1 \div 0.2. \qquad (8.8)$$

On the other hand, data on the dynamics of matter in galaxies, clusters and superclusters of galaxies indicated a high total density

$$\Omega > 0.2. \qquad (8.9)$$

The strongest lower bound on the cosmological density resulted from the analysis of the dynamics of formation of the large-scale structure and the effect of initial density fluctuations on the anisotropy of relic radiation.

1.2. The problem of self-consistent formation of structure of inhomogeneities and anisotropy of cosmic microwave background radiation

Soon after the discovery of cosmic microwave background radiation, it was stated (Silk 1968; Sunyaev, Zeldovich 1970) that the temperature of this radiation can not be strictly constant in all directions in the sky.

Fluctuations of density and velocity, whose evolution has led to the current structure of the Universe, had to leave 'footprints' in the cosmic microwave background radiation (CMB). The temperature of the CMB should vary, and the amplitude of temperature fluctuations is uniquely associated with initial fluctuations of density and velocity in the Universe. There were indications of the existence of these fluctuations at angular scales of the corresponding clusters of galaxies at

$$\frac{\delta T}{T} < 2 \cdot 10^{-5} \tag{8.10}$$

and at the level

$$\frac{\delta T}{T} \leq 10^{-5} \tag{8.11}$$

with angular dimensions

$$\theta > 1^{\circ}, \tag{8.12}$$

(Smoot et al 1992; Bennett et al 1996) fixing the amplitude of initial fluctuations of density and velocity at such a low level that the observed structure simply could not be formed in the Universe at low values of the average cosmological density. These data correspond to the values given by (8.9), or indeed even

$$\Omega \sim 1. \tag{8.13}$$

These contradictions were naturally eliminated in cosmological models with neutrinos with the mass

$$m_{\nu} \sim 30 \text{ eV} \tag{8.14}$$

(Szalay, Marx 1976; Zeldovich and Sunyaev; 1980; Doroshkevich et al 1980 a, b; Bisnovatyi-Kogan and Novikov 1980; Schramm, Steigman 1981; Zeldovich, Khlopov 1981).

Indeed, the observation of luminous matter, as well as an abundance of light elements, give only an estimate of the baryonic component of the total density, while the dynamic evaluation and arguments, considering the temperature fluctuations of the background radiation are associated with the total density, which in these models includes both the baryonic component and a component of the massive neutrinos.

Thus, cosmology of massive neutrinos has become a promising single solution of a number of old problems in astrophysics.

But the value of the cosmological models with massive neutrinos went far beyond a simple qualitative solutions to these problems.

Based on an analysis of the cosmological consequences of experimental evidence for the presence of the rest mass of the electron neutrino

$$14 < m_\nu < 46 \text{ eV} \tag{8.15}$$

(Lyubimov et al 1980), these models led during their development to irreversible changes in ideas about the period of structure formation. They have identified an inherent contradiction with the models of baryonic matter dominant in this period, making the existence of non-baryonic dark matter a necessary element of the theory of large-scale structure formation.

Comprehensive analysis of the data of the angular distribution of the thermal electromagnetic background and large-scale distribution of matter makes it possible in principle to obtain information about the physical nature of dark matter (Doroshkevich, Khlopov 1984).

Quantitative analysis of cosmological structures in models with massive neutrinos has led to a number of difficulties in comparing detailed predictions of these models with observational data on large-scale distribution of matter.

A quantitative difference was found between the estimates of the density of dark matter on the basis of the conditions of the formation of the structure in the observed isotropy of cosmic microwave background radiation and the estimate of the density of dark matter formed within the structure (Frank et al 1983; White et al 1983). This discrepancy could not be eliminated by simple replacement of massive neutrinos by some other weakly or superweakly interacting particles.

Measurements of the total cosmological density in the experiments BOOMERANG and WMAP in combination with an analysis of the evolution of the structure reject the idea of identity of the forms of dark matter dominating during the formation of the large-scale structure and in the present-day Universe. There is a widespread notion of 'dark energy. that dominates the present-day Universe as a form of matter with negative pressure. In its simplest form the role of this dark energy can be played by vacuum with positive energy density, which is a physical realization of the cosmological Λ-term in Einstein's equations. There are considered more sophisticated versions of the dark energy density, variable with time. The physical basis of this so-called 'quintessence' is given by the equation of state of the field with the hyperbolic potential $V(\varphi) \sim |\varphi|^{-n}$. This approach assumes the existence of both primordial dark matter and dark energy. Dark matter is dominant in the formation of the structure and then, at a redshift $z \lesssim 3$ the constant (or varying with time) density of dark energy begins to dominate the Universe. The required parameters of time-varying dark energy did not yet find a natural basis in the field theory, which makes another approach to the problem of time-varying dark energy deserving interest – its identification with the combination of cosmological term and uniform background of weakly interacting decay products of dark matter that formed the structure of the Universe. In other words, it is assumed that dark matter is multicomponent and its dominant component was unstable.

Analysis of the evolution of the large-scale structure of the Universe allows in this case to determine the mass and lifetime of the particles of dark matter. In the case of massive unstable neutrinos, we can define the parameters of hypothetical interactions that lead to the decay of the neutrino.

However, the problem of the physical nature of the dark matter, forming the structure of the Unvierse, has not yet found its final solution.

The choice of a candidate for the role of dark matter determines the physics of damping of density fluctuations, which determines the short-wave cutoff of the spectrum of these fluctuations.

In the simplest case of the thermal background of weakly interacting particles (e.g., massive neutrinos), the scale of short-wave damping R_c is determined by the mass of these particles

$$R_c \sim m^{-1} \tag{8.16}$$

The R_c scale determines the characteristic size of the first inhomogeneities. Depending on this scale, the dark matter is defined as hot, cold or warm.

The R_c scale, corresponding to superclusters or clusters of galaxies, is the hot dark matter (HDM), for example, relic hot gas of weakly interacting particles with mass

$$m \sim 30 \div 100 \text{ eV}. \tag{8.17}$$

In the case of the warm particles, such as the relic particles with mass

$$m \sim 1 \text{ keV}, \tag{8.18}$$

R_c corresponds to the scale of galaxies.

Inhomogeneities, formed by cold dark matter (CDM), have a much smaller R_c scale than the scale of galaxies. Candidates for the role of the CDM are weakly interacting particles with mass

$$m \gg 1 \text{ keV}, \tag{8.19}$$

such as the neutralino in supersymmetric models, or a heavy stable neutral lepton with mass

$$m > (1 \div 10) \text{ GeV}. \tag{8.20}$$

Invisible axions have a very small mass

$$m \sim 10^{-5} \div 10^{-3} \text{ eV}. \tag{8.21}$$

However, the scale of the damping of density fluctuations (Chapter 3) is as small as in the case of heavy particles with mass

$$m \gg 1 \text{ GeV}. \tag{8.22}$$

That is why the axions are also considered as candidates for the role of the CDM.

The ratio between the lifetime of a particle of dark matter τ and the age of the Universe t_U divides particles into stable,

$$\tau \gg t_U \tag{8.23}$$

and unstable

$$\tau < t_U. \tag{8.24}$$

The weakly interacting particles of dark matter, described above, represent a collisionless gas in which energy dissipation is a fairly lengthy process. Mirror and shadow matter (see Chapter 10) has internal interactions that lead to the dissipation rate comparable or even higher than the rate of dissipation in the baryonic matter. Consequently, there may be dissipating forms of dark matter which lead to the possible existence of compact objects of dark matter. This feature makes the interpretation of observational data even more complex.

1.3. Problem of physical consistency in the theory of large-scale structure formation

Another problem of the theory of cosmological structure formation is the spectrum of initial density fluctuations.

The most popular is the so-called 'flat' spectrum of fluctuations (Harrison 1967; Zeldovich 1970). Its benefits seem obvious, since it lacks a preferred scale. The amplitude of fluctuations in this spectrum is the same for all scales at a time when the size of the fluctuation is equal to the size of horizon.

The flat spectrum of density fluctuations follows from the simplest versions of inflationary scenario (Hawking 1982; Guth and Pi 1982; Starobinsky 1982). Such spectrum is also formed in the case of density fluctuations generated by the structure of long cosmic strings (Zeldovich, 1980).

However, the development of realistic models of multicomponent inflation which take into account several types of scalar fields (Kofman, Linde 1987; Sakharov, Khlopov 1993) and the development of the theory of cosmic strings taking into account the contribution of closed loops of these strings (Vilenkin 1985) and their disintegration into smaller loops (Smith, Vilenkin 1987), leads to more complex forms of the spectrum of initial fluctuations.

Moreover, a variety of effects of the theory of elementary particles in scenarios of the very early Universe (see Chapters 3 and 4) opens up new opportunities for physical interpretation of the phenomenology of adiabatic and entropy fluctuations and leads to a non-trivial pattern of physical phenomena, combining old ideas of the adiabatic and entropic theories.

The initial density fluctuations resulting from inflation (Chapters 3 and 4) are adiabatic and, it would seem, provide the fundamental support for the adiabatic theory. However, the evolution of these fluctuations in the CDM model leads to a hierarchical clustering characteristic of the entropy theory.

The realistic theory of cosmological structure formation is very probably a multicomponent theory. It can reflect all the main features of realistic models of elementary particles. This view is supported by both physical fundamentals of the cosmological models and the problems of simple one-parameter models of dark matter in reproducing the entire set of observational data.

Taking into account all these arguments, we can consider not only the adiabatic but also and entropy fluctuations, which may also find a physical basis in some scenarios of the very early Universe, based on predictions of models of elementary particles.

Entropy fluctuations arise naturally in inhomogeneous baryosynthesis which may be associated, for example, with the spatial dependence of the *CP*-violating phase which determines the magnitude of the baryon asymmetry of the Universe (Chapter 6). This can take place in the invisible axion models (Yoshimura 1983), or in a multicomponent model of spontaneous *CP*-violation (Kuzmin et al 1981). This may also occur in the supersymmetric models which ascribe the existence of baryon asymmetry of the Universe to the existence of the primordial condensate of scalar quarks (squarks) (Affleck, Dine 1985; Linde, 1985).

In the latter case, the superpotential does not depend on the field amplitude of scalar quarks, so that the field amplitude may be different in different regions of space. As a result, the spatial dependence of the baryon number appears.

Generalization of this approach is a model of spontaneous baryosynthesis (Dolgov 1992, 1996), in which the baryon asymmetry is formed due to creation of particles by the primordial scalar field. The presence of the 'valley' in the potential of this field leads to fluctuations in its amplitude in the inflationary stage, causing inhomogeneity of the resulting excess of baryons. This model became the basis of a recently quantitatively defined inflationary cosmological model with inhomogenous baryosynthesis (Khlopov et al 1999, 2000). In its extreme form, the inhomogeneity of baryosynthesis is reflected in the appearance of domains with an excess of antibaryons in the baryon asymmetric Universe (see Chapter 6).

The presence of the 'valley' of the potential of the scalar field that is not associated with baryosynthesis may be reflected in the emergence of strong primordial inhomogeneities. In the case of global $U(1)$ symmetry, violated initially spontaneously and then explicitly which is characteristic, in particular, of the model of the invisible axion (see Chapter 2), the presence of a 'valley' of the potential of the complex Higgs field after the phase transition with spontaneous symmetry breaking leads to the formation of the structure of cosmic strings. After the explicit symmetry breaking, this structure is transformed into an unstable structure of the 'walls bounded by strings', which is reflected in the spatial distribution of the energy density of coherent oscillations of the scalar field. Large-scale correlations in this distribution in the case of the invisible axion are called archioles (Sakharov, Khlopov, 1994a) and are discussed in Chapter 11.

If the spontaneous breaking of $U(1)$-symmetry occurs at the inflationary stage, the phase fluctuations of the complex Higgs field along the valley cover large spatial areas, so the phase transition with explicit symmetry breaking may result in the formation of closed walls whose collapse leads to the formation of clusters of black holes with masses reaching the masses of the nuclei of galaxies (Rubin et al 2001; Khlopov et al 2002).The existence of such clusters of primordial black holes (PBH) introduces a new element to the theory of galaxy formation.

These examples of the strong primordial inhomogeneities (archioles, supermassive PBH) do not contradict the global homogeneity and isotropy of the Universe, if the contribution of inhomogeneous components (ρ_i) to the total density is small: $\rho_i \ll \rho$. The strong inhomogeneity of this component is then consistent with the low inhomogeneity of the total density

$$\frac{\delta\rho}{\rho} \sim \left(\frac{\delta\rho_i}{\rho_i}\right)\left(\frac{\rho_i}{\rho}\right) \ll 1.$$

On the other hand, local effects of strong initial inhomogeneity of the density of dark matter and baryons can play an important role in the theory of the formation and evolution of the structure and require special consideration.

Thus, in recent years, cosmology has significantly extended its representation for both the possible physical nature of matter that forms the structure of cosmological inhomogeneities as well as the possible origin of the initial density fluctuations. Given the physical reasons arising from different parts of the theory of elementary particles, we can find in the theory of formation of the large-scale structure of the Universe the non-trivial reflection of the physical mechanisms of inflation, baryosynthesis and combinations of various forms of dark matter. Here, three central problems of modern cosmology are tied in a complex knot. The task of cosmoparticle physics is to untie this knot and find a physically self-consistent picture of galaxy formation.

2. Neutrino mass and large scale structure of Universe

Experimental indication of the existence of the electron neutrino mass (Lyubimov et al 1980) provoked an intense discussion of the cosmological role of massive neutrinos (Doroshkevich et al 1980a, b; Zeldovich, Khlopov 1981; Schramm, Steigman 1980; Sato, Takahara 1980; Bisnovatyi-Kogan, Novikov 1980, Bisnovatyi-Kogan et al 1980; Bond et al 1980; Klinkhammer, Norman 1981). It was of special interest to assign the existence of dark matter to massive neutrinos, as first proposed by Szalay and Marx (1976). Though the current data on neutrino oscillations exclude this possibility, the following detailed discussion of cosmological evolution of massive neutrinos provides useful tools for present-day studies of dark matter models.

2.1. Massive neutrinos in the Universe

We consider the cosmological evolution of massive neutrinos (Doroshkevich et al 1980a, b; Doroshkevich, Khlopov 1981).

As already noted in Chapter 3, at

$$T \gg m_e \tag{8.25}$$

neutrinos are in thermodynamic equilibrium with other particles – photons, electrons and positrons.

During this period, the mass of neutrinos may be neglected,

$$T \gg m_e \gg m_\nu, \tag{8.26}$$

and therefore

$$E_\nu = p_\nu, \tag{8.27}$$

where E_ν is the energy and p_ν the momentum of the neutrino.

According to Fermi statistics, the occupation number of ultra-relativistic neutrinos in the phase space

$$n = \left(\exp\left\{ \frac{p_\nu}{T} \right\} + 1 \right)^{-1} \tag{8.28}$$

with the chemical potential of neutrinos

$$\mu = 0. \tag{8.29}$$

In this case,

$$n_\nu = n_{\bar{\nu}}. \tag{8.30}$$

When

$$T \leqslant 1 \text{ MeV} \tag{8.31}$$

(Chapter 3) the neutrino gas is collisionless.

The momentum of the collisionless neutrinos decreases as

$$p \propto (1+z), \tag{8.32}$$

and its density in the phase space (occupation number) is conserved during cosmological expansion.

According to (8.28), this is equivalent to a decrease in neutrino temperature T in proportion

$$T \propto (1+z). \tag{8.33}$$

When

$$T \cong 0.1 m_e \tag{8.34}$$

the annihilation of electron–positron pairs takes place and this increases the temperature of the thermal background radiation relative to the temperature of neutrinos (Chapter 3)

$$T_\nu = \left(\frac{4}{11}\right)^{1/3} T_\gamma = 1.83 \cdot (1+z) K \tag{8.35}$$

at

$$T_\gamma < 0.1 m_e, \tag{8.36}$$

where

$$T_\gamma = 2.7 \cdot (1+z) K. \tag{8.37}$$

Distribution function (8.28) gives the concentration of the neutrino

$$N_\nu = \int n \frac{d^3 p}{(2\pi)^3} = 75 \cdot (1+z) \text{ cm}^{-3}. \tag{8.38}$$

It should be noted that the distribution function (8.28) is not an equilibrium one at

$$T_\nu \ll m_\nu, \tag{8.39}$$

since (8.27) is not valid in the non-relativistic case, and the ratio

$$p_\nu = \left(E_\nu^2 - m_\nu^2\right)^{1/2} \tag{8.40}$$

gives the density in phase space in the form of

$$n = \left(\exp\left\{\frac{\left(E_\nu^2 - m_\nu^2\right)^{1/2}}{T_\nu}\right\}+1\right)^{-1} \approx \left(\exp\left\{\frac{\left(2m_\nu E_k\right)^{1/2}}{T_\nu}\right\}+1\right)^{-1}, \tag{8.41}$$

where the kinetic energy of the neutrino

$$E_k = \frac{p_\nu^2}{2m_\nu} \approx E_\nu - m_\nu \ll m_\nu, \tag{8.42}$$

and

$$E_\nu + m_\nu \approx m_\nu \tag{8.43}$$

at

$$T_\nu < m_\nu. \tag{8.44}$$

Using equation (8.28), we obtain after averaging over phase space (Doroshkevich et al 1980a) the average kinetic energy

$$\langle E_k \rangle = \frac{\langle p_\nu^2 \rangle}{2m_\nu} = 6.47\frac{T_\nu^2}{m_\nu} \tag{8.45}$$

and average velocity

$$\langle v^{-2} \rangle = 0.385\frac{m_\nu^2}{T_\nu^2} \tag{8.46}$$

$$\sqrt{\langle v^2 \rangle} = 3.6\frac{T_\nu}{m_\nu} = 664\cdot(1+z)\,\text{km/s} \tag{8.47}$$

at $m_\nu =$ 30 eV. This average velocity corresponds to the temperature

$$T_{eq\nu} = \frac{m\langle v^2 \rangle}{2} = 5.5\cdot 10^{-5}(1+z)\,\text{K} \tag{8.48}$$

of the Maxwellian gas of neutrinos with mass (8.14).

The neutrinos become non-relativistic at a redshift

$$z = 4.5 \cdot 10^4 \left(\frac{m_\nu}{30\,\text{eV}} \right). \tag{8.49}$$

During this period the law of cosmological expansion changes. Non-relativistic neutrinos will soon begin to dominate the total cosmological density, and their dominance determines the law of cosmological expansion.

The approximate analytic expression for the equation of state of the Universe at the moment of transition from the RD stage to the MD stage can be derived on the basis of the distribution function of neutrinos (Doroshkevich, Khlopov, 1981).

Namely, we can find an approximation for the time dependence of energy and pressure of the gas of neutrinos, using the distribution function (8.28). This dependence has the form

$$\varepsilon_\nu = \varepsilon_{eq} x_\nu^4 \int_0^\infty \frac{x^3 \left(1 + \frac{9x_\nu^2}{x^2} \right)^{1/2}}{e^x + 1} \approx \frac{\varepsilon_{eq}}{\sqrt{2}} x_\nu^4 \left(1 + x_\nu^{-2} \right)^{1/2}, \tag{8.50}$$

$$p = \varepsilon_{eq} x_\nu^4 \int_0^\infty \frac{x^3 \left(1 + \frac{9x_\nu^2}{x^2} \right)^{-1/2}}{e^x + 1} \approx \frac{\varepsilon_{eq}}{\sqrt{2}} x_\nu^4 \left(1 + x_\nu^{-2} \right)^{-1/2}, \tag{8.51}$$

where

$$x_\nu = \frac{(1+z)}{(1+z_\nu)} \tag{8.52}$$

and ε_{eq} is the energy density of neutrinos at

$$z = z_\nu = 5.5 \cdot 10^4 \left(\frac{m_\nu}{30\text{ eV}} \right) \left(\frac{H_0}{50 \frac{\text{km}}{\text{s} \cdot \text{Mpc}}} \right)^2. \tag{8.53}$$

Here H_0 is the present-day value of the Hubble constant.

Equations (8.50) and (8.51) have the correct asymptotic behaviour as in the case

$$z \gg z_\nu, \tag{8.54}$$

corresponding to the equation of state

$$p = \frac{1}{3}\varepsilon,$$

and in the case

$$z \ll z_\nu \tag{8.55}$$

when the equation of state is

$$p \ll \varepsilon \tag{8.56}$$

which corresponds to the dust stage of cosmological expansion.

Using the approximate equation of state of gas of massive neutrinos, determined by the equations (8.50) and (8.51), we can find the law of the cosmological expansion at a time when massive neutrinos are non-relativistic and begin to dominate the Universe.

The transition from the relativistic law of expansion

$$a(t) \propto t^{1/2}$$

to the expansion law of the dust stage

$$a(t) \propto t^{2/3}$$

turns out to be rather long. This is slower than in the case of transition from the RD stage to the dominance of baryonic matter. The transition is even longer if there are several types of relativistic neutrinos in addition to the thermal electromagnetic background in the Universe, when the massive neutrino types are beginning to dominate the cosmological density.

The question of the mean free streaming of massive neutrinos is important for the evolution of density fluctuations.

When

$$z > z_\nu \tag{8.57}$$

particles are relativistic and their chaotic velocity is close to the speed of light, so the free streaming scale is close to the size of the horizon.

Neutrinos become non-relativistic at

$$z < z_\nu \tag{8.58}$$

and the free streaming scale is much smaller than the size of the horizon.

Thus, the maximum free streaming of neutrinos in an expanding Universe comes at a time

$$z \approx z_{\nu}, \tag{8.59}$$

when the particles become non-relativistic. The maximum length of the free streaming scale determines the most important parameter of the theory of gravitational instability in a gas of massive neutrinos: the scale of the cosmological structure R_c formed by massive neutrinos.

2.2. Gravitational instability of gas of massive neutrinos

Due to the isotropy and homogeneity of the unperturbed distribution of matter in the linear theory of gravitational instability it is convenient to use the Fourier representation for the fluctuations

$$\frac{\delta\rho}{\rho} = \frac{1}{(2\pi)^{3/2}} \int \delta(k,t) \cdot \exp\left(i\vec{k}\vec{r}\right) d^3k. \tag{8.60}$$

The time evolution is given by the transfer function $C(k, t, t_{\text{in}})$ relating the pertubation in the time period t to initial fluctuations in the period t_{in}. We have the relation

$$\delta\left(\vec{k},t\right) = \delta\left(\vec{k},t_{\text{in}}\right) \cdot C\left(k,t,t_{\text{in}}\right). \tag{8.61}$$

Initial density fluctuations $\delta(\vec{k},t_{\text{in}})$ are generated at the inflationary stage (Chapter 3) or by some other mechanism related to the physics of the very early Universe.

It is usually assumed that the phases are not correlated and the initial (stochastic) fluctuations of the density are determined by the spectrum

$$b^2(k) = \left\langle \delta^2\left(k,t_{\text{in}}\right) \right\rangle. \tag{8.62}$$

The widely used 'flat' spectrum (Zeldovich, 1972; Harrison, 1967), predicted in the simplest form of inflationary models, has the form

$$b^2(k) \propto k. \tag{8.63}$$

Consider the problem of the evolution of density fluctuations of massive neutrinos. Regardless of the physical mechanism of their origin, we assume that fluctuations of the metric, the density and velocity take place in the early Universe, and we consider their evolution.

Ultra-relativistic particles and radiation dominate in the RD stage of the evolution of the Big Bang Universe. At this stage both the evolution of the Universe as a whole and the development of density fluctuations are analyzed within a hydrodynamic model of the Universe with the equation of state of a relativistic fluid $p = \varepsilon/3$.

This model was studied by Lifshitz in 1946. The main results of his analysis were the following (see review by Shandarin et al 1983).

1) the pressure does not affect the evolution of fluctuations on large scales. At these scales, there is only one mode, growing with time, and another mode which eventually decays;

2) Gravity plays no role at small scales, and the density fluctuations are transformed into sound waves. The amplitude of these waves decreases slowly as the Universe expands in accordance with the theory of adiabatic invariants;

3) The boundary between large and small scales is the Jeans scale, corresponding to the distance traveled by the sound wave in an expanding Universe during time t

$$R_J = \frac{2ct}{\sqrt{3}}. \tag{8.64}$$

At a time when the temperature of radiation was much larger than the neutrino mass, neutrinos were relativistic, and this pattern remained almost unchanged. Of course, the gas of collisionless neutrinos can not be treated as a fluid with isotropic pressure, but it turned out (Khlopov 1999) that this distinction is not essential for the development of long-wave fluctuations.

Thus, in the early period, when the neutrinos are ultra-relativistic, the Jeans mass M_J in neutrino fluctuations was of the order of the masses under the cosmological horizon M_h

$$M_J \sim M_h = m_{\text{Pl}} \frac{t}{t_{\text{Pl}}}. \tag{8.65}$$

The Jeans mass increases linearly with time, and is the boundary between region I (Fig. 8.1), which corresponds to the growth of fluctuations in a gas of neutrinos, and the region II, in which there are no sound waves at small scales, and the fluctuations are damped due to free streaming. The boundary between the large and small scales is still close to M_h.

Damping of the neutrino density fluctuations (region II in Figure 8.1) is due to the free streaming of the weakly interacting neutrinos from areas with high density to areas with low density.

A detailed analysis of this problem is possible only on the basis of numerical calculations (Bond, Szalay 1982), but the asymptotic behavior for large k (small wavelengths)

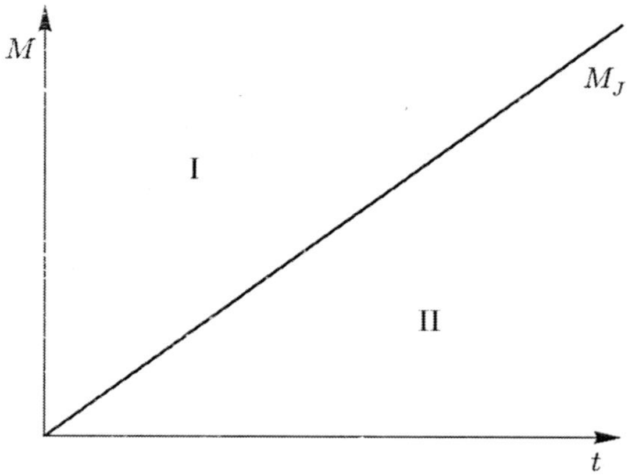

Fig. 8.1. Evolution of fluctuations of relativistic neutrinos.

$$kR_\nu \gg 1 \tag{8.66}$$

can be found analytically (Doroshkevich et al 1980a).

The distribution of neutrino momenta is given by the function of the Fermi–Dirac distribution (8.28) with zero chemical potential. Density fluctuations are associated with fluctuations of the distribution function. Adiabatic fluctuations correspond to the inhomogeneity of the density and average energy of neutrinos. The motion of the neutrino smoothes these inhomogeneities on the scale corresponding to the free streaming of the neutrino, so that fluctuations at the corresponding wavelengths are damped.

The problem of wave damping in a collisionless gas was considered in detail in the Big Bang Universe for the cases of plasma and neutral gas (Zeldovich and Novikov 1975).

The difference between these cases and the case of massive neutrinos because of the difference of the spectrum of neutrinos from the Maxwell distribution leads to a difference in the damping law.

Fluctuations in a gas of neutrinos with a wavelength much smaller than R_ν are damped mainly during the period t_1, z_1 when the neutrinos become non-relativistic (see below).

The condition

$$\lambda < R_\nu \tag{8.67}$$

in this period means that gravity can be ignored. Consequently, the problem is reduced to the trivial case of particle motion with constant velocity.

If the density in phase space is given by the function $n(\vec{p},\vec{x})$ at time t, then we obtain

$$n = n\left(\vec{p}, \vec{x} - \vec{v}\cdot(t - t_1)\right) \tag{8.68}$$

at an arbitrary moment t.

If the initial density in the phase space is given by

$$n = n_0(p) + \delta n_0(p)\cdot\exp\left(i\vec{k}\vec{x}\right) \tag{8.69}$$

when $t = 0$, in this case we obtain for the density fluctuations in the phase space

$$\delta n = \delta n_0(p)\cdot\exp\left\{i\vec{k}(\vec{x} - \vec{v}t)\right\}, \tag{8.70}$$

and density fluctuations are given by

$$\delta\rho(\vec{x},t) = \int \delta n(\vec{p},\vec{x},t)\,d^3p = \int \delta n_0(\vec{p})\cdot\exp\left(-i\vec{k}\vec{x}\right)d^3k. \tag{8.71}$$

Speed $\vec{v}$ should be viewed as a function of momentum $\vec{p}$

$$\vec{v} = \frac{c\vec{p}}{E} = c^2\vec{p}\left(\left(m_\nu c^2\right)^2 + p^2c^2\right)^{-1/2}. \tag{8.72}$$

In the case of an expanding Universe the product $\vec{k}\vec{v}t$ must be replaced by an integral

$$J = \int \vec{k}\frac{d\vec{x}}{dt}dt, \tag{8.73}$$

where $\vec{k}$ and $\vec{x}$ are expressed in terms of the coordinates in the companion volume

$$d\vec{l} = a\cdot d\vec{x}, \tag{8.74}$$

where a is the cosmological scale factor, so that

$$\frac{d\vec{x}}{dt} = \frac{\vec{v}}{a}. \tag{8.75}$$

Moreover, the momentum of free (test) particles decreases in inverse proportion to the scale factor a:

$$\vec{p} = \vec{p}_0\frac{a_0}{a}. \tag{8.76}$$

Therefore, the integral finally has the form

$$J = \text{const} \cdot \int k_\nu \frac{c^2 p_0 a_0}{a^2 \left(p_0 c^2 \frac{a_0^2}{a^2} + m_\nu^2 c^4 \right)^{1/2}} dt. \tag{8.77}$$

The relativistic phase integral can be reduced to

$$J = \text{const} \cdot \int k_\nu \frac{c}{a} dt, \tag{8.78}$$

and it converges at

$$t \to 0$$

because

$$a = a(t) \propto t^{1/2}.$$

In the non-relativistic stage we get

$$J = \text{const} \cdot \int k_\nu \frac{p_0 a_0}{a^2 m_\nu} dt. \tag{8.79}$$

Since

$$a = a(t) \propto t^{2/3}$$

at this stage, the integral converges at

$$t \to \infty.$$

Both estimates show that the integral must reach a maximum just in the transition from one stage to another.

Evaluation of the maximum value of the order of magnitude gives

$$J = k_\nu \frac{ct_1}{a(t_1)} \frac{p_0}{\langle p_0 \rangle} = \frac{k_\nu}{k_{\nu\,\max}} \frac{p_0}{\langle p_0 \rangle}, \tag{8.80}$$

where $k_{\nu\,\max}$ is the maximum wave vector corresponding to the minimum wavelength R_ν introduced above, p_0 is the momentum of the particles in an early ultra-relativistic stage $t_0 \ll t_1$, $\langle p_0 \rangle$ is the neutrino momentum averaged over the spectrum on the same stage, so that the ratio $p_0 / \langle p_0 \rangle$ does not depend on time. We have already used $\langle p_0 \rangle$ in calculations of t_1 and k_ν.

The law of damping of fluctuations depends on the type of function $\delta n_0(p)$.

If the fluctuations in the non-relativistic gas have the Maxwellian shape

$$\delta n \propto \exp\left\{-\frac{v^2}{\theta}\right\}, \tag{8.81}$$

then we obtain for a fixed time interval

$$\int \exp\left\{-\frac{v^2}{\theta} - ikvt\right\} dv \approx \exp\left\{-\frac{k^2 t^2 \theta^2}{4}\right\}. \tag{8.82}$$

But if the fluctuation includes particles at rest, so that

$$\delta n(\vec{p}) = \text{const} \cdot \delta(\vec{p}), \tag{8.83}$$

and such fluctuations are not damped.

Using the same arguments in the case of the invisible axion, we can easily find an explanation for the spectrum of cold dark matter for the primordial axions. Being coherent oscillations of the axion field, the fluctuations of the axion density correspond to the distribution of the particles of the type

$$\delta n \propto \delta(\vec{v}) f(\vec{x}), \tag{8.84}$$

for which there is no damping of short-wave fluctuations.

To derive the damping law, we must clarify the nature of the initial fluctuations that occurred at the stage when the wavelength was larger than the horizon. This issue is related to the mechanism of generation of density fluctuations which in the case of inflationary models is related to the parameters of the scalar field (Chapter 3).

In the case of inflationary models the adiabatic density fluctuations arise due to small deviations in the rate of expansion in different spatial regions which leads to a difference of temperature of these regions and fluctuations of the form

$$\delta n \propto \frac{\partial n}{\partial T} \exp\left(i\vec{k}\vec{x}\right). \tag{8.85}$$

To substantiate this claim, we consider (Doroshkevich et al 1980a) one-dimensional calculations of fluctuations in a relativistic gas, caused by gravity.

Consider the initial distribution of phase space density

$$n = n(p) = n\left(\frac{cp}{T}\right). \tag{8.86}$$

The effect of metric perturbations can be considered as an effect of the gravitational field. A change in the momentum of a relativistic particle being in a given metric or in a given field is proportional to the momentum. Therefore,

$$\frac{\partial n}{\partial t} = f(x)\, p \frac{\partial n\left(\frac{cp}{T}\right)}{\partial p}, \tag{8.87}$$

where

$$f(x) \propto \exp(ikx). \tag{8.88}$$

It is important to fulfill the following relation

$$p\frac{\partial n\left(\frac{cp}{T}\right)}{\partial p} = -T\frac{\partial n\left(\frac{cp}{T}\right)}{\partial T}. \tag{8.89}$$

If

$$\delta n \propto \frac{\partial n}{\partial T}, \tag{8.90}$$

we obtain the desired formula

$$n = n_0 + \delta n = n\left(p, T + \delta T \cdot \exp\left(i\vec{k}\vec{x}\right)\right). \tag{8.91}$$

The same is approximately true in the case of three spatial dimensions.

To evaluate the effect of damping, we consider

$$\delta\rho = \exp\{ikx\} \int \frac{d\left(\exp\left\{\frac{cp}{T}\right\} + 1\right)^{-1}}{dT} \exp\left(\frac{i\vec{k}\vec{p}R_v}{\langle p \rangle}\right) d^3p. \tag{8.92}$$

We obtain the decomposition

$$\left(\exp\left\{\frac{cp}{T}\right\}+1\right)^{-1}=\exp\left\{-\frac{cp}{T}\right\}-\exp\left\{-2\frac{cp}{T}\right\}+\dots, \tag{8.93}$$

where the first term dominates.

Integrating (8.92) under the assumption that

$$R_\nu k \gg 1, \tag{8.94}$$

we get

$$\left(\frac{\delta\rho}{\rho}\right)_\nu\Bigg|_{t\gg t_\nu}=\left(\frac{\delta\rho}{\rho}\right)_\nu\Bigg|_{\frac{kct}{a}=1}\cdot\left(\frac{k_{\max}}{k}\right)^4. \tag{8.95}$$

To find the predictions for the shortwave amplitude at large times

$$t\gg t_\nu, \tag{8.96}$$

we need to take into account that the shorter the wavelength, the later the growth of fluctuations starts at a given wavelength in the non-relativistic stage at t_1

$$t>t_1>t_\nu, \tag{8.97}$$

and the smaller the increase of this fluctuation in time t. This gives an additional suppression factor

$$(kR_\nu)^{-2}=\left(\frac{\lambda}{\lambda_J}\right)^2 \tag{8.98}$$

in the amplitude of the fluctuations.

On the other hand, only fluctuations with a wavelength

$$\lambda>ct \tag{8.99}$$

grow in the relativistic phase $t<t_\nu$, giving another suppression factor

$$\left(\frac{\delta\rho}{\rho}\right)_\nu\propto(kR_\nu)^{-2}. \tag{8.100}$$

Finally, the approximate law for the suppression of short-wavelength fluctuations for the wave numbers (8.94) is given by

$$\left(\frac{\delta\rho}{\rho}\right)_\nu=\frac{\delta\rho}{\rho}(R_\nu)\cdot(kR_\nu)^{-8} \tag{8.101}$$

so that the approximate formula for changing the initial amplitude of density fluctuations due to the damping can be interpolated as

$$\left(\frac{\delta\rho}{\rho}\right)_f = \left(\frac{\delta\rho}{\rho}\right)_{\text{in}} \left(1+k^2R_\nu^2\right)^{-4}. \tag{8.102}$$

Evolution of the density fluctuations drastically changes its character at a time when the neutrinos become non-relativistic.

This occurs when

$$t \sim t_\nu \tag{8.103}$$

at

$$T \sim T_\nu \sim 10^5\,\text{K}, \tag{8.104}$$

if the neutrino mass $m_\nu \sim 30$ eV.

It should be noted that the evolution of the Universe as a whole is determined only by the total density of neutrinos, so that option 1) when the mass of one type of neutrino is, for example, of the order of

$$m_\nu \sim 30\ \text{eV}, \tag{8.105}$$

and the mass of other types of neutrinos is small, and option 2) when the masses of all types of neutrinos are

$$m_1 \sim m_2 \sim m_3 \sim 10\ \text{eV}, \tag{8.106}$$

are equal.

But these two options differ in terms of the evolution of density fluctuations. Here, the heaviest neutrino plays a dominant role. It becomes non-relativistic before the others, thus causing a change in the evolution of fluctuations. The contribution of non-relativistic particles to cosmological density starts to dominate soon since the energy density of these particles decreases as they expand slower than the energy density of relativistic particles.

Evolution of relativistic particles is well described by the hydrodynamic model with the equation of state $p = 0$.

Changing the equation of state leads, on the one hand, to a change in the rate of cosmological expansion

$$a(t) \propto t^{2/3}, \tag{8.107}$$

and, on the other hand, to a change in the law of evolution of density fluctuations.

2.3. Formation of structure in the neutrino-dominated Universe

In the period prior to $t \sim t_\nu$, the evolutions of fluctuations in the neutrino, photon and baryon components were similar so that by the time (8.103) the amplitudes of all fluctuations in these components were the same.

However, after $t > t_\nu$, the evolution of fluctuations in these three components started to differ.

At large scale fluctuations in a gas of the neutrinos continue to grow, and at small scales they continue to decay, but the boundary is shifted toward smaller scales. All the fluctuations on the scales exceeding the Jeans scale at $t \sim t_\nu$ continue to grow

$$\frac{\delta\rho}{\rho} \propto (1+z)^{-1}. \tag{8.108}$$

The Jeans scale, corresponding to the period $t \sim t_\nu$ plays an important role in the problem of formation of the large-scale structure of the Universe, defining its many features (Szalay, Marx, 1976).

This scale can be expressed through the mass of the perturbed region, defined by neutrino mass and the Planck mass, which represent a combination of dimensional fundamental constants (Markov 1964; Szalay, Marx 1976, Bisnovatyi-Kogan, Novikov 1980)

$$M = \rho_{\text{tot}} R_\nu^3 = m_{\text{Pl}} \left(\frac{m_{\text{Pl}}}{m_\nu} \right)^2 . \tag{8.109}$$

More precisely, the appropriate length scale is expressed in terms of the value R_ν

$$R_\nu = 12.5(1+z)^{-1} \left(\frac{30 \text{ eV}}{m_\nu} \right) \text{Mpc} \tag{8.110}$$

(Doroshkevich, Khlopov 1981).

When $t > t_\nu$ the fluctuations in the photon and baryon components grow with the neutrino fluctuations (region III in Fig. 8.2) at scales greater than the corresponding Jeans length

$$R_J = \frac{2ct}{\sqrt{3}} \tag{8.111}$$

that is, when

$$M > M_J. \tag{8.112}$$

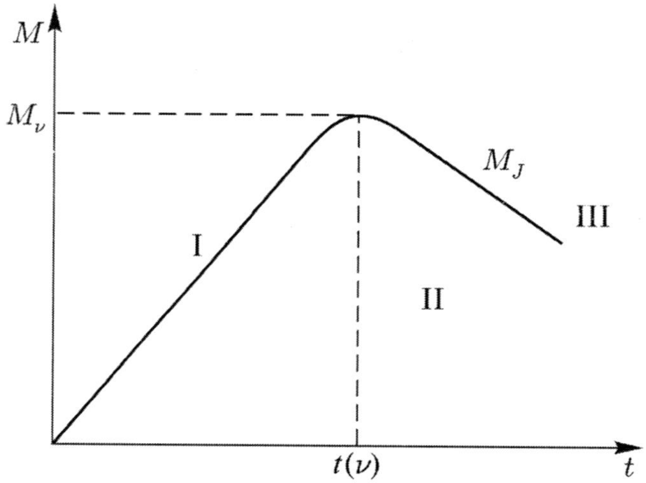

Fig. 8.2. Evolution of fluctuations of non-relativistic neutrinos.

At smaller scales, radiation pressure prevents the growth of fluctuations and, as before, the perturbations in the photon and baryon components are transformed into sound waves with constant amplitude (Bisnovatyi-Kogan et al 1980; Doroshkevich, Khlopov 1981).

Only after the recombination of hydrogen at

$$T < T_{\text{rec}} \cong 4.5 \cdot 10^3\,\text{K} \tag{8.113}$$

baryons cease to interact with radiation, and fluctuations in the baryonic component grow on all scales exceeding

$$M_b \approx 10^5\, M_\odot. \tag{8.114}$$

Value M_b corresponds to the Jeans mass, calculated for the internal pressure of the baryons.

However, the density fluctuations in the baryon and photon components must have in a period of recombination $t = t_{\text{rec}}$ on the scale

$$R_\nu \leqslant R \leqslant R_J\left(t_{\text{rec}}\right) \tag{8.115}$$

the amplitude much smaller than the amplitude of the fluctuations in the neutrinos in the same scale.

On a large scale

$$R > R_J\left(t_{\text{rec}}\right) \tag{8.116}$$

fluctuations in all three components grow at the same time and remain equal.

The difference of neutrino and baryon-and-photon fluctuations ensures that the fluctuations of the temperature of cosmic microwave background radiation in the neutrino-dominated Universe are small.

Freed from the influence of radiation pressure, the amplitude of the baryon fluctuations quickly reaches after recombination the amplitude of neutrino fluctuations so that a single mode of density fluctuations forms (Fig. 8.3).

The amplitude of fluctuations in the neutrinos and baryons do not match up to some time after recombination. But as was shown in (Doroshkevich et al 1980a), the differences between them disappear after recombination for several cosmological times.

After recombination of hydrogen, the radiation ceases to interact with matter and begins to pass almost freely through it, so that small-scale fluctuations of the temperature of the thermal background radiation transfer as if a snapshot of density and velocity fluctuations in the photon and baryon components during recombination (Silk, 1968; Sunyaev, Zeldovich, 1970; Doroshkevich, Khlopov, 1981). Thus, the smallness of the amplitude of the fluctuations of the CMB reflects the small amplitude of the fluctuations in the baryon and photon components.

Measurements of small-scale anisotropy of the cosmic microwave background radiation, carried out in recent years, confirm the existence of the so-called *Sakharov oscillations* in the diffraction structure of this anisotropy associated with the Doppler effect of the fluctuation of the velocity of matter during the recombination. The position of the first Doppler peak can thus determine the total density of the Universe,

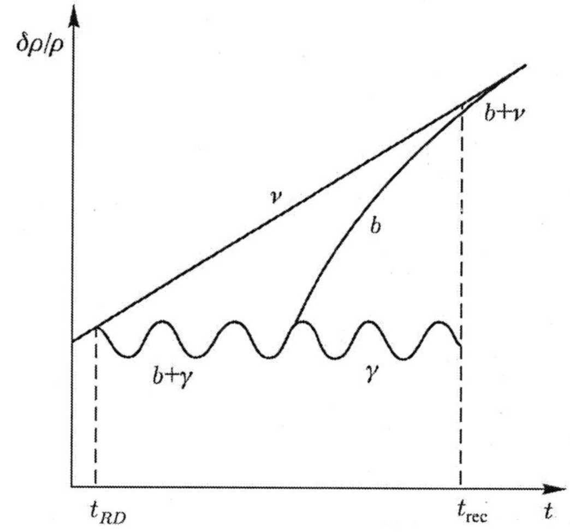

Fig. 8.3. Growth of the baryon density fluctuations in the neutrino-dominated Universe.

while the details of the diffraction structure refine other cosmological parameters, and, above all, the cosmological density of baryons. The results of the experiments and, in particular, BOOMERANG and WMAP experiments, lead to the total density equal to the critical, while the baryon density is no more than 0.05 times critical.

In cosmological models without non-baryonic dark matter (e.g., with massless neutrinos), the same fluctuations determine the amplitude of temperature fluctuations in the CMB and the formation of the large-scale structure of the Universe. In the neutrino-dominated Universe, the formation of this structure is associated with neutrino fluctuations, which begin to grow earlier and during recombination may become much larger than the fluctuations in the photon and baryon components at the same time, reflected in the small-scale CMB anisotropy.

Due to the presence of density fluctuations of massive neutrinos, even very small density fluctuations of matter can grow and form the structure of inhomogeneities.

Let us describe qualitatively the non-linear stage of evolution of fluctuations in the neutrino-dominated Universe.

As noted earlier, the gravitational interaction of massive neutrinos with matter leads to the unification of fluctuations with very high accuracy.

According to the non-linear theory of gravitational instability, the evolution of these fluctuations leads to the formation of 'pancakes', i.e., flat areas with high density and size which is determined by the scale of the damping of neutrino density fluctuations. Correlation of the phase of fluctuations in the neutrino and matter leads, in turn, to the simultaneous contraction of the baryonic matter. But the fate of these two components differs.

The compression of the gas component is stopped and the component is heated by the shock wave surrounding the compressed matter. This leads to the formation of a gas 'pancake' (Sunyaev and Zeldovich 1972; Doroshkevich et al 1980b).

The neutrino generates a collision-free 'pancake' (Doroshkevich et al 1980b, c) with the width 2–3 times greater that that of the gas 'pancake'. Therefore, the ratio of the densities of compressed matter and neutrinos is 2–3 times more than the average value in space.

Calculations of the photoionization of the intergalactic gas by the radiation of 'pancakes' (Doroshkevich et al 1978; Doroshkevich, Khlopov 1981) showed that in the absence of non-baryonic dark matter the number density of neutral hydrogen atoms can not be lower than

$$n_{\mathrm{HI}} \sim 3 \cdot 10^{-11} \ \mathrm{cm}^{-3} \qquad (8.117)$$

when $\Omega = 0.1$, while the observed spectra of distant quasars (Gunn, Peterson 1965) indicated that

$$n_{\text{H1}} < 10^{11}\ \text{cm}^{-3} \tag{8.118}$$

or even

$$n_{\text{H1}} < 10^{-12}\ \text{cm}^{-3}. \tag{8.119}$$

9

Physical nature of dark matter of the Universe

1. The structure of the Universe as a particle detector of dark matter

1.1. Constraints on unstable particles from the conditions of formation of the structure

To identify the relationship between the observational data and the parameters of the particles of dark matter, we turn to the description of gravitational instability, used in Chapter 4, where the formation of primordial black holes (PBH) was discussed.

This description is useful to distinguish explicitly the dependence of the parameters of the cosmological structure and anisotropy of the temperature of the thermal background from the parameters of the dark matter particles.

To illustrate this dependence, we recall the approach used in Chapter 4 to the problem of metric perturbations in the very early Universe.

The flat spectrum of metric perturbations corresponds to a constant amplitude at scales greater than the cosmological horizon. The density fluctuations on these scales grow like

$$\frac{\delta\rho}{\rho} \propto t. \tag{9.1}$$

The initial spectrum of density fluctuations was given in Chapter 8 in a period of the very early Universe (e.g., after inflation) and its evolution in time is determined by the transition function.

For simplicity, let us consider a simplified case, namely, we assume a constant spectrum of metric perturbations beyond the horizon and assume that the spectrum of the metric perturbations on the scale of interest to us is flat and is defined by a small dimensionless quantity δ

$$\delta \ll 1. \tag{9.2}$$

The flat spectrum means that at any given time the value $\delta\rho/\rho$ on the scale of the cosmological horizon coincides with δ.

Such a description does not account for factors arising from the change (or changes) of the equation of state.

However, these numerical factors are irrelevant to illustrate the main features of the relationship between the parameters of the dark matter particles and the observational data.

The evolution of fluctuations on scales smaller than the horizon is determined by the properties of the particles and the law of cosmological expansion.

We restrict our considerations to the class of weakly interacting particles.

In the context of the development of gravitational instability the primordial black holes also belong to this class.

Primordial magnetic monopoles (in the absence of the primordial magnetic field) can also be attributed to this class. Because of their large mass and low density of plasma in the RD stage, the plasma drag force after the electron–positron annihilation is negligible and, disregarding possible fluctuations of the magnetic charges, the monopoles can be viewed from the perspective of the issues under consideration as weakly interacting particles.

Thus, we consider the evolution of the density fluctuations of particles with mass m, with a frozen-out relative concentration

$$r = \frac{n_m}{n_\gamma} \tag{9.3}$$

and lifetime τ.

Consider first the sufficiently long-lived particles with

$$\tau > t_U, \tag{9.4}$$

where t_U is the age of the Universe.

Assuming in accordance with the simplest inflationary model (Chapter 3) and in accordance with the results of the BOOMERANG experiments that

$$\Omega = 1$$

and that the given particle dominates in the present-day cosmological density. This condition fixes the value of

$$(rm)_0 = 6.6\,\text{eV}\left(\frac{H_0}{50\dfrac{\text{km}}{\text{s}\cdot\text{Mpc}}}\right)^2\left(\frac{2.7\text{ K}}{T_0}\right)^3, \tag{9.5}$$

where H_0 and T_0 are the present-day Hubble constant and the temperature of the thermal background radiation, respectively.

From the condition that the total density is critical, we find that any form of stable particles must satisfy

$$(rm) < (rm)_0. \tag{9.6}$$

This condition can be weakened by a factor 2÷3 in the case of an upper limit on the total density which follows from the lower estimate of the age of the Universe.

In any case, the particles for which it is predicted that

$$(rm) \gg (rm)_0, \tag{9.7}$$

must be unstable and their lifetime must be less than the age of the Universe

$$\tau < t_U. \tag{9.8}$$

When

$$t < \tau \tag{9.9}$$

and

$$\tau > \frac{m_{\text{Pl}}}{(rm)^2}. \tag{9.10}$$

such particles dominate in the Universe.

After the decay of unstable particles, the relativistic products of their decay begin to dominate and experience the red shift in the stage of their dominance, so that when

$$t = t_U \tag{9.11}$$

the effective value *rm*, which is a measure of their contribution to the total density, does not exceed

$$(rm)_{\text{eff}} = rm\left(\frac{\tau}{t_U}\right)^{1/2}. \tag{9.12}$$

Thus, the upper limit on the total cosmological density does not exclude the unstable particles with

$$(rm) \gg (rm)_0$$

and lifetime

$$\tau < t_U\left(\frac{(rm)_0}{(rm)}\right)^2. \tag{9.13}$$

We now strengthen these limits by analyzing the effects of stable and unstable weakly interacting particles on the evolution of density fluctuations.

At a temperature

$$T \gg m$$

the particles with mass m are relativistic and density fluctuations decay by free streaming within the cosmological horizon.

At a time when

$$T \sim m$$

the particles become non-relativistic. The size of the horizon in this period is

$$R_m \sim \frac{m_{\text{Pl}}}{T^2} \sim \frac{m_{\text{Pl}}}{m^2}. \tag{9.14}$$

In today's Universe, this scale is

$$R_m^{\text{mod}} = R_m\left(t = t_U\right) = R_m \cdot \left(1 + z_m\right) = R_m \cdot \frac{m}{T_0} = \frac{m_{\text{Pl}}}{mT_0}, \tag{9.15}$$

where T_0 is the present-day temperature of the thermal electromagnetic background and z_m is the redshift corresponding to the period of

$$T \sim m.$$

Scale R_m determines the minimum scale of the inhomogeneities formed by the investigated particles.

Due to the weak (logarithmic or weak power-law) growth of inhomogeneities of the non-relativistic weakly interacting particles in a period when the temperature drops with the expansion from the value

$$T \sim m$$

to the value

$$T \sim rm \ll m \tag{9.16}$$

the density fluctuations on this scale grow slightly so that it is this scale which should determine the scale of the structure of inhomogeneities formed by the weakly interacting particles with mass m.

Thus, as was shown in Chapter 8 for the neutrino, there is an unambiguous relationship between the mass of weakly interacting particles and the scale of the cosmological structure created by these particles. We have

$$R_m \propto \frac{1}{m}. \tag{9.17}$$

On the basis of the adiabatic theory, the scale of the observed large-scale structure of the Universe may be related to the parameters of the spectrum of initial fluctuations, and we can determine the correlation length of fluctuations (Shandarin et al 1983).

The correlation function of the relative density fluctuations can be taken in the form of

$$\xi(r_{12}) = \frac{\langle(\rho(\vec{r}_1) - \langle\rho\rangle)\cdot(\rho(\vec{r}_2) - \langle\rho\rangle)\rangle}{\langle\rho\rangle^2} = \sigma^2\left(1 - \frac{r_{12}^2}{2r_c^2} + \ldots\right), \tag{9.18}$$

where $\langle\rho\rangle$ is the average cosmological density, σ is the dispersion of the fluctuations and r_c is the correlation length. On the basis of the observational data, the estimates of the correlation length (Doroshkevich 1983a, b) give

$$r_c \sim 10 \div 30 \text{ Mpc}. \tag{9.19}$$

Comparison of the values r_c and R_m^{mod} shows, taking into account the equality

$$R_m^{\text{mod}}(m) = r_c, \tag{9.20}$$

that the possible values of the mass m are in the interval (Doroshkevich, Khlopov 1984)

$$m = 40 \div 120 \text{ eV}. \tag{9.21}$$

Therefore, the scale of the observed large-scale structure of the Universe makes it possible, in principle, to determine the mass of weakly interacting particles that form the structure.

However, it is worth noting that all particles with fixed values rm also have the characteristic scale in the spectrum of their fluctuations, corresponding to the horizon at the moment t_m when the particles began to dominate the Universe. This scale is given by

$$R'^{\text{mod}} = \frac{m_{\text{Pl}}}{(rm)T_0}. \tag{9.22}$$

For any form of stable nonrelativistic particles that dominate in the dark matter of the present-day Universe this scale corresponds to a valid value r_c.

As discussed below, the presence of this characteristic scale can be regarded as evidence that the weakly interacting particles of any type may form some semblance of the observed structure (Melott et al1983; see the review in the book Khlopov 1999). This possibility can not be excluded on the basis of existing observational data. Moreover, it is this possibility that is regarded as the standard model of cold dark matter (CDM) of forming the large-scale structure of the Universe.

Our simplified approach allows us to relate the value δ with the predicted anisotropy of the cosmic microwave background radiation.

The value δ determines the amplitude of the fluctuations of density and velocity in the plasma, including fluctuations in the CMB temperature during the recombination period, so that there is a relation

$$\frac{\delta T}{T} = \gamma\delta < \left(\frac{\delta T}{T}\right)_{\max}^{\text{obs}}, \tag{9.23}$$

where the numerical factor

$$\gamma \sim 0.1 \div 1$$

is determined by the details of the processes that formed the fluctuations of the CMB temperature, as well as the conditions under which these processes take place.

On the other hand, the value δ determines the characteristic time of structure formation. Growth of fluctuations starting from the time t_m follows the law

$$\frac{\delta\rho}{\rho} \propto t^{2/3}.$$

The existence of the observed structure of the Universe corresponds to the growth of fluctuations to

$$\frac{\delta\rho}{\rho} \sim 1$$

and requires that

$$t_m \cdot \delta^{-3/2} < t_U. \tag{9.24}$$

We now consider the constraints that follow from the existence of the large-scale structure of the Universe, on the possible parameters of hypothetical weakly interacting massive metastable particles with lifetimes

$$\tau < t_U. \tag{9.25}$$

We assume that the particles decay only to weakly interacting particles, so that the constraints contained in the preceding chapters are not applicable.

If the value *rm* of these particles is much higher than the value (rm_0) given by equation (9.5), the presence of the relativistic decay products of the particles should lead to a strong slowing down of growth of fluctuations in the period determined by the values *rm* and τ.

Based, on the one hand, on the fact that the large-scale structure is formed and that, on the other hand, the thermal background is isotropic on the observed level, one can impose constraints on *rm* and τ, excluding the effect of inhibiting the growth of fluctuations, contrary to the relationship of these facts.

There are two possibilities.

In the first case, unstable particles decay before the non-relativistic particles with

$$rm = (rm)_0 \tag{9.26}$$

begin to dominate the Universe, that is, when

$$t_m > \tau. \tag{9.27}$$

This means that the considered unstable particles do not survive until the period of formation of the large-scale structure starts, and there should be more long-lived particles that form the large-scale structure.

For definiteness, we assume that particles of dark matter forming the structure, are stable with

$$rm = (rm)_0 .$$

Then, if the value rm of the unstable particles exceeds

$$rm = (rm)_0 \cdot \left(\frac{t_m}{\tau} \right)^{1/2}, \tag{9.28}$$

the stable particles with

$$rm = (rm)_0$$

do not dominate in the period

$$t = t_m,$$

when their density overtakes the density of the thermal background and primordial massless neutrinos, but start to dominate later, when their energy density is greater than the energy density of relativistic decay products of unstable particles, i.e., at the time t_1 determined from the equation

$$rm = (rm)_0 \cdot \left(\frac{\tau}{t_1} \right)^{1/2}. \tag{9.29}$$

In equation (9.24) we replace t_m by t_1 defined by the expression

$$t_1 = \left(\frac{rm}{(rm)_0} \right)^2 \tau > t_m. \tag{9.30}$$

The system of inequalities (9.23)–(9.24) leads to the constraint on the values of rm and τ

$$rm < (rm)_0 \left(\frac{t_U}{\tau} \right)^{1/2} \left[\frac{1}{\gamma} \left(\frac{\delta T}{T} \right)^{\text{obs}}_{\max} \right]^{3/4}. \tag{9.31}$$

This constraint is three orders of magnitude more stringent than the corresponding constraint resulting from lower estimates of the age of the Universe.

Now consider the second case

$$\tau > t_m. \tag{9.32}$$

In this case, the period of growth of fluctuations increases by factor $rm/(rm)_0$ if

$$rm > (rm)_0.$$

The growth of the characteristic time of evolution of the fluctuations follows from the following argument. Relativistic decay products of unstable particles should experience a red shift, which makes their energy density $rm/(rm)_0$ times lower so that their contribution to the total cosmological density is compatible with an upper limit on it and permits the stage of dominance in the Universe of stable particles with

$$rm = (rm)_0,$$

in which the effective growth of fluctuations takes place.

The existence of the large-scale structure is crucial in deriving the constraints on rm and τ in this case.

The dominance of the unstable particles before their decay at

$$t < \tau$$

begins in the Universe with a period

$$t'_m = \frac{m_{\text{Pl}}}{(rm)^2} < t_m = \frac{m_{\text{Pl}}}{(rm)_0^2} < \tau, \tag{9.33}$$

thus providing efficient growth of fluctuations on the scale of the minimum undamped scale of the fluctuations

$$R_{\min}^{\text{mod}} = \frac{m_{\text{Pl}}}{mT_0} \tag{9.34}$$

to the scale corresponding to the horizon at a time when the particles begin to dominate in the Universe

$$R_{\min}^{\text{mod}} = \frac{m_{\text{Pl}}}{(rm)T_0}. \tag{9.35}$$

The growth of fluctuations on this scale can ensure the formation of galaxies which apparently cannot be prevented by the intermediate RD stage of dominance of the relativistic decay products of unstable particles in the Universe.

However, the formation of the large-scale structure means that fluctuations on the scale of the structure grow to

$$\frac{\delta\rho}{\rho} \sim 1$$

and such growth can begin within the cosmological horizon just after the fluctuation at these scales will come 'out of' the horizon with the initial amplitude δ, i.e., after

$$t > \frac{r_c}{c \cdot (1 + z(t))}, \tag{9.36}$$

where r_c is the correlation radius of the structure, c is the speed of light and $z(t)$ is the red shift corresponding to the period when the scale of the fluctuations is comparable with the size of the horizon.

The growth of these large-scale fluctuations is suppressed at an intermediate RD stage of dominance in the Universe of relativistic decay products of unstable particles. The quite lengthy RD stage prevents the formation of the large-scale structure due to upper limits on the amplitude of δ following from the observed isotropy of the thermal electromagnetic background.

Thus, it is necessary to limit the possible duration of the intermediate RD stage, which leads to the following constraints on rm and τ

$$rm < \frac{t_U}{t_m} \delta^{3/2} (rm)_0 = (rm)_0 \frac{t_U}{t_m} \left[\frac{1}{\gamma} \left(\frac{\delta T}{T} \right)^{obs}_{max} \right]^{3/2}, \tag{9.37}$$

$$t_m < \tau < t_U. \tag{9.38}$$

The constraints (9.27) (9.31) and (9.37)–(9.38) are given in Fig. 9.1. These constraints do not involve detailed information about the properties of hypothetical particles. Only the possible contribution to the cosmological density of the investigated particles or their decay products is used.

In this sense, these constraints are as universal as constraints on weakly interacting particles resulting from constraints on the abundance of the primordial ^{4}He, on the age of the Universe or the upper limits on the concentration of PBHs in the Universe.

1.2. *Difficulties of simple models of dark matter*

Problems of the model of the neutrino-dominated Universe emerged after a detailed quantitative analysis of prediction of the simple model of dominance in the Universe of stable neutrinos with masses

$$m_\nu \sim 30 \text{ eV}.$$

On the one hand, this model predicts a clear cellular structure (Frank

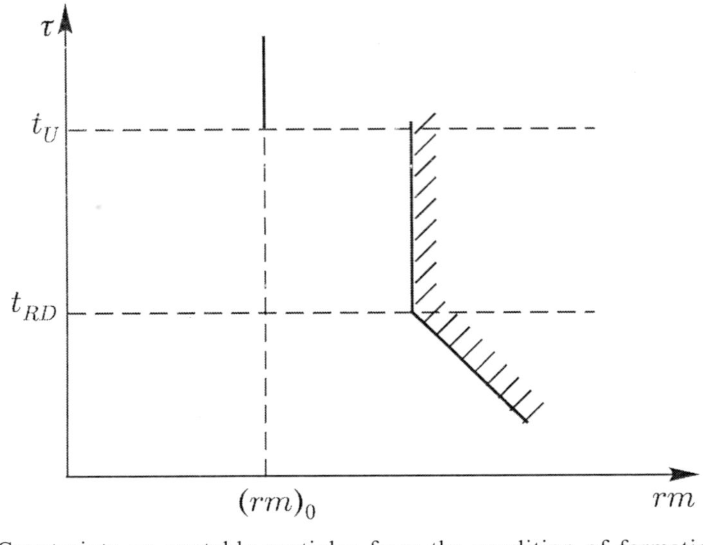

Fig. 9.1. Constraints on unstable particles from the condition of formation of the large-scale structure.

et al 1983; White et al 1983) which has been well confirmed both qualitatively (see review Shandarin et al 1983) and by a number of quantitative parameters. These parameters include:

I) evaluation of the contrast of structures (Shandarin et al 1983);

II) the results of cluster analysis and analysis of the methods of percolation theory (Shandarin et al 1983);

III) comparison of the scales predicted by the theory of scales with observational data (Doroshkevich 1983a, b);

IV) some features of the correlation relations (Shandarin et al 1983).

The predicted values of small-scale fluctuations in the CMB temperature

$$\frac{\delta T}{T} \approx (2 \div 3) \cdot 10^{-5} \qquad (9.39)$$

did not contradict the observational data at small angular scales that correspond to fluctuations on the scale of superclusters of galaxies (Shandarin et al 1983; Starobinsky 1983)

$$\frac{\delta T}{T} \approx (4 \div 6) \cdot 10^{-5}. \qquad (9.40)$$

However, the quantitative comparison of numerical models with observations revealed the following problems in the model of stable massive neutrinos (Frank et al 1983; White et al 1983; Doroshkevich and Khlopov 1984).

First of all, the advanced cell structure with

$$\Omega = 1$$

evolved so quickly that the formation of 'pancakes' (seen in distant QSO!) at redshifts

$$z \sim 4 \div 5 \qquad (9.41)$$

was incompatible with the preservation of the developed cellular structure at

$$z \sim 0.$$

Second, similar results followed from a study of the time dependence of the correlation functions obtained by numerical calculations.

Third, it proved impossible to explain on the basis of this model the observed parameters of clusters of galaxies – their masses, the central density and the fraction of matter which they contain.

All these difficulties were associated with too dense and too massive pancakes predicted in the neutrino-dominated Universe. These pancakes evolved rapidly and turned into a system of giant clusters of galaxies.

The simplest solution to these problems would consist in moving to a smaller cosmological density, for example,

$$\Omega \sim 0.3. \tag{9.42}$$

But this solution involved a shorter period of growth of fluctuations and structure formation with large initial amplitudes of density fluctuations δ, which contradicted the observed level of temperature fluctuations in the CMB.

In addition, the solution (9.42) must comply with a very specific type of inflationary models. And, finally, it clearly contradicted the results of BOOMERANG and WMAP experiments.

Melott et al (1983), Bond et al (1982) used the models with dominance in the Universe of heavy weakly interacting particles with mass

$$m \gg 30 \text{ eV} \tag{9.43}$$

or coherent oscillations of the scalar field, which have the same type of spectrum of density fluctuations.

In such models, the spectrum of density fluctuations is characterized by two scales.

The smaller scale corresponds to the horizon at a time when the particles become non-relativistic, and the larger scale is associated with the horizon at a much later period, when the particles begin to dominate in the Universe (Shandarin et al 1983; Bond et al 1982).

In this two-scale model the 'pancakes' correspond to the small scale, but there is some correlation in their distribution at distances corresponding to the large scale. These distances, as we mentioned above, in a cosmological model with critical density are quite close to the scale of the observed large-scale structure of the Universe. Such a model can lead to the formation of galaxies (and quasars) prior to formation of the large-scale structure, eliminating some of the problems of the neutrino-dominated Universe.

However, in the simplest version of this scenario of cold dark matter it was impossible to reproduce the clear structure for

$$z \sim 0.$$

The second problem was that the existence of a large number of galaxies in the 'voids' was inevitably predicted.

And finally, it was hardly possible to compare the observed and predicted densities of dark matter in galaxies and galaxy clusters, if the 'pancakes' formed at

$$z \sim 4 \div 5.$$

Enhanced versions of the CDM have been developed to resolve all these problems, using the hypothesis of so-called biasing.

It was thought that galaxies formed when the amplitude of the fluctuations exceeded the dispersion by a specific factor of 'biasing'. This hypothesis leads to a slight biasing in the distribution of total mass and luminous matter, but its physical basis is not clear.

To conclude the discussion on the simple scenario of dark matter, we consider the problem of the distribution of dark matter and luminosity on small scales.

The existence of inhomogeneities on the scale of galaxies is an obvious advantage of CDM models. Massive dark halos of galaxies provide the observed rotation curves of galaxies, and the evidence of their existence in the effects of gravitational lensing can be naturally explained in terms of CDM models.

On the other hand, the existence of such halos, especially in the case of dwarf galaxies, as well as the very formation of galaxies at high redshifts are difficult to explain in terms of models of hot dark matter (HDM).

Indeed, the suppression of small-scale fluctuations, which is so important for the formation of the large-scale structure of the Universe in HDM models, in the neutrino-dominated Universe, in particular, raises the problem of formation of small-scale objects of dark matter inside the neutrino 'pancakes'.

Moreover, this model has its own suppression of the baryon density fluctuations at scales smaller than the scale of superclusters. We will discuss the mechanism of this suppression because of its important application in the case of multicomponent dark matter.

We assume that there is a source of density fluctuations of the baryon charge, so that there are entropy fluctuations at the small scale.

We now consider the evolution of such fluctuations in the neutrino-dominated Universe.

Linearized evolution equations for small fluctuations of the baryon density can be easily derived (Doroshkevich et al 1980a) in the Newtonian approximation.

Before recombination in the Universe containing a mixture of massive neutrinos and baryons, electrons, and radiation, these equations are

$$\ddot{\delta}_\nu + \frac{4}{3}\frac{\dot{\delta}_\nu}{t} - \frac{2}{3}\frac{\delta_\nu}{t^2} = 0 \tag{9.44}$$

$$\ddot{\delta}_m + \frac{4}{3}\frac{\dot{\delta}_m}{t} + \frac{c^2k_1^2}{3}\left(\frac{t_1}{t}\right)^{4/3} = \frac{2}{3}\frac{\delta_\nu}{t^2}, \tag{9.45}$$

where we denoted

$$\delta_\nu \equiv \left(\frac{\delta\rho}{\rho}\right)_\nu,$$

and

$$\delta_m \equiv \left(\frac{\delta\rho}{\rho}\right)_b,$$

dots indicate derivatives with respect to time, c is the speed of light and k_1 is the wave vector of fluctuations at time t_1 when the massive neutrinos become non-relativistic.

Equations were derived under the following assumptions,

I) that the non-relativistic neutrinos dominate in the Universe during the period

$$z_\gamma < z < z_1, \tag{9.46}$$

where the redshift z_γ corresponds to the period of recombination,

II) that the cosmological model is close to the model of 'flat' Universe

III) we consider the scales smaller than the horizon and the gravitational interaction of baryons and radiation is assumed to be of no importance.

From equations (9.44) and (9.45) it follows that for

$$z_r < z < z_1$$

fluctuations in the neutrinos are growing

$$\delta_\nu = A\left(\frac{t}{t_1}\right)^{2/3}, \tag{9.47}$$

and the baryon density fluctuations evolve as

$$\delta_m = A + t^{-1/3}\left(C_1 \sin\left(\sqrt{3}k_1\left(\left(\frac{t}{t_1}\right)^{1/3}\right)\right) + C_2 \cos\left(\sqrt{3}k_1\left(\left(\frac{t}{t_1}\right)^{1/3}\right)\right)\right). \quad (9.48)$$

The constants C_1 and C_2 are easy to find from the initial condition at

$$t = t_1.$$

It is seen that the oscillating components slowly decay, and because of the presence of the growing mode in the neutrino gas the constant component forms

$$\delta_m = \frac{2A}{k_1^2 t_1^2}, \quad (9.49)$$

and growing with wavelength and reaching δ_ν at approximately the Jeans scale R_J.

After recombination, baryonic matter and radiation cease to interact, and the third term on the left side of the equation (9.45) can be discarded. Assuming the initial conditions at recombination t_γ

$$\delta_m\left(t_\gamma\right) = 0 \quad (9.50)$$

and

$$\dot{\delta}_m\left(t_\gamma\right) = 0, \quad (9.51)$$

we find that the fluctuations in the neutrino gas grow according to equation (9.47)

$$\delta_\nu = A\left(\frac{t}{t_1}\right)^{2/3},$$

and the baryon density fluctuations develop as

$$\delta_m = \delta_\nu\left[1 + 2\frac{t_\gamma}{t} - 3\left(\frac{t_\gamma}{t}\right)^{2/3}\right]. \quad (9.52)$$

If

$$t = t_\gamma\left(1 + \Delta\right), \quad (9.53)$$

then

$$\delta_m = \frac{8}{3}\delta_\nu \Delta^2 \quad (9.54)$$

and

$$\delta_m = \frac{1}{2}\delta_\nu \tag{9.55}$$

only at redshifts

$$z \sim 10. \tag{9.56}$$

When

$$z > z_1 \tag{9.57}$$

the entropy fluctuations are fluctuations of the baryonic charge against the uniform background radiation and neutrinos.
When

$$z < z_1 \tag{9.58}$$

baryon density fluctuations cause fluctuations in the density of neutrinos on the scales smaller than the horizon. Growth of fluctuations of neutrinos is given by

$$\ddot{\delta}_\nu + \frac{4}{3}\frac{\dot{\delta}_\nu}{t} - \frac{2}{3}\frac{\delta_\nu}{t^2} = \frac{2}{3}\frac{\Omega_m}{\Omega_\nu}\frac{\delta_m}{t^2}, \tag{9.59}$$

which has the solution

$$\delta_\nu = \frac{\Omega_m}{\Omega_\nu}\delta_m\left[-1 + \frac{3}{5}\left(\frac{t}{t_1}\right)^{2/3} + \frac{2}{5}\frac{t}{t_1}\right]. \tag{9.60}$$

For

$$\frac{\Omega_\nu}{\Omega_m} = 150$$

the amplitudes of fluctuations δ_m and δ_ν can grow only by 8 times to the present-day period, corresponding to

$$z = 0.$$

Neutrino fluctuations on scales

$$R < R_\nu \tag{9.61}$$

do not grow due to strong dissipation. After recombination, increase of δ_m is completely determined by the value δ_ν. For

$$\delta_\nu = 0 \tag{9.62}$$

the equation

$$\ddot{\delta}_\nu + \frac{4}{3}\frac{\dot{\delta}_\nu}{t} - \frac{2}{3}\frac{\Omega_m}{\Omega_\nu}\frac{\delta_m}{t^2} = 0 \tag{9.63}$$

has a solution corresponding to a decreasing mode

$$\delta_m \propto t^{-1/3} \tag{9.64}$$

and the weakly growing mode

$$\delta_m \propto t^n, \tag{9.65}$$

where the index n is given by

$$n = \frac{1}{6}\left(\sqrt{1 + 24\frac{\Omega_m}{\Omega_\nu}} - 1\right) \approx 2\frac{\Omega_m}{\Omega_\nu} \approx \frac{1}{75} \tag{9.66}$$

at

$$\frac{\Omega_m}{\Omega_\nu} = \frac{1}{150}.$$

1.3. Arguments in favour of unstable dark matter component

All the difficulties in the model of dominance in the Universe of the stable neutrinos were associated essentially with the contradiction between the condition of structure formation which assumed

$$\Omega \approx 1$$

and the condition for preserving the already formed structure that corresponded to the lower density of matter within a developed structure

$$\Omega < 0.5. \tag{9.67}$$

This contradiction could be interpreted as evidence that during a relatively short period early in its non-linear evolution the structure has lost much of its dark matter.

Formulation of the problem thus indicated a certain way of solving it. It became clear that this solution may be due to the instability of most of the dark matter that dominated the Universe during structure formation.

In fact, the same contradiction could be formulated within the framework of the simplest inflationary models such as the contradiction between the predicted critical density

$$\Omega = 1$$

and indications that the density within the observed inhomogeneities corresponded to the relation (9.67) (Gelmini et al 1984; Steigman, 1984; Turner et al 1984; Doroshkevich, Khlopov, 1984).

Thus, this issue became a problem of dark matter of two types. There should be dark matter that is dominant in the total density and homogeneously distributed, and non-dominant dark matter inside the inhomogeneities.

In the framework of the theory of the large-scale structure the existence of two contemporary forms of dark matter (homogeneous and inhomogeneous) reflects the existence of two stages in the evolution of dark matter. In the linear stage of growth of fluctuations the entire dark matter was distributed inhomogeneously, but in the early non-linear stage of the evolution of inhomogeneities a large part of dark matter must be located outside these inhomogeneities.

In the CDM scenario, all the dark matter is actually non-uniform, but its distribution differs from the distribution of visible matter because of the biasing. The dark matter, associated with galaxies, is only a fraction of dark matter.

However, most of today's dark matter might indeed be uniform and in this case (Doroshkevich, Khlopov 1984a, b) solution of both the problem of dark matter of the two types arising in inflationary models, and of problems of formation and preservation of the structure could be found on the basis of models of unstable dark matter (UDM). Then the condition for the formation of structure (see Section 1.1 of this chapter) imposes a lower limit on the lifetime of these particles, and the condition of slow non-linear evolution of the structure imposes an upper limit on this lifetime.

Thus, we could use the theory of formation of the large-scale structure to determine the desired properties of unstable particles that form the structure.

Indeed, the scale of the observed structure (its correlation length) determines the mass of particles of dark matter. Then conditions

I) the formation of structure and

II) slow non-linear evolution of the structure

fix the lifetime of unstable particles of dark matter.

As part of the UDM scenario we can naturally solve the problem of comparing inflationary models with observations. Here, the decay products of unstable particles form present-day uniform dark matter, and the stable particles of dark matter, giving the non-dominant contribution to the total density, are hidden inside the present-day mass inhomogeneities.

Methods of cosmoparticle physics make it possible, in principle, to determine the type of unstable particles of dark matter from the set of astronomical data. We illustrate this principal possibility on the example of the old Big Bang model for the case when the fermions are the relic unstable particles.

If the particles are in equilibrium with the hot plasma in the early Universe, they decouple the sooner the weaker their interaction. The sooner the particles decouple from the rest of the particles in the plasma, the larger particles still remain in equilibrium and the lower the relative concentration of relic particles r is (Chapter 3).

If the initial concentration of particles in the Universe is not determined by their decoupling but is generated by some non-equilibrium processes, the concentration of frozen-out fermions can only be less than the equilibrium concentration of relativistic particles, taken at the time of decoupling.

The bosons can be in the coherent state, as is the case of an invisible axion or inflaton.

In the case of fermions, the maximum value rm at a fixed m corresponds to decoupling of relativistic particles from the equilibrium with hot plasma in the early Universe.

This argument follows naturally from the fact that non-equilibrium processes lead to an increase in specific entropy, and hence an increase in the number density of thermal photons.

Also, since any non-equilibrium process that occurs after decoupling of the particles also leads to an increase in specific entropy (and the number density of photons), the maximum concentration of frozen-out fermions is defined in the Big Bang model by the number of types of particles that were in equilibrium in the decoupling period (Chapter 3).

On the other hand, in models of unstable particles, their values rm may exceed $(rm)_0$ that corresponds to

$$\Omega_{DM} = 0.25$$

for the stable particles.

The higher the value rm, the earlier the non-relativistic dominance in the Universe starts and the longer is the period of growth of small initial fluctuations. For large rm the structure can be form from fluctuations with small amplitudes δ, as compared with the case of stable particles.

The ratio (9.23)

$$\frac{\delta T}{T} = \gamma\delta$$

establishes the correspondence between the amplitude δ and the amplitude of temperature fluctuations of the thermal electromagnetic background ($\delta T/T$). From this relationship on the basis of the observational data on CMB anisotropy we obtain an estimate on the value δ and, consequently, the lower limit on rm.

Looking at this lower limit on rm together with an estimation of mass m from the correlation length of the structure, we obtain a lower limit on r which can be compared with the maximum possible value of r, estimated for the decoupled particles in thermal equilibrium. Thus, the value r can in principle be fixed and thereby we can obtain some indirect information about the strength of interaction of dark matter particles.

The condition of substantial growth of large-scale fluctuations, derived in part 1.1, leads to serious difficulties in models of the unstable neutrino of the type Gelmini et al (1984) and Davis et al (1981).

For example, in the model proposed by Gelmini et al (1984) the mass of unstable neutrinos with a relative concentration

$$r = \frac{3}{11}$$

was taken to be

$$m_\nu = 550 \text{ eV}, \tag{9.68}$$

that gives the value

$$rm = 150 \text{ eV}. \tag{9.69}$$

Arguments of section 1.1 allow us to indicate that the formation of the large-scale structure in this model is virtually impossible.

A simple argument presented here can be modified for more

sophisticated cosmological scenarios and cosmological consequences of particle physics.

Extension of a simple inflation model to multi-component scenarios, taking into account a set of Higgs fields that determine the hierarchy of symmetry breaking, can lead to a very specific spectrum of density fluctuations (Linde 1984; Kofman, Linde 1987), if the phase transition, corresponding to symmetry breaking, occurred in the inflationary stage. This spectrum is determined by the parameters of the potentials of the scalar field, deviating greatly from the 'flat' Harrison–Zeldovich spectrum which has no characteristic scale.

The rather complicated 'hidden sector' of the particle theory forms in the case of particles of the shadow or mirror matter, having by no means weak interaction among themselves but ultraweak interaction with the ordinary particles. Existence of sufficiently strong internal interactions alters the ratio between the scale of the decay of fluctuations and the parameters of the particles of dark matter.

For example, in the case of mirror matter (Chapter 10) the existence of mirror photons and mirror electromagnetism leads to the same damping scale in the mirror matter as for the ordinary baryonic matter. In the same manner, the damping scale of fluctuations of mirror matter is determined by the recombination of mirror electrons and nuclei into the mirror atoms, rather than by the masses of mirror particles.

The invisible axions (Chapter 2) give another example of a non-trivial relation between the scale of damping and the mass of particles of dark matter. Despite the very small axion mass, the scale of damping of the density of axions is very small. The reason for this is that in the coherent axion field oscillations this scale is determined by multiplying the mass of the axion and the amplitude of these oscillations.

Thus, the above arguments, in general, constrain certain specific combinations of the parameters of dark matter physics.

It should be noted that data from observations of non-thermal electromagnetic background in different ranges (Chapter 2 and compare with Zeldovich and Novikov 1975; Dolgov, Zeldovich 1980), impose rather stringent constraints on the possible present-day energy density of electromagnetic radiation from the decay of dark matter. The upper limit is given by

$$\rho_\gamma < 10^{-9} \rho_{cr} \tag{9.70}$$

in X-ray and gamma bands,

$$\rho_\gamma < 10^{-6} \rho_{cr} \tag{9.71}$$

in the infrared and optical ranges and

$$\rho_\gamma < 10^{-11} \rho_{cr} \tag{9.72}$$

in the radio range.

These data lead to a constraint on the relative probability of photon decays of dark matter particles.

This constraint was discussed in connection with the possible decay of the neutrino

$$\nu_H \to \nu_L + \gamma$$

in the present-day Universe (Zeldovich, Khlopov 1981, Petkov 1977; Shabalin 1980, Aliev, Vysotsky 1981; Berezhiani et al 1987).

Rough estimates of the possible upper limits on the branching ratio of photon decays can be obtained from the upper limits given by the relations (9.70)–(9.72), multiplying them by a factor taking into account the reduction in the relative contribution of radiation to the total density due to dust expansion after the decay of dark matter. These upper limits provide additional information on the possible physical nature of dark matter particles and their interactions.

In the class of unstable weakly interacting particles the problem of the formation and evolution of the structure can find a solution that attracts weakly interacting particles with mass

$$30 < m < 100 \text{ eV} \tag{9.73}$$

and lifetime

$$10^8 < \tau < 10^9 \text{ years} \tag{9.74}$$

with the dominant decay into weakly interacting particles.

In the group of the known particles such particles could be massive neutrinos, if there were a new type of neutrino interaction which led to decay of neutrinos corresponding to the conditions (9.73) and (9.74). The simplest version of the model of structure formation by unstable dark matter – the cosmological model of the unstable neutrino was developed in (Doroshkevich, Khlopov 1984; Doroshkevich, Khlopov 1984; Doroshkevich et al 1985, 1987, 1988, 1989).

The neutrino mass (9.73), assumed in this model, was not in conflict with the results of direct experimental measurements of neutrino mass (see Hagiwara et al 2002). However, the existence of neutrino oscillations in experiments SNO (SNO collab., Ahmad et al 2002), KAMLAND (KAMLAND collab., Eguchi et al 2003) and SuperKamiokande (SuperKamiokande collab., Kajita, Totsuka, 2001) and the estimates of allowable difference of neutrino masses arising from these data, combined with the direct experimental upper limit on the mass of the electron neutrino, exclude the mass range (9.73) for the known neutrinos.

Nevertheless, the unstable dark matter can be physically justified in the models with the hierarchy of symmetry breakings, as we demonstrate in the gauge theory of broken symmetry of generations of quarks and leptons discussed in Chapter 11. This gives good reason to treat it with seriousness in the models of dominance of vacuum energy density in the present-day Universe.

The set of arguments in favour of the dominance of vacuum energy density such as the difference between the estimates of the total density and the density inside inhomogeneities, the evidence in favour of slowing down evolution of the structure, the problem of relating the total and baryon densities in rich clusters of galaxies is the same as the argument in favour of unstable dark matter. In essence, the difference in these two models is due to the fact that in the case of unstable dark matter the *absolute* slowdown of the evolution of the structure takes place after decay of unstable particles, and the model of the vacuum energy density is characterized by the *relative* slowdown of the rate of this evolution due to the acceleration of general expansion of the Universe. Arguments in favour of the present-day regime of accelerated expansion of the Universe are connected with the analysis of observational data on distant supernovae of type I, on the measured high value of the Hubble constant ($H_0 > 50$) and lower limits on the age of the Universe. This set of data confirms the present-day regime of accelerated expansion, excluding unstable dark matter as an alternative to the vacuum energy density dominance but does not exclude the possibility of its existence and the important dynamic role during the formation of the large-scale structure of the Universe.

2. Dark matter in the galaxy

2.1. Condensation of dark matter in galaxies

In the stage of formation of galaxies massive neutrinos, neutralinos, axions, and most other forms of dark matter represent the collisionless gas. They can interact with matter only gravitationally. Therefore, there was no energy dissipation due to radiation in the collisionless gas of dark matter particles, as was the case with ordinary baryonic matter.

Nevertheless, during the motion of dark matter particles in the non-stationary gravitational field of the baryonic matter shrinking due to dissipation of energy by radiation, there is an effective mechanism of energy dissipation operating also for collisionless particles. Compression of ordinary matter leads to the condensation of collisionless dark matter which leads to a significant increase in the density of dark matter in the galaxy compared to its mean cosmological density (Zeldovich et al 1980).

Let us consider the mechanism proposed in (Zeldovich et al 1980) and consistently developed by Fargion et al (1995) in more detail. For the sake of simplicity, we consider collisionless dark matter as a gas of massive neutrinos.

When ordinary matter (baryons) are compressed due to loss of energy as a result of energy dissipation, the particles of dark matter (neutrinos) move in a potential that varies with time.

Since the energy of particles moving in a time-dependent potential is not conserved, the neutrinos also lose energy and thus increase their density. Consider the simplest case of radial motion of particles whose density does not depend on the radius in a contracting system of baryons. If $\rho_\nu(t)$ and $\rho_b(t)$ are the density of neutrinos and baryons, respecitvely, neutrino motion is described by the equation

$$\frac{d^2r}{dt^2} = -\omega(t)^2 r, \tag{9.75}$$

where

$$\omega^2(t) = \frac{4\pi G}{3}\left(\rho_\nu(t) + \rho_b(t)\right). \tag{9.76}$$

Let baryons slowly increase their density. For slow changes of ω the amplitude of oscillations can be found from an adiabatic invariant

$$\frac{E}{\omega} = \omega R^2 = \text{const}, \tag{9.77}$$

where E is the energy and R is the amplitude of the oscillations.

With increasing ω the amplitude of neutrino oscillations decreases and their density increases, respectively, according to the equation

$$\frac{\rho_\nu(t)}{\rho_\nu(0)} = \left(\frac{R(0)}{R(t)}\right)^3 = \left(\frac{\omega(t)}{\omega(0)}\right)^{3/2} = \left(\frac{\rho_\nu(t)+\rho_b(t)}{\rho_\nu(0)+\rho_b(0)}\right)^{3/4}. \tag{9.78}$$

Because of the radiative energy losses the baryon density can grow so that the condition is valid

$$\rho_b(t) \gg \rho_\nu(t). \tag{9.79}$$

Then we obtain from (9.78) that the density of neutrinos increases with time as

$$\frac{\rho_\nu(t)}{\rho_\nu(0)} = \left(\frac{\rho_b(t)}{\rho_\nu(0) + \rho_b(0)} \right)^{3/4}. \tag{9.80}$$

Analytical treatment described above was backed in (Zeldovich et al 1980) by numerical analysis, confirming the law of condensation (9.80).

This process of collisionless gas condensation can occur in any of the collapsing systems of ordinary matter, if compression is supported by self-gravitation at all stages.

It is believed that this condition does not correspond to the initial stage of formation of an object smaller than a globular cluster in which the decisive role is played by the development of thermal instability, and the compression of the system is provided by the effects of external pressure of hot gas (Chapter 10).

The discussed mechanism should be effective in the process of galaxy formation but, apparently, does not not work during the formation of globular clusters and smaller celestial objects (such as stars).

If massive neutrinos (or any other form of collisionless dark matter) dominate the cosmological density, then such a mechanism explains the formation of the massive halo of galaxies from dark matter particles. It has been used, for example, in (Doroshkevich et al 1980b) to explain, in the scenario of neutrino-dominated Universe, why massive neutrinos remain at the periphery of galaxies and do not give a tangible contribution to the density of the central parts of galaxies.

The assumption that the hypothetical particles dominate in the galactic halo can be used to estimate the distribution of the particles, without reference to the dynamic model of halo formation. Using the model of the distribution of mass in the halo, we can simulate the distribution of particles in the galaxy, and calculate the effects of weak annihilation, as was done in (Silk, Srednicky 1984) for supersymmetric particles.

However, because of the universality of the mechanism of condensation (9.80) the mechanism can also be used for the case of a small contribution of hypothetical particles to the total density, thus providing a reasonable estimate of the expected distribution of particles in the Galaxy and their possible effects, even if they do not play any significant role in the dynamics of halo formation.

2.2. Annihilation of weakly interacting massive particles in the Galaxy

There are many theories of the origin of cosmic rays associated with active galactic nuclei, supernova explosions, the processes in the magnetospheres of pulsars and the various mechanisms in galactic magnetic fields (Ginzburg et al 1990; Hill et al 1987).

Cosmoarcheology opens up new possible mechanisms for the generation of cosmic rays and gamma background, coupled with the cosmological consequences of particle theory which lead to sources of energetic particles in the Universe.

It offers a new perspective on the relationship between cosmic ray physics, ultrahigh-energy physics and cosmology, leading to exotic sources of cosmic rays and gamma background, and discovering new aspects of experimental physics of cosmic rays and gamma astronomy.

Intensive study of cosmoarcheological chains linking the data on cosmic rays and gamma background with the ability to test models of inflation, baryosynthesis and dark matter, as well as the consequences of different cosmological GUT models, supersymmetry, and theoretical mechanisms of *CP*-violation and the neutrino mass was carried out in the project ASTRODAMUS (Khlopov et al 1996a,b,c).

One of the effects of new physics in cosmic rays is associated with the effect of weak annihilation of weakly interacting massive particles (WIMP) in the Galaxy. Consider the example of the effect of heavy massive stable neutrinos (Zeldovich et al 1980; Fargion et al 1995).

Condensation of heavy neutrinos in the Galaxy increases the rate of annihilation of neutrinos, causing the generation of cosmic rays. The most stringent limit on the mass of heavy neutrinos can be obtained by considering the electron (and positron) component of cosmic rays which is most sensitive to sources of cosmic rays.

The flux of relativistic electrons and positrons from the annihilation of neutrinos in the Galaxy is given by

$$J = T_e \frac{dn_\nu}{dt} \frac{c\delta}{8\pi} \, \mathrm{cm}^{-2}\mathrm{s}^{-1}\,\mathrm{ster}^{-1}, \tag{9.81}$$

where

$$\frac{dn_\nu}{dt} = n_\nu^2 \sigma_\nu v \tag{9.82}$$

is the annihilation rate of neutrinos per unit volume of the Galaxy, σ_ν is the annihilation cross section of neutrinos in the Galaxy

$$T_e \approx 10^7 \text{ years} \tag{9.83}$$

is the lifetime of cosmic rays in the Galaxy, c is the speed of light, δ is the number of relativistic electrons and positrons produced in a single act of annihilation of neutrinos.

In order to get a limit on the mass of heavy neutrinos, we consider the annihilation of heavy neutrinos in the e^+e^- pair in the Galaxy. Because the heavy neutrinos in the Galaxy are non-relativistic, the ultra-relativistic electrons in the annihilation process

$$\nu\bar{\nu} \to e^+e^- \tag{9.84}$$

will be born monochromatic with the energy of the order

$$E \approx m$$

so that even for

$$m = 1 \text{ TeV}$$

the scatter of energy in the source will be

$$\Delta E \ll 1 \text{ TeV}.$$

Electrons and positrons can also be produced in secondary processes of decay of μ-, τ-leptons and quarks, but these processes contribute only to the soft part of the spectrum of cosmic rays and so we will not consider them.

The annihilation cross section for the process (9.84) in the Galaxy can be rewritten as

$$\sigma_\nu v \approx 2.9 s M_W^2 D_Z 10^{-36} \text{cm}^{-2}\,\text{s}^{-1}, \tag{9.85}$$

where D_Z is the propagator of the Z^0-boson given by the expression

$$D_Z s \propto \frac{s}{\left(s - M_Z^2\right)^2 + \Gamma^2 M_Z^2}, \tag{9.86}$$

Γ is the width of the Z^0-boson, and the s-invariant has the meaning of the square of the total energy in the system of the centre of inertia.

Therefore, we have the flux of cosmic electrons

$$J_e \approx 10^{13} n^2 s M_W^2 D_Z \text{ cm}^{-2}\text{s}^{-1}\text{ster}^{-1}. \tag{9.87}$$

The experimental energy spectrum of cosmic electrons (Lang, 1974) integrated over the energy resolution ΔE of detector was given by

$$J^{\text{exp}} \approx 1.16 \cdot 10^{-2} \left(\frac{E}{1 \text{ GeV}} \right)^{-2.6} \left(\frac{\Delta E}{1 \text{ GeV}} \right) \text{cm}^{-2}\text{s}^{-1}\text{ster}^{-1}, \tag{9.88}$$

where

$$3 < E < 300 \text{ GeV}$$

and

$$\Delta E \ll E.$$

By comparing (9.87) and (9.88) it was found in (Fargion et al 1985) find that the existence of heavy neutrinos is forbidden in the mass range

$$60 < m < 115 \text{ GeV}. \tag{9.89}$$

The fact that there was no constraint in the mass range 44÷60 GeV was due to the smallness of the relic density (cross-section of annihilation of neutrinos is very large near the Z^0-boson peak). The constraint in the area of large neutrino mass is a consequence of reduction of the flux (9.87)

$$J_e \propto m^{-8} \tag{9.90}$$

compared with the experimental flux (9.88)

$$J_{\text{exp}} \propto m^{-2.6}.$$

In fact, the constraint (9.89) can serve only to illustrate the method which removed from the number of candidates for the role of dark matter particles of the Galaxy the massive neutrino ν_τ with a large magnetic momentum (Khlopov et al 1996d). As they move in the

Galaxy realistic analysis takes into account the energy loss due to synchrotron radiation. More detailed analysis also takes into account the inhomogeneity of the distribution of dark matter in the Galaxy, as well as the relation of diffusion in space with the lifetime of the electron of a given energy in the Galaxy.

2.3. Search for WIMP in the Galaxy on the basis of their direct and indirect effects

Let us consider the above indirect information about the WIMP together with experimental data on the direct search for the effects of elastic scattering of WIMP on nuclei in the underground experiments in order to obtain constraints on the mass of a stable neutrino of 4th generation, following the results by (Fargion et al 1998, Golubkov et al 1999).

For the sake of definiteness, consider the standard electroweak model, which includes one additional generation of heavy neutrinos ν and a heavy charged lepton L in the standard doublet $SU(2)_L$. In order to guarantee the stability of the heavy neutrino, we assume that its mass does not exceed the mass of the respective charged lepton M_L

$$m < M_L$$

and that the heavy neutrino is a Dirac particle.

It is known that the current experimental results do not exclude the existence of heavy Dirac neutrinos with the mass

$$m > \frac{M_Z}{2}, \tag{9.91}$$

where M_Z is the mass of the Z^0-boson.

In the early Universe at high temperatures such heavy neutrinos should have been in thermal equilibrium with other types of particles. With the fall in temperature, heavy neutrinos become non-relativistic at

$$T \sim m$$

and their concentration rapidly decreased exponentially. With further expansion of the Universe the temperature was lowered to the values of the freezing-out temperature T_f, and the weak processes were too slow to maintain equilibrium of neutrinos with other particles. As a consequence, the number density of heavy neutrinos is given for their frozen-out concentration in the Universe

$$n = \frac{6.6 \cdot 10^3}{\sqrt{n_*}} \left(\frac{m_p}{m_{\text{Pl}}} \right) \left(\frac{m_p}{m} \right) \left(\frac{\rho_{\text{cr}}}{10^{-29} h^2 \frac{\text{g}}{\text{cm}^3}} \right) \left[\int_0^{x_f} dx\, m_p^2 (\sigma v) \right]^{-1} \text{cm}^{-3}, \tag{9.92}$$

where

$$\rho_{\text{cr}} = 1.879 \cdot 10^{-29} h^2 \frac{\text{g}}{\text{cm}^3} \tag{9.93}$$

is the critical density of the Universe, h is the dimensionless Hubble constant normalized as

$$h = \frac{H}{100 \frac{\text{km}}{\text{s} \cdot \text{Mpc}}}, \tag{9.94}$$

m_{Pl} is the Planck mass, m_p is the mass of the proton;

$$x = \frac{T}{m}; \; x_f = \frac{T_f}{m}; \tag{9.95}$$

σ is the annihilation cross section, v is the relative velocity of the neutrino–antineutrino pair in its centre of mass, g_* is the effective number of relativistic degrees of freedom when

$$T = T_f.$$

For the bosons the contribution to g_* is equal to 1, and for fermions 7/8. The freezing-out temperature is calculated by iteration from the equation

$$x_f^{-1} = \ln \left(\frac{0.0955 m_{\text{Pl}} \sigma m \sqrt{x_f}}{\sqrt{g_*}} \right) \tag{9.96}$$

and in this case is

$$T_f \approx \frac{m}{30}. \tag{9.97}$$

In general, these processes can lead to the annihilation of heavy neutrinos in the Universe

$$\nu\bar{\nu} \rightarrow f\bar{f}, W^+W^-, ZZ, ZH, HH.$$

However, (see Fargion et al 1997 and references therein), the dominant process is

$$\nu\bar{\nu} \rightarrow f\bar{f} \tag{9.98}$$

below the threshold of production for the W^+W^- pair and

$$\nu\bar{\nu} \rightarrow W^+W^- \tag{9.99}$$

above the threshold of production of W^+W^-.

Figure 9.2 shows the dependence of the cosmological density

$$\rho = 2mn$$

of the heavy neutrinos as a function of neutrino mass. In the area

$$m \sim \frac{m_Z}{2}$$

the density of the neutrinos is extremely low because of the enormous value of the annihilation cross section near the pole of the Z^0-boson. When the neutrino mass increases, the annihilation cross section decreases and this leads to an increase in the concentration of the neutrino which reaches its maximum at

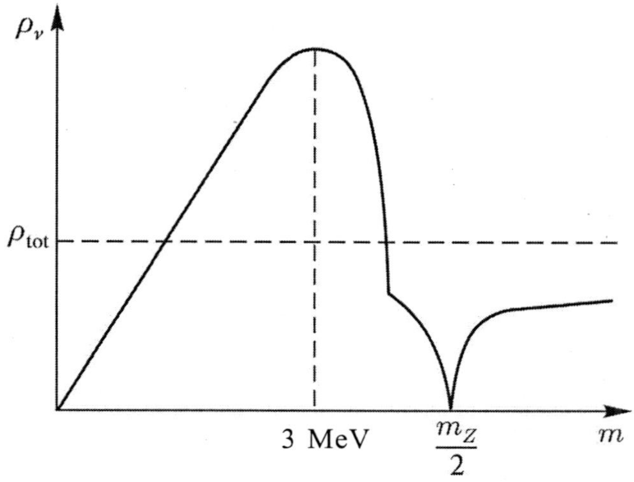

Fig. 9.2. Dependence of the density of neutrinos in the Universe on the neutrino mass.

$$m \approx 100 \text{ GeV}.$$

When

$$m > m_W$$

the annihilation channel into the W^+W^- pair opens and quickly becomes dominant, and because its cross section increases as m^2 the present-day density of the neutrino decreases again with the mass of neutrinos.

As we see from Fig. 9.2, the neutrino density in the allowable mass range of 4-th neutrino is small compared to the critical density. However, condensation of the neutrino, considered in section 2.1, leads to a significant increase in the density of neutrinos in the Galaxy, compared with the average cosmological density. For the central density of neutrinos, we obtain

$$n_{0G} \approx n \left(\frac{\rho_{0G}}{\rho_b} \right)^{3/4}, \quad (9.100)$$

where

$$\rho_{0G} \approx 10^{-20} \frac{\text{g}}{\text{cm}^3} \quad (9.101)$$

is the central density of matter in the galaxy and ρ_b is the density of baryonic matter in the Universe.

Usually, it is assumed that the density of the dark matter halo in the Galaxy decreases with distance from the centre according to the law

$$\rho(r) = \frac{\rho_0}{1 + \left(\frac{r}{a} \right)^2}, \quad (9.102)$$

where

$$2 < a < 10 \text{ kpc} \quad (9.103)$$

is the radius of the halo core.

We assume that the density of neutrinos also follows distribution (9.102). Taking the minimum value for the halo core

$$a = 2 \text{ kpc}$$

and substituting in (9.102) the distance from the Sun to the centre of the Galaxy

$$r_{\text{Sun}} \approx 8.5 \text{ kpc}, \tag{9.104}$$

we obtain that the density in the solar neighbourhood is estimated by a factor of about 19 compared with the central density given by expression (9.100). Therefore, taking into account (9.100)–(9.104), we find that the density of heavy neutrinos in the vicinity of the Sun satisfies the condition

$$n_{\text{Sun}} > 3.3 \cdot 10^6 n. \tag{9.105}$$

It should be noted that the value (9.105) is an order of magnitude higher than the value derived from estimates of the density of dark matter in the vicinity of the Solar system under the assumption of direct proportionality of the local density of 4th generation neutrinos to their contribution to the cosmological density. However, despite the seeming transparency and credibility of this assumption (Dolgov,2002), its conclusion is not perfect even in the case of dominance of CDM in the dark matter of the galaxy. Problems of the CDM models (see, for example, Salucci 2003) in explaining the observed density profile of dark matter of the galaxies might reflect the need for some other, non-trivial forms of dark matter. In this case, the identity of the spatial distribution of 4th generation neutrinos (Dolgov 2002) and the dominant dark matter of the Galaxy, and the subsequently estimated identity of the local density of 4th generation neutrinos are, strictly speaking, incorrect. With this caveat in mind, we apply the above results to the experimental data on the direct search for the effects of elastic scattering of WIMPs on nuclei.

Neutrino scattering on nuclei occurs through the exchange of the Z^0-boson and, therefore, the axial interaction which leads only to small spin-dependent effects can be neglected. In the non-relativistic limit, the cross section of neutrino coherent scattering on a nucleus is given by

$$\sigma_{\text{el}} \approx \frac{m^2 M^2}{2\pi\left(m+M\right)} G_F^2 Y^2 K^2, \tag{9.106}$$

where

$$K = N - \left(1 - \sin^2 \theta_W\right) Z \tag{9.107}$$

N, Z is the number of neutrons and protons, respectively, M is the mass of the nucleus, G_F is the Fermi constant, Y is the average hypercharge ($Y \sim 1$). The nuclear degrees of freedom can be taken into account by the nuclear form factor.

From a comparison with the results of underground experiments CDMS with a Ge detector (see the review Khlopov 1999; Cline 2003) it was found that the existence of massive neutrinos is forbidden in the mass range

$$60 < E < 290 \text{ GeV}. \tag{9.108}$$

It should be noted that the constraint (9.108) represented the minimum area excluded by the underground experiments with the Ge detector, because this result was obtained with a careful evaluation of possible values of the astrophysical and cosmological parameters. But even such a cautious approach led to greater constraint on the mass for the heavy neutrinos than the ratio (9.89) which was derived from analysis of the electron spectrum in cosmic rays. The further results of CDMS experiment together with the results of other underground experiments (XENON100, first of all) strengthen these constraints, being accompanied by the indirect searches for dark matter in cosmic ray experiments PAMELA and FERMI.

If there is a Higgs boson, the limit (9.108) does not exclude the possibility that the heavy neutrinos have the mass in the range

$$|m_H - m| < \Gamma_H, \tag{9.109}$$

where Γ_H is the width of the Higgs meson, because in this case the annihilation in the s-channel

$$\nu\bar{\nu} \to H \to \dots \tag{9.110}$$

during freezing-out can lead to a significant decrease in the concentration of relic neutrinos and hence the number density of neutrinos in the Galaxy.

The underground search for dark matter by its annual modulation effects in large NaI(Tl) detectors gives a positive result with more and more higher level of significance in experiments DAMA/NaI and DAMA/LIBRA (Bernabei et al 2000, 2003, 2010).

In the combination with constraints from XENON100 experiment this positive result cannot be interpreted in the terms of the 4th generation neutrino with mass

$$m > 45 \text{ GeV}, \tag{9.111}$$

but it can find explanation in the framework of composite dark matter scenarios (Khlopov et al 2010), implying in particular existence of stable u_4-quarks of the 4th generation (Khlopov 2006).

The theoretical analysis of the contribution of radiative effects of the quarks and leptons of the 4th generation to the precision-measured parameters of the standard model allows for such a possibility at the mass of the 4th generation neutrino in the region of 50 GeV (Maltoni et al 2000; Novikov et al 2002).

In the experiments with cosmic rays in AMS spectrometer, which is in operation at the International Space Station, additional check of the physical nature of dark matter is possible.

Features in the spectrum of cosmic positrons related to the diffusion and slowing down of monochromatic positrons with energies above 45 GeV would give a clear sign of annihilation in the galactic halo of Dirac neutrinos because the direct annihilation of Majorana fermions, such as neutralinos in the electron–positron pair in the Galaxy is strongly suppressed (Golubkov et al 1999). Constraints on such feature from the experiment PAMELA together with constraints from the data on cosmic antiprotons and gamma background put further limits on the mass of 4th neutrino.

Another indirect effect of the existence of primordial massive stable neutrinos in the Galaxy is that they are captured and stored in the central parts of the Earth and the Sun (Belotsky et al 2001, Belotsky et al 2002). With accumulation the concentration of trapped neutrinos and antineutrinos increases and the probability of their annihilation also increases. The matter of the Sun and the Earth is opaque to all products of this annihilation, with the exception of the three known neutrino types. Therefore, the main manifestation of the annihilation of massive neutrinos in the Earth and the Sun is the flux of neutrinos from the annihilation products. Unlike all other discussed types of the WIMP (e.g., neutralinos), for which a similar effect in the annihilation in the Earth and the Sun must also be the case of generation of the neutrinos is manifested in a noticeable relative probability of annihilation in the channel $\nu_e\bar{\nu}_e$ that generates monochromatic neutrinos of all three known types of energy $E_\nu = mc^2$. For Majorana particles such as neutralinos, this annihilation channel is strongly suppressed. If the equilibrium is reached between the capture and annihiliation rates, the total flux of neutrinos from annihilation is directly determined by the total rate of capture of heavy neutrinos by the Earth.

In the capture effect of the 4th generation neutrinos in the Earth their 'slow component' in the Solar system could play a crucial role. As was shown in (Damour, Krauss, 1999) the interaction of galactic WIMPs with the matter of the Sun can lead to their capture in the

Solar system. This mechanism of formation of the 'slow component' of the WIMPs is that the WIMPs lose energy in the scattering with matter on the periphery of the Sun and move to the finite trajectories in the Solar system. In this case, the perturbing effect of the planets deflects this path, preventing its subsequent intersection with the surface of the Sun. Estimate of this effect in the case of 4th generation neutrino (Belotsky et al 2002) leads to the flux of the slow component in the vicinity of the Earth, comparable to the neutrino flux in the Galaxy. Because of an order of magnitude lower velocity of the slow component of the 4th generation neutrino (~30 km/s) compared with the galactic 4th generation neutrinos (~220 km/s), the particles of the slow component could be trapped by the Earth much more efficiently, enhancing the effect of annihiliation of the 4th generation neutrino with the mass 50÷80 GeV by a factor of 4÷100. This enhancement could increase the sensitivity of the data of underground neutrino observatories to the expected flux of monochromatic neutrinos. Such an increase of sensitivity assumes, however, sufficiently long lifetime of slow component in the Solar system, what may not be the case due to perturbing effects of planets. The trapped neutrinos also form a 'neutrino core' inside the Earth, the direct effects of which apparently are hard to observe.

Detailed test of the hypothesis of 4th generation neutrinos in astrophysical data must inevitably take into account the possible inhomogeneity of distribution of neutrinos in the Galaxy, as well as the heterogeneous nature of the information on this distribution, derived from the results of various astrophysical research studies (Belotsky, Khlopov 2003). In fact, the direct search for WIMPs examines the fluxes of cosmic (mostly galactic) particles penetrating the Earth at the present time, whereas the monochromatic fluxes of neutrinos from annihilation in the Earth result from the accumulation of the 4th generation neutrinos for all the lifetime of the Solar system. Gamma rays from the annihilation of 4th generation neutrinos in the galactic halo reach the Earth directly from the place of annihilation, while the cosmic antiprotons and positrons come only after long-term diffusion in the galactic magnetic fields. In the case of positrons, the effects of energy loss during this diffusion are also important.

Search for the effects of massive cosmic neutrinos can be supplemented by searches for such neutrinos in accelerators (Fargion et al 1996). The reaction

$$e^+e^- \to \nu\bar{\nu}\gamma \qquad (9.112)$$

makes it possible to analyze the range of the neutrino masses

$$m \sim \frac{M_Z}{2}, \tag{9.113}$$

for which the astrophysical studies of the neutrino are difficult because of the low density of neutrinos.

If heavy neutrinos exist, there may arise an interesting effect of hadronless bremsstrahlung of the birth of the Higgs boson in accelerators

$$e^+e^- \to ZH \to Z\nu\bar{\nu} \to l^+l^-\nu\bar{\nu} \tag{9.114}$$

and this mode can be dominant. In fact, if the mass of the Higgs boson is less than 160 GeV, the channel of its decay into a pair of 4th generation neutrino–antineutrino dominates, and the Higgs boson turns out to be 'invisible', what can significantly affect the strategy of its experimental search (Belotsky et al 2003)

Thus, comprehensive studies, which include underground experiments, accelerator searches, and astrophysics research, may clarify the physical nature of even small components of multicomponent dark matter.

The important physical meaning of existence of such a small component of dark matter as the cosmic neutrino of the 4th generation, reveals the connection between this hypothesis with the phenomenology of superstring theory, identified in (Belotsky et al 2000; Belotsky, Khlopov 2001; Khlopov, Shibaev, 2002; Belotsky et al 2008). In the phenomenology of the heterotic string the GUT symmetry has a rank (i.e. the number of conserved charges) higher than the rank of the symmetry of the standard model. On the other hand, the Euler characteristic of the topology of the compactified six dimensions defines in this approach the number of generations of quarks and leptons, which can be both 3 and 4. The difference in the ranks of the symmetry groups of grand unification and the standard model implies the existence of at least one new conserved charge, which may be associated with quarks and leptons of the fourth generation. This may explain the Dirac nature of the mass of the 4th generation neutrino and stability of this massive neutrinos. If a new conserved charge is the gauge charge and the neutrino is the lightest fermion of the 4th generation, the 4th generation neutrino, having a new charge, like electric, must be absolutely stable, just as is stable the lightest electrically charged particle – electron. When the mass of the 4th generation neutrino is ~50 GeV the occurrence of the new interaction does not affect the kinetics of freezing-out of the 4th generation neutrino, discussed above, since the annihilation cross section of 4th generation neutrino dominates the Z-resonant channel. However, the slow annihilation of neutrinos and antineutrinos in the Galaxy, the Sun or Earth is significantly enhanced by the new long-range effect which adds to the probability of processes the factor $\sim 2\pi\alpha_y c/v$, completely analogous to the well-known Coulomb factor $\sim 2\pi\alpha c/v$ in the processes with slow charged particles, first considered by Sommerfeld (1931) and Sakharov (1948). Here α_y is a dimensionless constant of the new y-interaction, analogous to the electromagnetic fine structure constant. Such 'Sommerfeld–Sakharov enhancement' increases by more than an order of magnitude the expected fluxes of products of annihilation of the 4th generation neutrinos in the Galaxy. It is curious that the effect of 'Sommerfeld–Sakharov enhancement' does not lead to any apparent increase in the flux of monochromatic neutrinos from the annihilation of 4th generation neutrinos in the Earth. The increase in the annihilation rate accelerates the rate of establishment of kinetic equilibrium between the capture of neutrinos and their annihilation. Since such an equilibrium must be established also in the absence of Sommerfeld–Sakharov enhancement (Belotsky et al 2002), the inclusion of this enhancement does not change the estimation of the monochromatic neutrino flux, increasing the reliability of calculation of the equilibrium conditions.

Thus, testing the hypothesis of the 4th generation neutrino opens an approach to the study of the phenomenology of string theory. Confirmation of this hypothesis and the existence of the new interaction will create opportunities whose value can not be overestimated.

In fact, the new interaction should be common to all quarks and leptons of the 4th generation. In this case, the lightest quark of the 4th generation, being unstable against decay to the fourth neutrino, can have has a lifetime exceeding the age of the Universe. According to (Khlopov 2006; Belotsky et al 2009), if it is the u_4-quark with the +2/3 charge, and there is an excess of $\bar{u}_4$ in the Universe, stable $(\bar{u}_4\bar{u}_4\bar{u}_4)$ state with charge −2 can bind with primordial helium in dark atoms of O-helium (OHe), which can be a dominant form of dark matter in the Universe. OHe scenario of composite dark matter can be implemented also in AC model, based on the almost commutative geometry, or in the models of walking technicolor (models of new non-abelian color with gauge constant which is not running, but walking – changes much slower with the momentum transfer).

These nuclear interacting dark atoms implement the scenario of warmer than cold dark matter and can shed new light on the puzzles of dark matter searches. Slowed down in the terrestrial matter OHe cannot cause nuclear recoil, expected for WIMPs in underground detectors, but can lead to annual modulations in energy release due to radiative capture of OHe by nuclei. Annual modulations in the local concentration of OHe in the underground detectors, caused by the orbital motion of the Earth, follow from the rapid adjustment of this concentration to the incoming flux of the cosmic OHe, determined by the equilibrium between this flux and OHe diffusion to the center of Earth.

If the bound state of OHe with nuclei exists in DAMA/NaI and DAMA/LIBRA detectors with a few keV energy, annual modulations in the radiative capture to this level can explain results of these DAMA experiments. Such level may not exist at all or can have different energy in detectors with different chemical content, what can explain the controversy in the results underground experiments. This example shows that the solution of the puzzles of the dark matter searches can have very nontrivial implementations. Finally, we note that the condensation in the Galaxy might be also important for direct search for the light neutrinos and axions.

The light neutrinos with the mass

$$m \sim 30 \text{ eV}$$

and the velocity dispersion in the Galaxy ~300 km/s have a wavelength of order

$$\lambda \sim \frac{\hbar}{mv} \sim 6 \cdot 10^{-4} \text{cm}. \quad (9.115)$$

As a result of the interaction of the weak vector current, the galactic neutrinos can scatter coherently at solids with sizes

$$a \leqslant \lambda.$$

Momentum transfer in such a scattering can lead to a small acceleration, which, in principle, could be found (Shvartsman et al 1982). For realistic values of ordinary neutrino masses (less than 1 eV) such acceleration is below the sensitivity of the current experimental methods.

The existence of the photon interaction with the axion $a\gamma\gamma$ leads to the possibility of axion–photon conversion in the RF cavity (Sikivie 1983). This principle was used in the axion halo- and helioscopes for the experimental search for the axion flux, respectively, from the galactic halo and the Sun (see Sikivie 2010 for review and references). Astrophyscial probes for axions imply effects of axion-photon conversion in magnetic fields.

Shadow and mirror matter is a form of dark matter that is very difficult to detect in the direct experimental searches. Astronomy is a unique source of information about the existence and possible properties of these forms of dark matter, which we discuss in the next chapter.

10

Mirror world in the Universe

1. Mirror particles

1.1. Equivalence of left and right in the nature

If we put aside the mirror world of Lewis Carroll's 'Alice in Wonderland', the fundamental reasons for the existence of mirror partners were first revealed in the same article by Lee and Yang (Lee, Yang 1956), in which they questioned the conservation of parity in weak interaction. This study brought them a Nobel Prize.

In quantum field theory, the transformation of *P*-parity, i.e. inversion of spatial coordinates

$$P:\ \left(t,\vec{x}\right)\rightarrow\left(t,-\vec{x}\right), \tag{10.1}$$

leads to the transformation of the field operator

$$P:\ \Psi\left(t,\vec{x}\right)\rightarrow\Psi^{P}\left(t,\vec{x}\right). \tag{10.2}$$

Prior to the study by Lee and Yang it was assumed that the parity is conserved in all the fundamental interactions of elementary particles. This means that the inversion of the coordinate axes leads to the transformation of the field of a particle into a different field, describing the particles which also exist in nature.

For example, a left-handed polarized state of the field with nonzero spin are converted into the right-handed polarized states of the same field, so that

$$\Psi_{L}^{P}\left(t,\vec{x}\right)=\Psi_{R}\left(t,\vec{x}\right) \tag{10.3}$$

and conversely, the right-handed state of the field converted to the left-handed state.

The field operator corresponding to a particle with zero spin can be invariant with respect to P-transformation, so that as a result of the P-inversion of the spatial coordinates

$$\Psi^{P}\left(t,-\vec{x}\right)\rightarrow+\Psi\left(t,\vec{x}\right). \tag{10.4}$$

These fields correspond to scalar particles.

The field operator of a particle with zero spin can change its sign under the P-transformation

$$\Psi^{P}\left(t,-\vec{x}\right)\rightarrow-\Psi\left(t,\vec{x}\right) \tag{10.5}$$

and then it corresponds to a pseudoscalar particle.

Conservation of parity means that the Lagrangian is invariant under P-inversion, i.e. is a scalar.

P-inversion can be regarded as a reflection in the mirror. Until 1956, it was assumed that the mirror image of a process with any fundamental particle leads either to the same process, or to some other process also existing in nature.

Violation of parity in weak interactions was the beginning of the study of processes in which this fundamental rule is violated.

Thus, the neutrino that is born in beta decay has only one polarization. P-transformation leads to the neutrino state which does not participate in weak interactions.

In the beta decay of polarized nuclei, electrons are emitted mainly in the direction of nuclear spin. Mirror reflection of this process leads to a process with the preferred direction of electron emission opposite to the nuclear polarization which is not observed in Nature.

In the conclusion of their article, Lee and Yang (1956) examined the obvious theoretical problem associated with the violation of parity. Non-conservation of parity leads to non-equivalence of left- and right-handed oriented systems of coordinates.

Indeed, the P-transformation of the coordinate system, which describes the P-violating process, corresponds to the transition from left-handed to right-handed coordinate system, or a mirror reflection of the process. As a result, because of P-non-conservation this transformation leads to a process that does not exist in nature (see Fig. 10.1). On the other hand, the existence of the preferred coordinate system means that empty space-time has some preferred orientation. The only difference between left- and right-handed coordinate systems is associated with the direction of rotation along the shortest arc to match the X axis with the Y axis around a fixed Z axis. It seems

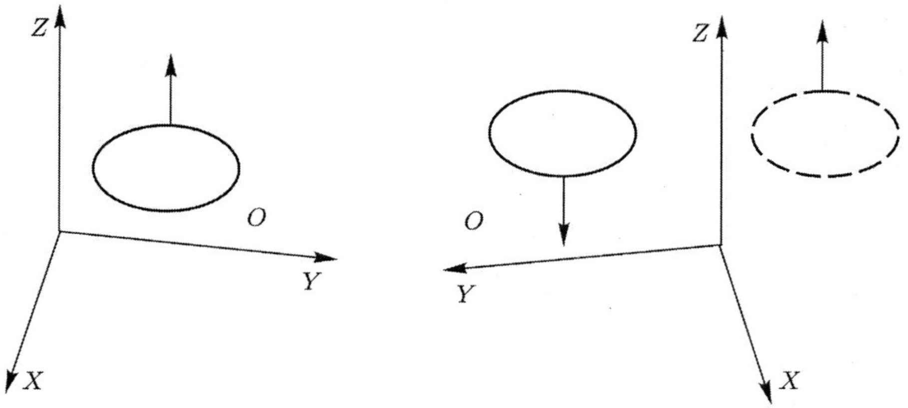

Fig. 10.1. Mirror reflection of processes with violation of P-parity leads to processes that do not exist in nature.

doubtful that God had any reason to select a preferred direction for the rotation in empty space.

To restore the equivalence of right and left, Lee and Yang (1956) suggested that the mirror partners should exist for all known particles. In this case, the P-inversion should be accompanied by interchanging the ordinary particles and their mirror partners.

If P-parity has the opposite sign in the mirror world, the combination of reflection and exchange of the mirror partners will not lead to any change in the description of the original process and its mirror image.

Economical choice of the mirror partners based on the idea of CP-invariance was then proposed. Here the already found partners of particles of ordinary matter, i.e., antiparticles, were identified with mirror twins.

The antiparticles, it seemed, showed the basic property of mirror particles – their strict symmetry with the corresponding ordinary partners. The absolute values of all their fundamental parameters (mass, charge, lifetimes, the relative probabilities of different decay channels) seemed to be strictly equal to the same parameters of the particles.

Indeed, the CP-transformation matches the observed left-handed state of the neutrinos with the right-handed state of antineutrinos

$$CP: \quad \nu_L \to \bar{\nu}_R, \tag{10.6}$$

and since the right-handed antineutrino is also born in weak interactions, the symmetry between left and right is restored.

Likewise, conversion of P-parity matches the β^+-decay of antinuclei with the β^--decay of polarized nuclei. Theory of P-violating weak interactions predicts the opposite direction of the preferred positron emission compared with the emission of electrons.

Consequently, the combination of mirror reflection together with interchange of the particles and antiparticles seemingly supports the equivalence of left-handed and right-handed coordinate systems.

However, the discovery of CP-violation showed the wrong choice of selecting antiparticles as mirror partners, and once again raised the question of finding the true set of mirror particles and their expected properties (Kobzarev et al 1966).

It turned out that when the mirror particles are not identified with antiparticles, they can not participate in the same interactions as ordinary particles.

The strict symmetry between the ordinary and mirror electrons leads to serious problems in atomic physics, if ordinary and mirror worlds take part in the same electromagnetic interaction. We should in this case expect doubling of atomic states because of the additional degrees of freedom associated with the mirror electron states (Kobzarev et al 1966).

A similar problem arises in hadron physics, if the strong interaction of ordinary and mirror particles is identical. Because of mirror particles, there is doubling of some hadronic states, for example, there should be doubling of the states of the neutral pion.

The discovery of the intermediate $W^\pm$- and Z^0-bosons in weak interactions and measurements of their total width excluded the common weak interaction between ordinary and mirror particles. Thus, not only ordinary fundamental particles of matter but also gauge bosons, carrying their interaction, should have their mirror partners.

1.2. Fractons and Alice strings

The easiest way to incorporate mirror particles in the model of elementary particles is to add to $SU(2)\otimes U(1)\otimes SU(3)_c$ – the gauge symmetry of the standard model – the same symmetry related to the mirror particles.

It must be assumed that the mirror particles correspond to the trivial representation of the ordinary gauge group and, conversely, the ordinary state transform as scalars under the transformations of the mirror gauge group.

This means that particles of a certain mirrority (i.e., ordinary or mirror) are sterile with respect to interactions of the other mirrority.

Thus, the mirror electron interacts with the mirror photon like the ordinary electron interacts with ordinary photons. However, the mirror electron does not interact with ordinary photons, like the ordinary electron 'does not feel' the mirror photon.

One might expect that states related to a non-trivial representation of the gauge group of both mirrorities must comply with the hypothetical particles with very specific properties.

Indeed, the state of the X-lepton, corresponding to the representation of the group

$$\left[SU(2)\otimes U(1)\otimes SU(3)_c\right]_O\otimes\left[SU(2)\otimes U(1)\otimes SU(3)_c\right]_M$$

$$X\ \left(2,\frac{1}{3},1;1,1,3\right), \qquad (10.7)$$

has the properties of ordinary leptons and mirror quarks. It is a doublet with respect of the usual electroweak gauge group and a triplet with respect to mirror colour transformations.

Due to the confinement of the mirror colour this state should be linked with the other mirror colour states in composite particles, which gives the state looking like an anomalous lepton.

The bound states of X-leptons and mirror partners of ordinary quarks q_M, for example, baryon-like states (Xq_Mq_M) behave in normal electromagnetic processes as colourless particles with fractional charges. Such particles are called *fractons*.

Fractons also arise as bound states of quarks that corresponds to the representation of the gauge group

$$\left[SU(2)\otimes U(1)\otimes SU(3)_c\right]_O\otimes\left[SU(2)\otimes U(1)\otimes SU(3)_c\right]_M$$

$$X\ \left(1,1,3;2,\frac{1}{3},1\right). \qquad (10.8)$$

X-quarks have the fractional mirror electric charge and are neutral with respect to the usual electromagnetic interaction.

The bound state of X-quarks with ordinary quarks in baryon-like or meson-like hadrons is a fractionally charged colorless particle, such as baryon (Xud) with electric charge $+1/3$.

In the early Universe, X-quarks (or X-leptons) should have been born, frozen-out and form fractons during the QCD confinement. The specific feature of the fractons as compared with free quarks (Zeldovich et al 1967) is that the primordial concentration of the fractons may significantly decrease in the present-day astrophysical objects, say in the Earth, due to the recombination processes

$$(Xud)+(\bar{X}u)\rightarrow(X\bar{X})+\text{ordinary hadrons} \tag{10.9}$$

followed by decay of X-quarkonium

$$(X\bar{X})\rightarrow \text{ ordinary hadrons} \tag{10.10}$$

or

$$(X\bar{X})\rightarrow \text{ mirror particles.} \tag{10.11}$$

The recombination of X-quarks (10.9)–(10.11) is caused by 'Coulomb' attraction of the uncompensated mirror electric charge of X-quarks and antiquarks, which reduces the abundance of fractons and makes their existence compatible with the negative results of searches for fractionally charged particles (Khlopov 1981).

In the framework of Grand Unification the gauge symmetry

$$[SU(2)\otimes U(1)\otimes SU(3)_c]_O\otimes[SU(2)\otimes U(1)\otimes SU(3)_c]_M$$

of the ordinary and mirror particles is incorporated into a single group symmetry of GUT – G_{OM}.

Breaking of G_{OM} leads to the separation of ordinary and mirror sectors of particles, subject to strict discrete symmetry between them. Topology of G_{OM} symmetry breaking under this condition is reduced to the case of $U(1)$ symmetry breaking and ensures the existence of topological solutions of the type of strings.

We can illustrate this strict topological result (Schwartz, Tyupkin 1982) by the following simple considerations.

Breaking of G_{OM} divides the Higgs multiplet of the GUT group into the sets of Higgs fields H_O and the H_M corresponding to the subgroups of separated symmetries of the ordinary and mirror particles. Discrete symmetry between the multiplets H_O and H_M makes these two sets of fields equivalent to the real and imaginary parts of a complex field, which reduces the analysis to the case of broken $U(1)$ symmetry. Then a topological argument is valid describing the existence of solutions of the type of strings in the case of broken non-Abelian symmetry.

Schwartz (1982) found that these solutions have the property to change the mirrority of the object as a result of the object going around the string along the closed path. Schwartz noted that Alice, the heroine of the book by L. Carroll, could come 'through looking glass' (to the mirror world) passing around this string, and called such solutions Alice strings.

Phase transition in the early Universe, in which the GUT symmetry G_{OM} is broken to the symmetry with separate sectors of the ordinary and mirror particles should lead to the structure of cosmic strings.

The evolution of this structure leads to straightening of long strings under the horizon and the collapse of small string loops. As a result of this evolution, the contribution of the structure of strings to the total density has a time dependence

$$\rho_s \sim \frac{\mu}{t^2}, \tag{10.12}$$

where μ is the mass per unit length of the string associated with the scale of symmetry G_{OM} breaking as

$$\mu \sim \Lambda^2. \tag{10.13}$$

From equations (10.12) and (10.13) we find that the relative contribution of strings to the total cosmological density is given by

$$\Omega_s = \frac{\rho_s}{\rho_{\rm cr}} \sim G\mu \sim \left(\frac{\Lambda}{m_{\rm Pl}}\right)^2. \tag{10.14}$$

Because of negative pressure, corresponding to a false vacuum inside the string, the space-time properties of cosmic strings are very different from those properties of the one-dimensional structure of ordinary matter. General relativity predicts a zero gravitational potential outside the string, but a wedge-shaped 'cutout' appears in a flat space-time along the string and its cross-linking leads to a particular motion of particles under the action of the string. The angular size of the wedge-shaped cutout is of the order of

$$\delta \sim G\mu \sim \left(\frac{\Lambda}{m_{\rm Pl}}\right)^2. \tag{10.15}$$

Because of the cutout in space-time along the string, its intersection with the line of sight to an astronomical object leads to the effect of gravitational lensing. Instead of one observed object there appear two images whose angular coordinates are separated by the value (10.15).

Motion of the string leaves the wave trace in passage through matter. The matter perturbed by the string starts to move in a wave trace to the plane of motion of the string with velocity v_W

$$\frac{v_W}{c} \sim (G\mu)^{1/2} \sim \frac{\Lambda}{m_{\rm Pl}}. \quad (10.16)$$

It is interesting to note that the results of rigorous calculations of this phenomenon on the basis of general relativity coincide with the formal Newtonian approach to the calculation of a one-dimensional material string with linear density μ moving with the speed of light (Gasilov et al 1985).

Alice strings are typical cosmic strings and should have all of the above properties. However, the change of the relative mirrority of objects around the Alice strings adds new dimensions to its observable effects.

An object defined as ordinary by going around the string on the left is defined as the mirror object when traversing the string on the right.

Therefore, when the Alice string intersects the line of sight the mirrority of the object relative to the observer changes. Normal stars become mirror and invisible, and the mirror stars on the contrary become ordinary and visible.

Both mirror and ordinary photons originate from the objects with mixed mirrority. The Alice string, crossing the line of sight, changes the mirrority of the radiation. Radiation of the mirror photon to the left of the string is the emission of ordinary photons on the right and vice versa.

For uncorrelated processes of ordinary and mirror radiation, the Alice string. crossing the line of sight, leads to exchange of mirrority, which can be seen as the abnormally rapid change in luminosity. Blinnikov and Khlopov (1982) noted that this effect may explain a 100 second change of the luminosity in X-rays from the object QSO 1525+227, which occurred in 1981 in the Einstein X-ray laboratory.

The effect of the gravitational lens on the Alice string gives rise to two images of different mirrority. The radiation of the second image is mirror relative to that the first image.

If ordinary and mirror radiation of the source is not correlated, the usual identification of the gravitational lens effect on the string by comparing the two spectra can not be correct in the case of Alice strings (Sazhin, Khlopov 1989).

Accounting for various mirror images can revive the interpretation in terms of the effect of the gravitational lens on the string for the broad pair Q1146 +111 B, C, and offer such interpretation for the 'twin galaxies' in the field of 0249-186 or double quasar GL Q2345 +007 (Oknyansky 1999).

Development of a model of mirror particles allows for the possibility of violation of strict symmetry between the ordinary and mirror particles.

A shadow world forms in this case. Strict correspondence between the object and its image in a plane mirror is transformed in this case into a bizarre distortion of the size and shape of its shadow.

Cosmoparticle physics is designed to develop a system of tests for both the mirror and shadow matter. The cosmological evolution of the mirror and shadow matter is an important part of this analysis.

Foundations of the theory of cosmological evolution of the mirror world, presented in this chapter, have received extensive development in the works (Ignatiev, Volkas 2000a, 2000b, 2003; Foot 2003; Foot et al 2001; Berezhiani et al 2001a, 2001b, 1996; Bento, Berezhiani,2002; Blinnikov 1996; Khlopov 1999, Okun 2007; Ciarcellutti 2010 and references therein).

2. Mirror particles in early Universe

2.1. Inflation and the constraints on the mirror domain structure

In the model of chaotic inflation (Linde 1984), the initial amplitudes of ordinary and mirror inflatons may be different, which leads to the formation of a domain structure in the distribution of ordinary and mirror matter (Dubrovich, Khlopov 1989).

In areas where the amplitude of the ordinary inflaton is higher the ordinary particles start to dominate after inflation, and the admixture of mirror particles should be exponentially small. Conversely, the dominance of the mirror inflaton leads to low density of ordinary particles after inflation.

Since the area of inflation tends to be much greater than the observable part of the present-day Universe, the mirror asymmetric inflation corresponds, as a result of the ordinary inflaton, to the exponentially small density of mirror matter under the present-day cosmological horizon. In the opposite case, the inflation caused by the mirror inflaton, leads to an exponentially small density of ordinary matter, which is obviously impossible in the observed region of the Universe today.

If the inflaton does not have definite mirrority and equal amounts of ordinary and mirror particles are born after inflation, the domain structure can form due to random local asymmetry of the amplitudes of the ordinary and mirror scalar fields at different periods after general inflation, such as phase transitions.

The scale of such a domain structure is determined by the specific parameters of fields that define the process in question, and it can be much smaller than the scale of the modern cosmological horizon (Dubrovich, Khlopov 1989; Khlopov et al 1989, 1991).

In the case of such small-scale structure, its scale should be either less than the size of the horizon in the period of nucleosynthesis or greater than the size of the supercluster. These restrictions follow from the analysis of the effect of the mirror domain structure on the physical processes in the early Universe after nucleosynthesis (Dubrovich, Khlopov 1989).

The mirror domain looks like a void for the ordinary particles. In the RD stage, radiation pressure should have squeezed the ordinary matter in the domain that would lead to collision of subrelativistic shock waves.

Nuclear collisions in the shock wave would affect the abundance of light elements, so the condition for the absence of overproduction of D, ^{3}He, ^{6}Li, ^{7}Li and ^{7}Be imposes stringent constraints on strong shock waves.

Energy release even in the collision of weak shock waves would distort the shape of the Planck spectrum of the cosmic microwave background radiation, which allows us to strengthen constraints on the domain structure up to the scale of superclusters of galaxies.

The case of the acceptable small-scale structure corresponds to the mass

$$M \ll M_{\odot} \tag{10.17}$$

and practically does not differ in their cosmological features from the case of the initially homogeneous mixture of ordinary and mirror particles.

The acceptable large-scale mirror domains corresponding to the masses

$$M \gg 10^{16} M_{\odot} \tag{10.18}$$

would look like giant voids in the distribution of ordinary matter and in a very special case it could lead to the so-called 'island' model of the Universe (Dolgov et al 1987).

The observed level of anisotropy of the thermal electromagnetic background excludes the case in which the outer boundary of the present-day mirror domain is outside the cosmological horizon. This eliminates the structure of these domains (Dubrovich, Khlopov 1989) on scales

$$l_h(t_{\text{rec}}) \cdot \sqrt{1+z_{\text{rec}}} < l < l_h(t_U), \tag{10.19}$$

where $l_h(t_{rec})$ is the size of the horizon during recombination at a redshift of $z = z_{rec}$ and $l_h(t_U)$ is the size of the present-day horizon.

It should be noted that these constraints on the domain structure are based on averaging the effects in the Universe under the assumption of symmetric distribution of mirror and ordinary domains. In the asymmetric case where ordinary matter is dominant within the present-day cosmological horizon, there may be isolated mirror domains, if the average effect of these domains in the observable Universe is not in contradiction with the upper limit.

2.2. Inhomogeneous baryosynthesis and island distribution of mirror baryons

The need for a special set of mirror partners which do not coincide with antiparticles follows from the fact that in the presence of *CP*-violation in the world of ordinary particles, the equivalence between the left- and right-handed coordinate systems must be restored. For mirror partners, the *CP*-violating effects are equal in magnitude but have opposite sign in comparison with ordinary particles. In particular, this leads to the opposite sign of *CP*-violating effects in baryosynthesis in the mirror world.

Thus, the generation of baryon excess of the ordinary particles formally corresponds to the generation of antibaryon excess for mirror particles, providing a strict symmetry between the baryosynthesis processes in the ordinary and mirror worlds.

However, since the sign of the baryon number for the mirror particles is not observable, we will refer to it as 'baryon' excess in the case of both mirrorities.

Since the evolution of the ordinary and mirror matter is symmetric, the local process of generation of baryon excess in the early Universe lead to the simultaneous production of equal baryon excess in the ordinary and mirror matter. In the absence of the domain structure the ordinary and mirror baryons are born in the Universe with equal local density.

In the presence of the domain structure the scale of domains and the average densities of ordinary and mirror baryons in the domains should be equal.

If the generation of baryon excess is not associated with *CP*-violating local processes in which the baryon number in the ordinary and mirror matter is not conserved, in principle, at any scale there may be the interesting possibility of 'entropy' fluctuations of density in the relative distribution of the excess of ordinary and mirror baryons (Khlopov et al 1989 1991).

In particular, the mechanism of baryosynthesis in supersymmetric models of GUT (Affleck, Dine 1985; Linde 1985) relates the cosmological baryon asymmetry with the existence of the primordial condensate of supersymmetric scalar quarks and leptons. Even in the case of strict symmetry in the superpotentials of ordinary and mirror particles, this mechanism ensures the inhomogeneity of distribution of excess of ordinary and mirror baryons.

Unlike the mirror domains discussed in the previous paragraph, the inhomogeneity of the baryon excess represents local suppression of baryons of different mirrority. The energy density of relativistic mirror and ordinary particles in these areas is the same.

In the RD stage, the existence of regions with a strong dominance of mirror baryons over ordinary baryons (or vice versa) does not lead to any strong dynamic effects and it can not be excluded by the observational data.

In such areas called baryon islands with fixed mirrority (Khlopov et al 1989, 1991), astronomical objects can be formed on any scale, up to the scale of the present-day horizon.

2.3. Nucleosynthesis and the mirror world

It should be noted that in all cases, except for very large mirror domains, the relativistic mirror particles are present in the same amount as the ordinary relativistic particles during nucleosynthesis.

This means that after the first second of the expansion when freezing-out of the ratio of the number of protons and neutrons in ordinary matter took place, the total density should take into account the contribution of the mirror photons, mirror electron–positron pairs, right-handed neutrinos and left-handed antineutrinos should be taken into account in the total density.

Such a doubling of the relativistic particle species during the period of nucleosynthesis should lead to an increase in the primordial abundance of normal ^{4}He (Blinnikov, Khlopov 1980, 1982,1983; Khlopov, Chechetkin 1987; Carlson, Glashow 1987)

$$Y_{\text{prim}} \geqslant 28\%. \qquad (10.20)$$

Taking into account the widespread accepted upper limit on the primordial helium abundance, $Y < 25\%$ (see Chapters 3 and 5), we conclude (Carlson, Glashow 1987) that a homogeneous mixture of ordinary and mirror matter is ruled out by the observations.

Taking this fact into account, we should either consider the asymmetric cosmological solutions for mirror matter (Berezhiani et al

1996), or assume that the symmetry between the ordinary and mirror matter has been broken (Senjanovic et al 1984), or that shadow matter should be considered as the most realistic case (Kolb et al 1985; Khlopov 1999).

Even in this case, the subsequent discussion of the symmetric case of mirror matter is of interest because its quantitative definiteness is very useful for identifying qualitative astronomical effects that characterize even the most general asymmetric case.

It should be noted that the observed abundance of ^{4}He (see Chapter 5 and references therein) is

$$Y_{\text{obs}} = (28 \pm 12)\% \qquad (10.21)$$

which in itself does not contradict the predictions of the mirror world. At the same time, long-term observations of radio-recombination lines in the region *HII* together with an analysis of the ionization structure of these regions (Ershov et al 1988; Tsivilev 1991; Gulyaev et al 1997) permitted uncertainties in the value of the primordial abundance of helium in the range (23.8÷28.4)%, which also did not exclude the case of the symmetric mirror world. However, the refinement of these data (Tsivilev 2002), reducing the uncertainty, makes this case less likely.

In fact, we can not directly observe the primordial helium, and in the absence of direct methods we must use reasonable extrapolation of observational data to their pregalactic values. No matter how justified such an extrapolation is, it is model-dependent so that in the absence of model-independent results, the question of the existence of homogeneously mixed ordinary and mirror matter should not be considered definitively settled.

On the other hand, it would be naive to expect the realization of this ideal strictly symmetric case in nature, as well as the systematic errors in the determination of primordial helium abundance can hardly admit this case.

3. The formation of astronomical objects from mirror matter

With all these caveats in the last section, we will follow in our analysis the overall scenario of cosmological evolution of mirror and ordinary matter homogeneously mixed during the formation of galaxies (Blinnikov, Khlopov 1980, 1982, 1983; Khlopov et al 1989, 1991).

This scenario, since its earliest version, presupposes the equality of the cosmological density of mirror and ordinary matter and accounts for the existence of dark matter in the form of massive neutrinos (Blinnikov, Khlopov 1980, 1982, 1983) or in any other form (Khlopov et al 1991), dominant in the Universe during the formation of galaxies and determining together with dark energy the main part of the present-day cosmological density. This assumption together with inflation and baryosynthesis discussed in the previous section, converts the analysis in the framework of a realistic inflationary model with baryosynthesis and dark matter/energy.

The scenario assumes the dominance in the RD stage of ordinary and mirror radiation of the same density together with relativistic light neutrinos with a small admixture of ordinary and mirror baryons with equal densities. The scenario can naturally take into account the existence of a small (in the RD stage) admixture of nonrelativistic ordinary and mirror particles with equal densities, such as ordinary and mirror neutralinos, axions, and so on.

It should be noted that the equal densities of mirror and ordinary particles during nucleosynthesis contradict the widely accepted assumptions about the abundance of primordial ^{4}He and can hardly fit the observational data even with the account for possible systematic errors in the determination of this abundance.

So, taking this assumption into account, we must find a mechanism that would suppress the contribution of mirror partners to the cosmological density in the first second of the expansion. A possible implementation of such a mechanism was proposed in (Berezhiani et al 2001a).

However, this has almost no influence on the further discussion that examines the evolution of mirror baryons whose contribution to the cosmological density during nucleosynthesis is negligible in the case of strict symmetry between the ordinary and mirror particles.

At the end of the RD stage the non-relativistic dark matter particles, starting to dominate in the Universe, contribute to the development of gravitational instability within the cosmological horizon.

The specific choice of the model of formation of the large-scale structure, as well as the form of dark matter defining the structure, are not essential for the further consideration of the effects of mirror baryonic matter.

The mirror baryons with an average density equal to the density of ordinary baryons form a small admixture of non-relativistic matter, taking part in the overall development of gravitational instability. The character of this development depends on the form of dark matter that dominates the cosmological density.

The influence of mirror matter on the large-scale structure is possible only in case of large-scale mirror domains corresponding to the mass scale

$$M > 10^{16} M_{\odot} \tag{10.22}$$

or the large-island distribution of baryons.

In this case, the scale of inhomogeneities in the mirror baryons can be arbitrary so that 'pure' mirror objects can form at any scale. Given this possibility, we further consider the mirror objects of all possible scales.

It should be noted that the large-scale 'island' distribution of baryons island can produce a physical mechanism of 'biasing' in the distribution of luminous and dark matter. The mirror baryonic islands in this case should look like voids in which no galaxies formed from ordinary matter exist.

However, the problems of rapid evolution of the structure inherent seemingly to all models of formation of the structure by stable dark matter (Doroshkevich et al 1989) are preserved in this case.

To be specific, following (Blinnikov, Khlopov 1980, 1983; Khlopov et al 1989, Khlopov 1991), we consider the 'pancake' scenario of structure formation.

The main features of this scenario were stored in a model of unstable dark matter, more realistically reproducing the observational data.

Moreover, as we shall see, the main features of the formation of mirror baryonic objects are also inherent in any other models of dark matter, such as the CDM scenario.

In the absence of an island or a domain structure in the distribution of baryons, the mirror and ordinary baryons with dark matter form a common non-linear flattened structure (pancake). Shock waves form at the boundary of the ordinary and mirror matter.

The cooled gas is compressed by shock waves due to ordinary and mirror electromagnetic radiation, which leads to fragmentation of the pancake.

Fragmentation of the ordinary and mirror matter in the pancake takes place independently as the ordinary gas pressure does not affect the mirror matter, and vice versa, while the gravitational potential of the fragments is too small for the gravitational capture of the fragments of another mirrority.

This is crucial in explaining the separation of mirror and normal astronomical objects on the scales on which thermal instability plays an essential role. Homogeneously mixed ordinary and mirror baryonic matter forms mainly 'pure' stellar objects with a definite mirrority up to the scale of globular clusters.

Let us consider in more detail this fundamental feature of the evolution of homogeneously mixed ordinary and mirror gases.

3.1. Separation of ordinary and mirror matter

The mass of fragments or ordinary and mirror gases is determined (Doroshkevich et al 1978; Blinnikov, Khlopov 1980, 1983) by the scale of the developing thermal instability (Field 1965; Doroshkevich, Zeldovich 1981)

$$l_{\text{th}} \cong \left(\frac{\lambda T}{\varepsilon}\right)^{1/2}, \tag{10.23}$$

where the electronic conductivity is

$$\lambda = 10^{-6} T^{5/2} \frac{\text{erg}}{\text{cm}\cdot\text{s}\cdot\text{K}}, \tag{10.24}$$

where T is temperature, and radiation energy losses are equal

$$\varepsilon = 10^{-27} n^2 T^{1/2} \frac{\text{erg}}{\text{cm}^3\cdot\text{s}}. \tag{10.25}$$

The mass of the fragment is given by

$$M_{\text{th}} \cong m_p n l_{\text{th}}^3 \cong 5\cdot 10^9 M_{\odot} \left(\frac{T}{10^6\,\text{K}}\right)^{13/2} \left(\frac{10^{-15}\,\frac{\text{erg}}{\text{cm}^3}}{P}\right)^2, \tag{10.26}$$

where the pressure of the shock front $P = nkT$ is a universal function of the momentum of the 'pancake' and is given by

$$P = 4\cdot 10^{-15}\left(1+z_0\right)^4 \frac{\text{erg}}{\text{cm}^3} \tag{10.27}$$

for $\Omega = 0.3$ (Doroshkevich et al 1978).

Taking the red shift during the formation of a pancake equal to

$$z_0 \sim 5 \div 2, \tag{10.28}$$

we get the mass of the ordinary and mirror fragments of the order

$$M_{\text{th}} \sim 10^2 \div 10^5 M_{\odot}, \tag{10.29}$$

respectively, since the temperature of the hot gas is always close to

$$T = 10^6 \text{K}. \tag{10.30}$$

Similar estimates of the masses of fragments of ordinary gas were obtained (Doroshkevich 1980) for the different mechanisms of fragmentation.

Since the ordinary gas pressure has no effect on clouds of mirror matter, and vice versa, there is no reason for mixing of the clouds of the ordinary and mirror gases, if such mixing is not set by the initial conditions of density fluctuations. The strict equality of the initial conditions of fragmentation of the ordinary and mirror matter seems, however, rather artificial.

Clouds of the ordinary and mirror matter gas may merge with clouds of the same mirrority, to increase their mass, gravitationally bind and compress, forming clusters of stars.

Then, the joint merger of clouds of the ordinary and mirror gas with the mass of the order of

$$M \sim 10^6 \div 10^9 M_\odot \tag{10.31}$$

takes place and a large gravitationally bound system forms. At this stage, in principle, possible gravitational capture of the ordinary fragments by the mirror fragments, and vice versa, may take place. Let us estimate the probability of such capture (Blinnikov, Khlopov 1983).

Let a protocluster with mass M and characteristic radius R split into fragments with mass M/N and size r. Even if we assume that the ordinary and mirror fragments are at rest and the mean distance between them is of the order of r, the characteristic time of their merger is of the order of several free-fall times, which is given by

$$t_{ff} \sim \left(\frac{r^3}{Gm} \right)^{1/2} \tag{10.32}$$

and this value is approximately

$$t \sim (G\rho)^{-1/2}, \tag{10.33}$$

where $\rho \sim M/R^3$ which is close to the hydrodynamic time scale of the proto-clusters. Thus, in a characteristic time, as estimated above, the Maxwell velocity distribution of clouds is established with a dispersion of the order of the virial velocity of the fragments in the cluster, which is given by

$$v_0^2 \sim \frac{GM}{R}. \tag{10.34}$$

This means that the probability of low relative velocity of clouds is small.

The total probability w that a fragment of matter will be captured by a fragment of mirror matter is determined by their proximity in the phase space, that is by the product of the probability w_s that the fragment of mirror matter will be close (at distance r) to a fragment of ordinary matter, and the probability w_v that the relative velocity of a fragment of mirror matter would be less than the escape velocity for a fragment of ordinary matter, which corresponds to the relative velocity of fragments being

$$v < v_1 \cong \left(\frac{Gm}{r}\right)^{1/2}. \tag{10.35}$$

It should be noted that the probability of a merger of two clouds of ordinary matter is determined by their size and rate of energy loss.

The total probability of gravitational capture of a fragment of mirror matter by a cloud of ordinary matter is

$$w = w_s \cdot w_v, \tag{10.36}$$

where

$$w_s = 1 - \left(1 - \frac{r^3}{R^3}\right)^N \tag{10.37}$$

and

$$w_v = v_0^{-3} \int_0^{v_1} v^2 \exp\left(\frac{v^2}{v_0^2}\right) dv \cong \left(\frac{v_1}{v_0}\right)^3 \cong \left(\frac{mR}{rM}\right)^{3/2} \cong \left(\frac{R}{Nr}\right)^{3/2} \tag{10.38}$$

(Blinnikov, Khlopov, 1983a).

When $N \gg 1$ the maximum probability (10.36) is formally reached in the earlier stages of compression at

$$r \sim \frac{R}{N^{1/3}}. \tag{10.39}$$

Then

$$w \cong \frac{1}{N}. \tag{10.40}$$

Since $N \sim M/n$, we obtain for the mass of the protocluster

$$M \sim 10^{14} M_{\odot}, \tag{10.41}$$

and for the mass of the fragments

$$m \sim 10^{6} \div 10^{9} M_{\odot} \tag{10.42}$$

the upper limit on the probability of the merger of the ordinary and mirror clouds, which is equal

$$w \leqslant \frac{m}{M} \sim 10^{-8} \div 10^{-5}. \tag{10.43}$$

The probability of merger of the clouds of the same mirrority, that is, of two clouds of ordinary matter or two fragments of mirror matter is determined by the cross section of their collision

$$\sigma_c \cong 4\pi r^2 \left(1 + \frac{v_1^2}{v_0^2}\right). \tag{10.44}$$

Therefore, the characteristic collision time for one fragment is of the order

$$t_c \sim \frac{R^3}{N\sigma_c v_0}. \tag{10.45}$$

At the beginning of contraction, when r is large, this characteristic time is shorter than the free fall time in a cluster, which corresponds to

$$t_c < \frac{R}{v_0}. \tag{10.46}$$

Thus, only one of N fragments could have a comparable amount of ordinary and mirror matter which must form stars during the subsequent fragmentation.

The admixture of mirror (M) matter in the remaining fragments of ordinary (O) matter (and vice versa) is provided by uniform filling of the pancake with the O + M gas. The relative fraction attributable to this admixture is a small quantity

$$f \sim \frac{\langle \rho \rangle}{\rho_i}, \tag{10.47}$$

which is the ratio of the average gas density in the pancake (or later in the galaxy) and the average density inside the object in question.

At the stage of thermal instability the pressure of hot gas with temperature $T \sim 10^6$ K is nearly equal to the pressure of cold gas with temperature $T \sim 10^4$ K, so that the relative fraction attributable to the admixture of matter with different mirrority is given by

$$f \sim \frac{\langle \rho \rangle}{\rho_i} \cong \frac{10^4}{10^6} = 10^{-2}. \tag{10.48}$$

This part freely (apart from the effects of accretion, discussed below) passes through the inhomogeneities and is not involved in the processes of their evolution, including the processes of star formation. At smaller scales the density contrasts increase, respectively reducing the fraction of the mirror admixture. Thus, the average density of the Galaxy is

$$\langle \rho \rangle \cong 10^{-24} \, \mathrm{g/cm^3}, \tag{10.49}$$

and in the volume of the Solar system, the free flux of mirror matter (taken inside the Sun) provides an admixture not exceeding

$$M_{\mathrm{M}} \leqslant 10^{21} \mathrm{g}, \tag{10.50}$$

and the proportion of mirror matter in the Earth according to the estimates (Blinnikov, Khlopov 1983) should be less than

$$f \sim 10^{-24} \tag{10.51}$$

that seems almost impossible to detect.

Therefore, in future consideration the separation of ordinary and mirror matter within the individual fragments with mass

$$M \sim 10^6 \div 10^9 M_{\odot}, \tag{10.52}$$

will mean that the average probability that the fraction of admixture of matter with different mirrority in these fragments is greater than

1%, is not exceeding

$$w \leqslant 10^{-5}. \tag{10.53}$$

The largest single fragments with the mass

$$M \sim 10^{9} M_{\odot} \tag{10.54}$$

apparently form dwarf galaxies. The admixture of ordinary matter in the mirror dwarf galaxies should not exceed 1% and can be detected only in the effect of a gravitational lens.

On the other hand, in the case of the hot dark matter scenario, the admixture of mirror matter and hot dark matter (massive neutrinos) in the dwarf galaxies from ordinary matter must be small and, thus, they practically do not contain dark matter.

The cold dark matter scenario predicts small-scale inhomogeneities of dark matter which may lead to the presence of dark matter in dwarf galaxies as well as the comparable amounts of ordinary and mirror matter in such galaxies.

On the scale

$$M < 10^{6} M_{\odot} \tag{10.55}$$

the estimates of mixing of ordinary and mirror matter mentioned in the above scenario for the case of hot dark matter, are completely valid in the scenario of the cold dark matter.

The ordinary and mirror fragments corresponding to these small scales, form, together with dark matter (hot or cold), the normal and giant galaxies. Dissipation processes play a very important role in establishing a balance in the gas components.

Because the dissipation laws are strictly symmetrical in the ordinary and mirror gases, and condensation of the already formed stellar systems is determined by a common gravitational interaction, we should then expect that the ordinary and mirror matter, almost completely separated on the scale of single fragments, have the same distribution on the scale of galaxies.

The almost complete separation of the ordinary and mirror matter on the scale (10.55) leads to the formation of stars and stellar systems constructed from the matter with definite mirrority (not considering here a small admixture due to accretion and the specific conditions under which the mixing is enhanced – see below). With the same caveats, in general, the formation of binary stars like (mirror-star) - (ordinary star) or of stars from a comparable amount of mirror and ordinary matter is strongly suppressed.

Indeed, such suppression can be estimated under normal conditions in the galaxy (Blinnikov, Khlopov 1983). This estimate is significantly increased in some specific cases, such as molecular clouds or globular clusters, which will be discussed in Section 4.2.

The process of capturing the stars in a binary system requires either a triple collision, or a double collision with a strong tidal dissipation. At the normal number density of stars in the Galaxy, which is equal to

$$n_s \sim 0.1\ \mathrm{pc}^{-3} \tag{10.56}$$

these processes cannot take place. They should be considered only in dense star clusters or galactic nuclei. Blinnikov and Khlopov (1983), following (Fabian et al 1975; Press, Teukolsky 1977) estimated the rate of formation of binary systems due to tidal dissipation.

In a cluster with radius R, which contains solar-type stars, the rate of formation of binary systems of ordinary stars is

$$\dot{N}_{\mathrm{bin}} \cong 3\left(\frac{N}{10^5}\right)^{1.4}\left(\frac{R}{5\mathrm{pc}}\right)^{-2.4}\left(10^9\ \mathrm{years}\right)^{-1}. \tag{10.57}$$

In clusters of ordinary stars, the rate of formation of a double system of the (ordinary) – (mirror) star type is multiplied by a factor which takes into account a small fraction of mirror matter and is usually evaluated as very small.

For example, if an admixture of mirror matter is even of the order of 1%, the rate of formation of such binary systems is

$$\dot{N}_{\mathrm{bin}}\left(\mathrm{O}-\mathrm{M}\right) \sim 0.3\cdot\left(10^{10}\ \mathrm{years}\right)^{-1} \tag{10.58}$$

(Blinnikov, Khlopov, 1983).

In galactic nuclei where the amount of ordinary and mirror matter is expected to be comparable, the number of binary systems of the (ordinary) – (mirror) star type must be equal to the number of binary systems of the (ordinary star)–(ordinary star) type but individual stars are not observable there.

Because of the expected strict symmetry between the ordinary and mirror interactions, the evolution of mirror stars should take place in the same fashion as for the ordinary stars of the same mass. As a result of such evolution a mirror star should form a mirror white dwarf, a mirror neutron star, a black hole (whose properties are obviously not dependent on the physical nature of the matter from which it was formed), or it should be completely destroyed.

On the other hand the average densities of ordinary and mirror matter (for example, ordinary and mirror gas) in galaxies should be equal. Strict symmetry is also expected in the distribution on types of objects and in the mass and velocity distribution of these objects.

In particular, the strict symmetry of the distributions should lead to the existence of non-luminous compact objects in the spherical component of our Galaxy with the density of the order of baryonic density. Such objects (MACHO) were observed in the galactic halo by the effect of microlensing, and we should seriously consider the interpretation of such objects based on the hypothesis of the existence of a mirror world (Blinnikov 1996; Khlopov 1999).

A single disk with the comparable amount of mirror and ordinary matter should form at comparable orbital angular momenta of galactic ordinary and mirror components. Separate ordinary and mirror disks apparently cannot form (Blinnikov, Khlopov 1983). Their combined attraction should lead to a merger into a single disk. The average density of dark matter (mirror matter) in the disk must be equal to the density of visible matter.

The question of the existence of this 'local dark matter' has a rather long history (Oort 1965; Gliese 1977). Observational estimates (Hill et al 1979; Jones 1972; Balakirev 1976; House, Kilkenny 1980) indicating that the density of dark matter in the vicinity of the solar system should not exceed 30% of the density of visible matter are, as it turns out (Blinnikov, Khlopov 1983), insufficiently accurate to exclude local dark matter of the order of the density of visible matter. This possibility was supported by a series of articles (Bahcall 1984; Bahcall et al 1992), where the total local density near the Solar system was estimated as

$$\frac{\rho_{\text{tot}}}{\rho_v} \sim 2.3, \qquad (10.59)$$

where ρ_v is the mean density of objects from visible matter in the local part of the disk components of the Galaxy.

Note that the formally symmetrical mirror world implies that

$$\frac{\rho_{\text{tot}}}{\rho_v} = 2, \qquad (10.60)$$

but with the account for the ordinary normal stars with low luminosity, such as brown dwarfs, and appropriate mirror stars, as well as a small contribution to the density of the disk of dark matter particles, forming a massive galactic halo,

$$\frac{\rho_{dm}}{\rho_v} \lesssim 0.1 \qquad (10.61)$$

the results of Bahcall, provided they are correct (see critical notes by Kuijken and Gilmore 1989, 1991), would form a serious argument in favour of the mirror world. However, the implementation of the HIPPARCHOS program, greatly increasing the statistical reliability of the data on the local dark matter, preclude these results, indicating that its density is less than 50% of that the visible matter.

It should be emphasized that although the local dark matter, if it exists, provides a rather small, of the order of the baryon density, contribution to the total density

$$\Omega_{ldm} \sim \Omega_b \leqslant 0.05 \qquad (10.62)$$

and plays no significant dynamical role in the cosmological evolution, its physical meaning is difficult to overestimate.

The reason for this is that weakly interacting particles, considered as candidates for cosmological dark matter are a collisionless gas, which condenses into a visible part of the galaxy according to the law (Chapter 9)

$$\frac{\rho_{dm}(t)}{\rho_{dm}(0)} \propto \left(\frac{\rho_b(t)}{\rho_b(0)}\right)^{3/4}, \qquad (10.63)$$

so that no more than 10% of the density of visible matter can be provided by particles of dark matter halos. Consequently, the local dark matter with the density comparable to the baryon density, must have a non-trivial physical nature. It must have an effective mechanism for dissipation as it is the case for baryonic matter.

The simplest solution is to try to explain the existence of local dark matter by some low-luminosity objects, such as brown dwarfs. If this solution does not pass, we are faced with the problem of the existence of dissipated dark matter, such as shadow or mirror matter. This leads to the dark matter particles that interact weakly with ordinary matter, but have a much stronger and even long-range interaction between them.

3.2. Accretion of gas on astronomical objects with different mirrority

The sufficiently pure separation of the ordinary and mirror matter in astronomical objects with mass

$$M < 10^6 M_\odot \tag{10.64}$$

makes the process of accretion of ordinary gas on the mirror objects (and vice versa) most important when mixing of ordinary and mirror matter in these objects (Blinnikov, Khlopov 1983).

According to the classical accretion theory, the gravitating body with mass M and velocity v traps matter at a distance

$$R_A \sim \frac{2GM}{v^2}. \tag{10.65}$$

For subsonic motion, the velocity v must be replaced by the speed of sound v_s.

From the standard formula for the mass flow

$$\dot{M} = \pi R^2 \rho v \tag{10.66}$$

we find that for constant boundary conditions the mass absorbed during time t is expressed by the ratio

$$\frac{\Delta M}{M} \cong 10^{-5} \left(\frac{M}{M_\odot}\right)\left(\frac{10\ \text{km/s}}{v}\right)^3 \left(\frac{\rho}{10^{-24}\ \text{g/cm}^3}\right)\left(\frac{t}{10^{10}\ \text{years}}\right). \tag{10.67}$$

Thus, even for a large globular cluster with the mass

$$M \sim 10^6 M_\odot, \tag{10.68}$$

and the velocity of movement

$$v \sim 100 \frac{\text{km}}{\text{s}}, \tag{10.69}$$

the amount of accreting matter is negligible.

The same is true for dwarf galaxies, having masses

$$M \sim 10^9 M_\odot \tag{10.70}$$

and velocities

$$v \sim 10^2 \div 10^3\ \text{km/s}. \tag{10.71}$$

Because of the very low density gas in galaxy clusters which is about

$$\rho \sim 10^{-27}\,\text{g/cm}^3, \tag{10.72}$$

we obtain from (10.67) that

$$\frac{\Delta M}{M} \sim 10^{-5} \div 10^{-2}. \tag{10.73}$$

In fact, equation (10.67) is strictly true only for accretion onto the star. In the case of star clusters and galaxies, it gives only an upper limit on the mass fraction of accreted matter. The reason for this is that the normal velocity v of such objects

$$v \geqslant v_1 \tag{10.74}$$

exceeds the escape velocity v_1, so that the capture radius

$$R_A \leqslant r \tag{10.75}$$

is smaller than the radius of the cluster r. This makes the gravitational potential quite inefficient for capture. If the radius of accretion onto the mirror star is small compared to the radius of capture

$$r \ll R_A, \tag{10.76}$$

the fall of ordinary matter in such a star would lead to the luminosity of the order

$$L \sim \frac{G\dot{M}M}{r}. \tag{10.77}$$

If the density of the accreting material is of the order of (10.72), this luminosity is very low even in the case of accretion onto the neutron star, that is such a star can be seen only from a relatively small distance of a few tens of parsecs.

During the evolution of mirror stars of the order or

$$t \sim 10^{10}\ \text{years}, \tag{10.78}$$

and for the radius of the mirror neutron star

$$r \sim 10^6\ \text{cm} \tag{10.79}$$

with the mass of the order of the mass of the Sun the mass of ordinary matter contained in it is estimated as

$$\Delta M \sim 10^{26} \div 10^{28}\ \mathrm{g} \tag{10.80}$$

(Blinnikov, Khlopov,= 1983).

When the accreting ordinary matter loses angular momentum it is compressed. In accordance with the equilibrium conditions, compression in the gravitational field of a mirror star takes place up to densities

$$\rho \sim 10^{9} \div 10^{10}\ \mathrm{g/cm^{3}} \tag{10.81}$$

inside the mirror neutron star (10.79).

The magnetic fields, frozen in the accreting matter, experience rapid change when the matter reaches the centre of an optically opaque central part of nucleus of ordinary matter in the mirror star. V. Shvartsman estimated size of such a nucleus

$$r_{cO} \sim 10^{5}\ \mathrm{cm}, \tag{10.82}$$

and the characteristic time of rapid change in luminosity in the photosphere of the nucleus of ordinary matter because of rearrangement of the magnetic fields, should be

$$\Delta t \sim \frac{r_{cO}}{c} \sim 3\cdot 10^{-6}\ \mathrm{s}. \tag{10.83}$$

As pointed out by Shvartsman, we can use methods for searching the effects of rapid variations in the MANIA experiment (Shvartsman, 1977) and to separate the effect of a mirror neutron star from the similar effects of isolated neutron stars and black holes.

The idea was based on an estimate of the gravitational radius of a black hole with the mass of a few solar masses, and the radius of the surface of an ordinary neutron star, which must be an order of magnitude greater than the radius of the photosphere of the nucleus of ordinary matter in the mirror neutron star. In the case of the mirror-neutron star this results in a very short variation time and this can not be explained by similar effects from a single black hole or an ordinary neutron star.

For a larger radius of the mirror star its ordinary luminosity due to accretion of ordinary matter correspondingly decreases, and accreting ordinary matter cools down and forms objects such as planets or comets in a potential well of the mirror stars, if its potential does not destroy such objects by tidal forces. Indirect effects of such a mirror star can be seen only at a very short distance form the Solar system, less than 1000 astronomical units. Because of the very small probability of formation of a binary (ordinary star) – (mirror star) system, the presence of such a mirror star in the close vicinity of the Sun is very improbable.

The most stringent constraints on the existence of a nearby invisible companion of the Sun of any physical nature follow from the spatial distribution of pulsars with anomalously low rates of deceleration. Acceleration of the Solar system in the direction of an unseen companion would have shifted the distribution to the blue edge, thus causing its anisotropy. Using the observational data for the distribution of pulsars with measured rates of deceleration in the celestial sphere, Blinnikov and Khlopov (1983) derived a restriction on any invisible companion with a mass M at a distance d from the Sun

$$M \leqslant 1 M_{\odot} \left(\frac{d}{1000 \text{ a.m.u.}} \right)^2 . \quad (10.84)$$

According to equation (10.67) at an average density of mirror matter in the galactic disk of the order

$$\rho \sim 10^{-24} \text{ g/cm}^3 \quad (10.85)$$

and the velocity of the Sun relative to the disk components of the order

$$v \sim 10 \div 20 \text{ km/s} \quad (10.86)$$

the Sun during its lifetime can capture approximately

$$\Delta M \sim 10^{27} \div 10^{28} \text{ g} \quad (10.87)$$

of mirror matter. Some of the mirror matter can have angular momentum and can form planets orbiting inside the Sun near its surface where the gravitational force is maximal. Blinnikov and Khlopov (1980) estimated that if such a planet has a mass

$$M_{\text{M}} \sim 10^{26} \text{ g} \quad (10.88)$$

and moves at a depth of about

$$h = 2 \cdot 10^4 \text{ km} \quad (10.89)$$

beneath the Solar surface, then this planet can cause non-radial oscillations of the Sun with the observed period

$$T = 160 \text{ min} \tag{10.90}$$

(North et al 1976; Severnyi et al 1979).

The existence of such a planet, consisting of mirror matter, is consistent with observations of the shift of Mercury's orbit. Such a planet can be detected using high-precision gravimetric measurements.

The specific angular momentum, which is required for the formation of a planet (it is an order of magnitude smaller than the angular momentum of Mercury) could be due to fluctuations of the order of 1% of the mirror gas at the accretion radius

$$R_A \sim 10^{14} \text{ cm}. \tag{10.91}$$

The mirror planet can be formed outside the Sun, and then descend along a spiral inside the Sun due to tidal friction.

Consider, following Blinnikov and Khlopov (Blinnikov and Khlopov 1983, Blinnikov and Khlopov 1983), the effects caused by movement of the mirror planet (M-planet) in the solar interior.

According to the standard solar model, the temperature inside the Sun at a depth of (10.89) is about

$$T \sim 10^5 \text{ K}, \tag{10.92}$$

so that the speed of sound is

$$v_s \cong 3 \cdot 10^6 \text{cm/s}. \tag{10.93}$$

The velocity of the M-planet in a circular orbit is $v = 4.4 \cdot 10^7$ cm/s, so that its accretion radius is $r_A \sim 10^4$ cm

If the radius of the planet is

$$r \geqslant 10^8 \text{ cm}, \tag{10.94}$$

we have

$$\frac{r}{r_A} \gg \frac{v}{v_s}, \tag{10.95}$$

and a rather unusual case of accretion (Blinnikov, Khlopov 1980).

For

$$r > r_A \tag{10.96}$$

the substance of the Sun is not captured by the mirror planet and if the condition (10.95) is fulfilled, the rate of perturbation of the incoming gas Δv everywhere is much less than the speed of sound,

$$\Delta v \ll v_s. \tag{10.97}$$

This means that even if the accretion shock wave forms, it will be weak and heating of the solar matter will be negligible, so that the corresponding energy release per unit time is of the order

$$\dot{E} \sim \dot{M}\frac{(\Delta v)^3}{v_s}, \tag{10.98}$$

where

$$\dot{M} \sim \pi r^2 \rho v \tag{10.99}$$

is the mass of the matter of the Sun which will be perturbed per unit time (but not captured!) by the mirror planet.

Thus, the mirror planet should cause only weak gravitational and weak acoustic effects.

It is interesting to note that the search for a mirror planet inside the Earth are quite groundless. The probability of the existence of the primordial mirror matter in the Earth is negligible and the accretion of mirror matter can not lead to its preservation in the Earth, because the velocity of mirror matter near the Earth is determined by the gravity of the Sun and can not be less than

$$v_{oE} = \sqrt{2}v_E = 42 \text{ km/s}, \tag{10.100}$$

where the velocity of the orbital motion of the Earth is

$$v_E \cong 30 \text{ km/s}. \tag{10.101}$$

Thus, the velocity of mirror matter, accreted by the Solar system near the Earth, is higher than the escape velocity for the Earth and no capture of mirror matter by the Earth takes place.

4. Observational physics of mirror matter

Based on the scheme of the evolution of mirror matter, discussed above, and taking into account the possible effects of 'island' distribution of baryons, we consider the observable effects of mirror matter predicted in various astronomical scales.

From the above consideration, it is clear that all possible observational effects of mirror matter should be due solely to gravitational interaction with ordinary matter. Taking into account the possibility of the island distribution of mirror baryons it is evident that any types of mirror objects can exist.

In the most general terms one can distinguish two types of effects. This is a case of pure gravitational interaction and the situations in which the gravitational effect causes the gas dynamic effects.

In the first case, the effects of mirror matter are manifested in the peculiar velocities of objects. The effects of this type can be called 'kinematic'. It is clear that they mostly occur in cases where an ordinary object is situated in the gravitational field of some mirror configuration with a much greater mass.

In the second case, we consider the effects arising due to the gravitational effects of various types of object on the gas of different mirrority.

We discuss in more detail some examples of interaction between different objects with different mirrority which can be investigated by experimental means.

4.1. Galaxies and clusters of galaxies with specific mirrority

In the case of an island-like distribution of baryons in galaxies or clusters of galaxies, these astronomical objects can have a certain mirrority, i.e. they consist mainly of either ordinary or mirror matter. Possible admixtures of ordinary matter in galaxies or of mirror matter in ordinary galaxies can arise either because of small local mirror asymmetry in the baryosynthesis processes (see Section 2.2.) or due to accretion of intergalactic gas on such objects.

There are the following observed effects of the existence of mirror galaxies and clusters of galaxies.

I. Capture of normal galaxies by a cluster of mirror galaxies can lead to the formation of objects with large peculiar velocities or of small groups of galaxies with velocities that lead to an anomalous virial paradox, i.e., with the velocity dispersion of up to

$$\left\langle v^2 \right\rangle^{1/2} \sim (1 \div 2) \cdot 10^3 \text{ km/s}, \tag{10.102}$$

which is typical of dense rich clusters of galaxies (see Khlopov et al 1991 and references therein).

The peculiar velocity components of massive galaxies can be found in principle at the level of

$$v_{\text{pec}} \leqslant 10^3 \text{ km/s}, \tag{10.103}$$

by the methods proposed by Zeldovich and Sunyaev (1982) to measure the peculiar velocities of galaxy clusters, i.e. to measure the distortions of the blackbody background radiation from the scattering of electrons in the gaseous halo of the galaxy.

The probability of capture of galaxies by rich clusters is quite high.

Indeed, suppose that the dissipation of energy required for capture occurs at the distances between the centers of galaxies of the order of the diameter of the galaxy d. Then the clusters containing N galaxies and having diameter D, will capture the background galaxies with a probability

$$w = n\pi \cdot d^2 \cdot \frac{2}{3} D = 4N\left(\frac{d}{D}\right)^2 \cong 0.01 \div 1. \tag{10.104}$$

Here

$$n = \frac{N}{\frac{\pi}{6} D^3} \tag{10.105}$$

is the number density of galaxies in the cluster. Numerical estimates are given for the rich clusters with

$$N = 10^3 \div 10^4 \tag{10.106}$$

galaxies, and for the ratio of diameters of galaxies and clusters of galaxies which is equal to

$$\frac{d}{D} = 10^{-3} \div 10^{-2}. \tag{10.107}$$

II. In the above process of capturing galaxies from ordinary matter by a mirror cluster of galaxies a significant amount of gas is inevitably lost. As a result, ‘poor’ clusters of galaxies formed in the potential of the mirror cluster can have a significant amount of intergalactic gas that fills the region with the size typical for rich clusters.

The quantity and hence the density of intergalactic gas should be a factor of

$$k = \alpha \frac{N_{\text{M}}}{N_{\text{O}}} \tag{10.108}$$

smaller than in rich clusters. Here N_M is the number of mirror galaxies in a rich cluster, and N_O is the number of ordinary galaxies captured by the rich cluster. The numerical factor $\alpha > 1$ takes into account that in contrast to the capture by the rich cluster of galaxies with the same mirrority the cooling flow (see work Khlopov et al 1991 and references therein) does not arise and the gas does not leave the cluster.

Thus, when

$$\frac{N_O}{N_M} \approx 10^{-2} \tag{10.109}$$

we can expect that the suppression factor is equal to

$$k^{-1} \approx 0.03 \tag{10.110}$$

and the emission measure of the intergalactic gas will be

$$k^{-2} \approx 10^{-3} \div 10^{-2} \tag{10.111}$$

times less than in the case of rich clusters of the galaxies.

The X-ray telescopes can be used to search for QSO up to the redshift of $z = 5 \div 10$ redshift and the intergalactic gas at redshifts $z = 1 \div 3$ in rich clusters with redshifts up to $z = 2 \div 4$.

Consequently, in the case considered here, the intergalactic gas can be observed at redshifts

$$z = 1 \div 3. \tag{10.112}$$

Observations of the hot intergalactic gas with no apparent rich cluster of galaxies can be a strong argument for the existence of mirror (shadow) matter.

III. Gravitational interaction of the ordinary and mirror galaxies leads to a distortion of their shape. The shape of the ordinary galaxies should be distorted without any visible reason for it. Development of numerical methods for calculating the tidal interactions between galaxies (see Toomre, Toomre 1972; Khlopov et al 1991 and references therein) may provide a solution for the inverse problem of determining the parameters of the body, causing distortion, on the basis of the form of distorted galaxies.

In contrast to the pertubations caused by single black holes with the same mass, the perturbations from the mirror galaxies are not associated with the observed effects of accretion onto such a black hole. The corresponding effects of accretion on the active galactic nucleus of the mirror galaxy will be suppressed by the ratio of the mass of the galactic nucleus M_{nuc} to the total mass of the galaxy M_{gal}, which is of the order

$$\frac{M_{nuc}}{M_{gal}} \leqslant 10^{-2} \div 10^{-4} \tag{10.113}$$

for active galactic nuclei.

IV. Gravitational perturbation caused by the mirror galaxy can trigger a burst of star formation in ordinary gas-rich galaxies, or in the protogalactic cloud of an ordinary gas. As a result, the ordinary galaxy will be observed as irregular.

Phenomena III and IV can also be caused by non-luminous clusters of gravitationally bound dark matter, which could be formed due to biasing in the distribution of baryonic and dark matter.

However, the collisionless gas of particles of dark matter can not form dense inhomogeneities typical for mirror matter in which a dissipation mechanism operates. So, we should consider the effects caused by the mirror matter of intermediate density which differ from the effects of both black holes and diffuse weakly interacting dark matter.

V. If the activity of galactic nuclei is determined by the existence of black holes with masses of order

$$M_{BH} \sim 10^{6} \div 10^{10} M_{\odot}, \tag{10.114}$$

then the symmetry between the ordinary and mirror matter should indicate the existence of black holes in the nuclei of the mirror galaxies. In this case, mirror galaxies with active nuclei can be seen as isolated supermassive black holes and their observable properties will be determined by the amount of ordinary matter in their vicinity.

As stated in (Blinnikov, Khlopov 1983), the presence of mirror matter in the mixed galaxies (with ordinary and mirror matter) makes it easy to explain the origin of close binary systems of supermassive black holes in galactic nuclei.

VI. Massive mirror galaxies can cause a gravitational lens effect without any optically visible source of this effect.

VII. Fast movements of the mirror masses can be a source of gravitational waves, without any kind of observable effects. They can be distinguished from the supermassive black holes, which can also be a source of gravitational radiation, on the basis of the effects of accretion onto such black holes.

VIII. The presence of an ordinary gas in a mirror galaxy leads to formation of single clouds or low-mass galaxies with a large internal velocity dispersion, i.e., with a large internal virial paradox.

Intergalactic gas clouds can be considered as candidates for such formation. In a number of neighboring clusters of galaxies there is cold gas in the form of massive clouds of HI with the mass

$$M > 10^6 M_\odot \tag{10.115}$$

(Haynes 1979) and the sizes that are typical for galaxies

$$r \sim 20 \div 25 \text{ kpc.} \tag{10.116}$$

The most massive of all known highly ionized clouds was discovered (see the references in Khlopov et al 1991) in the G11 group of galaxies in the constellation Leo near the galaxy M96.

This cloud has a size not less than

$$r \geqslant 30 \div 100 \text{ kpc.} \tag{10.117}$$

The mass of highly ionized hydrogen in it is evaluated as

$$M > 10^9 M_\odot \tag{10.118}$$

and the concentration

$$n = 4 \cdot 10^9 h_{50} \text{ cm}^{-3}, \tag{10.119}$$

where the dimensionless Hubble constant is defined as

$$h_{50} = \frac{H}{50 \dfrac{\text{km}}{\text{s} \cdot \text{Mpc}}}. \tag{10.120}$$

The surface luminosity of the cloud in the optical range is less than 30 stellar magnitudes per unit of arcsec. Such clouds can affect the absorption spectra of QSO.

4.2. Mirror matter on the scale of globular clusters

Now consider the case of separation of mirror and normal matter at the scale of globular clusters.

Globular clusters are among the oldest astronomical objects that formed possibly before the formation of galaxies from the inhomogeneities

$$M > 10^6 M_{\odot}. \qquad (10.121)$$

They are objects of a definite mirrority, even if initially consisted of homogeneously mixed ordinary and mirror matter (see §3).

I. Capture of ordinary stars by the mirror globular cluster may lead to the formation of an opened cluster of ordinary stars with a long lifetime and strong virial paradox.

The best chance to form such objects by the capture of ordinary stars have mirror globular clusters, moving near the plane of the galactic orbits with not a very large eccentricity. But even better opportunity to create an opened cluster in the gravitational field of a globular mirror cluster occurs when the opened cluster is formed from ordinary gas, captured by the mirror globular cluster in the period of separation of matter of different mirrority.

Part of the gas with different mirrority in this case is about

$$f \approx 10^{-2} \qquad (10.122)$$

(Blinnikov, Khlopov 1983, see § 3).

The lifetime of normal opened clusters with respect to decay (Wielen 1971)

$$t_{dc} = (1 \div 3) \cdot 10^8 \text{ years}, \qquad (10.123)$$

whereas the age of such clusters, forming in a potential-type mirror globular cluster may be as large as

$$t_{dc} \approx 10^{10} \text{ years}. \qquad (10.124)$$

There are several old opened clusters in the Galaxy with a sufficiently long lifetime.

For NGC 188 the age is estimated as (Vandenberg 1983; Janes, Demarque 1983)

$$t_{dc} = (5 \div 10) \cdot 10^9 \text{ years}, \qquad (10.125)$$

For M67 the age is estimated as (Nissen et al 1987)

$$t_{dc} = (5 \pm 0.5) \cdot 10^9 \text{ years}, \qquad (10.126)$$

The age of NGC 752 is estimated as (Twarog 1983)

$$t_{dc} \approx 2 \cdot 10^8 \text{ years.} \tag{10.127}$$

For NGC 2243 and Melloffe 66 their age is estimated at (Gratton 1982)

$$t_{dc} \approx 6 \cdot 10^9 \text{ years.} \tag{10.128}$$

The stars captured by the mirror globular clusters may have a different age in contrast to the ordinary opened clusters.

II. The capture of ordinary stars by a mirror globular cluster or of a mirror star by an ordinary globular cluster may be followed by the formation of similar closed binary systems with components of different mirrority, O–M binary systems. The existence of the non-relativistic invisible companion is a specific feature of such systems (see below).

Mixed (O–M) globular clusters can form in cases where clouds of gas of ordinary and mirror matter, generating globular clusters, are close in phase space (see §3).

In addition to the explicit virial paradox in compact mixed O–M globular clusters with radius R and containing N stars over time t tidal scattering may results in the formation of binary systems as a result of a large concentration of stars. For the O–M clusters with $N = 2 \cdot 10^5$ stars, the number of binaries that form within the radius $R = 5$ pc during 10^{10} years is

$$N_{\text{bin}} \cong 60 \tag{10.129}$$

and nearly half of them will be mixed.

III. As was shown in (Khlopov et al 1989, 1991), the effects of mirror matter in ordinary molecular gas clouds are significantly enhanced when compared to the general case considered in section 3.1.

Giant molecular clouds are most abundant among the massive objects in galaxies. With the mass of order of the mass of a globular cluster, i.e.

$$M = 10^4 \div 10^6 M_\odot, \tag{10.130}$$

these clouds are an order of magnitude more numerous than the globular clusters in the Galaxy (see Khlopov et al 1991 and references therein). Their size is also an order of magnitude larger than the size of a globular cluster.

Consequently, in a galaxy containing a comparable amount of ordinary and mirror matter, the interpenetration of the ordinary and mirror clouds into each other must occur quite frequently.

Molecular clouds contain a lot of internal inhomogeneities, i.e. areas with high density and close to the development of gravitational instability.

The structure of star forming regions shows that even relatively small perturbations can cause the process of star formation in molecular clouds.

The usual source of bursts of star formation are shock waves so that the youngest objects are found in the outer layer of molecular clouds (see Khlopov et al 1991 and references therein).

Passage of a massive body through a molecular cloud may also initiate the process of star formation. These star-forming regions can be found by the spatial distribution of gravitational perturbations, and they can occupy a significant part of the cloud.

In the case of comparable amounts of ordinary and mirror matter in the Galaxy which we consider here, such pertubations may be caused, above all, by a mirror molecular cloud, and with the probability of an order of magnitude smaller by a mirror globular cluster.

Because molecular clouds collide with low relative velocities $v \sim 10$ km/s and are strongly dissipating objects, the collision of ordinary and mirror clouds can form a giant molecular cloud with mixed mirrority, with the probability of star formation with mixed mirrority and of binary O–M systems in this cloud being considerably higher.

Repeated gravitational interactions of stars with inhomogeneities within molecular clouds may lead to the capture of some stars by molecular clouds, greatly increasing the effects of accretion of interstellar gas on these stars. This leads to a higher rate of accretion of gas onto stars with different mirrority, that is, O-gas on M-stars or M-gas on O-stars.

4.3. Effects of mirror matter at the scale of stars

In section 3 we have already noted some effects of existence of mirror matter in the stellar scale. We present here a somewhat more systematic approach to such effects which was proposed in (Khlopov et al 1989, 1991).

I. Accretion of intergalactic gas on the stars with different mirrority leads to the admixture of matter with another mirrority in the stars.

Estimates based on analysis of the processes of accretion, presented in section 3, give the following values for the admixture of ordinary matter to the mirror stars (and, conversely, for admixture of M-gas to the O-star) $\Delta M \sim 10^{-7} \div 10^{-6} M_{\odot}$.

If such an admixture of mirror matter in the Sun forms a mirror planet near the Solar surface, this may explain the source of non-radial solar oscillations with a period $T = 160$ min.

V. Shvartsman pointed to the possibility of observing accretion of interstellar ordinary gas on a single mirror neutral star on the basis of the effects of rapid variations of luminosity (see Section 3.2 for more detail).

II. Rotation of an ordinary neutron star and mirror matter, captured by the star, around a common center of mass leads to periodic variations in the pulsar period due to the Doppler effect.

In contrast to the same effect of difficult to observe planets of ordinary matter the variations produced by the mirror matter can have a period corresponding to the orbits that are so close that the formation of normal planets, and even more so their survival after a supernova explosion is not possible.

The search for such effects require precise 'timing' of pulsars in the characteristic time intervals shorter than several hours.

III. A protoplanetary disk without a young star in the centre should form as a result of the accretion of ordinary gas and dust on the mirror stars in interstellar space.

Such disks can be found by the radio line of the CO molecule – 2.6 mm.

The mass of the central configuration can be determined by the Doppler effect, it will be strongly contrary to the low luminosity of the central body.

Accretion in areas with a high density of gas of the other mirrority and the formation of binary systems of mixed mirrority provide several new features in search for mirror matter.

As stated in paragraph 4.2, the O–M binary systems can form inside a globular cluster also during star formation in the course of the mutual penetration of giant ordinary and mirror molecular clouds.

IV. Star formation or the evolution of the mixed binary O–M systems may lead to the formation of a mixed O–M star, containing a comparable amount of ordinary and mirror matter.

For ordinary matter in stars of mixed mirrority the relationship between the main stellar parameters (mass, radius, colour, effective temperature, etc.) can vary greatly.

In particular, these stars can occupy an unusual position in the Hertzsprung–Russell colour-luminosity diagram.

A supernova explosion in the mirror component of the mixed O–M star is followed by the rearrangement of its gravitational field. This inevitably leads to the restructuring of the star and a simultaneous change in the properties of an optically visible object. As regards the star, its size should increase and the surface temperature should decrease.

V. In close binary O–M systems, accretion of ordinary matter in a potential well of the mirror companion should lead to the formation of an accretion disk with no apparent centre of accretion.

In the case of a non-relativistic mirror star when the potential has a flattened shape of the minimum a thick accretion disk or a spherical configuration of the accreting ordinary matter should form around the invisible mirror companion. The configuration that is characteristic of highly non-spherical stars forms when the mass exchange rate is typical of close binary stars.

One of the promising methods for studying such objects is to analyze the linear polarization of their radiation (Khlopov et al 1991 and references therein).

In the case of the low mass ordinary companion in such a binary system, the typical disk around the mirror companion should remain fairly cool and look like a faint infrared source with a mass sharply contradicting the value of the 'disk mass' which is determined by the Doppler effect and is equal to the mass of the mirror star.

In the broad O–M binary systems, the effects of accretion onto a non-degenerate mirror star may be absent. Such stars can be detected as invisible companions of an ordinary massive star.

Estimation of the proportion of O–M systems amongst the binary stars can be found in the existing catalogs of spectral binary systems.

Catalogues (Batlen et al 1978; Pedoussaut et al 1984, 1988) include the physical parameters and orbital elements for about 1500 spectral binary systems.

Given that the theoretical upper limit on the mass of the neutron star is

$$M_{ns} < 3M_{\odot} \tag{10.131}$$

(Baym, Pethick, 1979) one could put a constraint in the analysis

$$M_{inv} > 3M_{\odot} \tag{10.132}$$

on the mass of the second invisible companion, in order to eliminate the systems of faint white dwarfs and neutron stars.

45 binary systems in the catalogues (Batlen et al 1978; Pedoussaut et al 1984, 1988) satisfy this criterion, since the mass function for these binary systems is

$$f(M) > 3M_{\odot}. \tag{10.133}$$

Among these 45 binary systems, only six do not have lines corresponding to the second companion. Taking into account the possibility that the second companion is a faint, ordinary star or black hole, Khlopov et al (1989) estimated the proportion of O–M binary systems

$$\alpha < \frac{6}{45} = 13\%. \tag{10.134}$$

VI. Consider (Khlopov et al 1989; 1991) now the properties of these six binary systems and the ability to distinguish between black holes and massive mirror stars as invisible companions.

Parameters of the six spectral binary systems are given in the catalogue (Batlen et al 1978; Pedoussaut et al 1984, 1988).

These binary systems do not contain the lines belonging to the second companion.

The mass function of these binary systems satisfies (10.133).

From the definition of the mass function

$$f(M) = \frac{M_2 \sin^3 i}{\left(1 + \frac{M_1}{M_2}\right)}, \tag{10.135}$$

where i is the inclination of the orbit, we find that

$$M = f(M)\frac{\left(1 + \frac{M_1}{M_2}\right)}{\sin^3 i} > f(M) > 3M_{\odot} \tag{10.136}$$

So in both cases, the second companion cannot be a white dwarf or neutron star.

If we consistently analyze this system, we definitely exclude as a second companion ordinary stars at the stage of ignition of thermonuclear combustion in the centre of the star, which means that the system has either a black hole or a mirror star.

The main criterion for choosing between these two options is the total rate of energy release in a binary system.

In case of the black hole, this rate should be several orders of magnitude higher than in the case of a mirror star.

Taking into account the possible effect of screening of radiation by an accretion disk, it is important to consider the integrated luminosity of the entire spectrum.

Moreover, accretion onto black holes should result in very rapid changes in luminosity with the minimum characteristic time of order

$$t \sim \frac{r_g}{c}. \tag{10.137}$$

In this context, the AO620-00 binary system, which is an X-ray nova with millisecond bursts, should be rather considered as a system with a black hole, rather than with a mirror star.

In the W Cru and V600 Her binary systems, lower estimates of the mass of the second invisible companion give, respectively, 16 and 32 solar masses.

With such a mass of the second companion, if it is neither a black hole nor a mirror star, its luminosity must be higher than that of the visible companion.

VII. The influx of ordinary matter to a mirror white dwarf or mirror neutron star can lead to the formation of a dense area of ordinary matter in the centre of the accretion disk having the size of the mirror star or its central nucleus. The latter is an order of magnitude smaller than the size of the entire star.

The observed properties of such a dense area are close to the corresponding properties of ordinary degenerate stars with some quantitative differences, such as perhaps the smaller size and higher temperature.

In addition, in the case of a mirror white dwarf we will see in its place a mostly hydrogen-containing ordinary object created by the accretion of ordinary matter.

Detonation in such objects can lead to phenomena such as the nova explosion with some possible quantitative differences.

In the case of a mirror neutron star, there may be bursts with quantitative parameters that may differ from the usual case.

An ordinary neutron star, which is located in a binary system with a non-relativistic mirror star, would be visible as a radio pulsar paired with an invisible component.

Such binary systems can be distinguished from ordinary relativistic binary systems of stars by the rate of change of the binary system orbit, which will be many orders of magnitude higher than the rate due to emission of gravitational waves in the ordinary binary system.

The rapid evolution of the orbit of the O–M binary system can occur due to

a) Movement of the apsidal line because of the finite size of the non-relativistic mirror companion;

b) Accretion of mirror matter from the non-relativistic mirror companion on the observed neutron O-star;

c) Loss of mass of the system because of its non-conservation during mass exchange, and so on.

11

Cosmoparticle physics of horizontal unification

1. Physical grounds of horizontal unification

1.1. Symmetry of generations of fermions

The problem of generations (families) of fermions remains one of the central problems of elementary particle physics.

The standard gauge model $SU(3)\otimes SU(2)\otimes U(1)$, as well as its 'vertical' extensions, based on gauge groups of unification $SU(5)$, $SU(10)$, and so on, are built into one fermion generation. These models do not contain any deep physical causes of the mass hierarchy between generations of fermions and the observed smallness of mixing angles.

In these models, Yukawa couplings are arbitrary and should initially be set for each fermion separately to reproduce the experimental data for their masses and mixing angles.

The equality between the quark–lepton generations

$$\begin{pmatrix} u \\ d \\ \nu_e \\ e \end{pmatrix};\begin{pmatrix} c \\ s \\ \nu_\mu \\ \mu \end{pmatrix};\begin{pmatrix} t \\ b \\ \nu_\tau \\ \tau \end{pmatrix}$$

with respect to strong and electroweak interactions largely presupposes the existence of 'horizontal' symmetry between the generations.

It is reasonable to consider the concept of local symmetry $SU(3)_H$, which was first proposed by Chkareuli (1980).

Under the action of this symmetry the left-handed polarized quarks and lepton components transform as $SU(3)_H$ triplets and the right-polarized – as antitriplets. Their mass term transforms as

$$3 \otimes 3 = 6 \oplus \overline{3} \qquad (11.1)$$

and, therefore, can only form as a result of horizontal symmetry breaking.

This approach can be trivially extended to the case of n generations, assuming the proper $SU(n)$ symmetry. For three generations, the choice of horizontal symmetry $SU(3)_H$ is the only possible choice because the orthogonal and vector-like gauge groups can not provide different representations for the left- and right-handed fermion states.

In the above approach, the hypothesis that the structure of the mass matrix is determined by the structure of horizontal symmetry breaking, i.e., the structure of the vacuum expectation values of horizontal scalars carrying the $SU(3)_H$ breaking is justified.

The mass hierarchy between generations is related to the hypothesis of a hierarchy of such symmetry breaking. This hypothesis is called - the hypothesis of horizontal hierarchy (HHH) (Berezhiani, Chkareuli 1982, 1983; Berezhiani 1983).

The simplest implementation of HHH requires the introduction of additional superheavy fermions which acquire their masses directly from the horizontal scalars. Ordinary quark and lepton masses are introduced by their 'see-saw' mixing (Berezhiani 1983) with these heavy fermions.

The concept of Grand Unification (GUT) provides another argument in favour of the chiral horizontal symmetry G. In the GUT models, the left-polarized quarks and leptons are in the same irreducible representation of GUT groups as the antiparticles of their right-components. Under $G_{\mathrm{GUT}} \otimes G_H$ symmetry, the left and right components must transform as adjoint representations of the group G_H, then G_H symmetry must be chiral symmetry.

It is hoped that the development of superstring theory will lead to the complete unification of the horizontal and vertical symmetries based on a single fundamental symmetry G which includes $G_{\mathrm{GUT}} \otimes G_H$. The most developed simple version of a realistic model of strings with the gauge group $E_8 \otimes E'_8$ (Candelas et al 1985; Witten 1985) does not allow for the inclusion of horizontal symmetry. However, such integration is possible within a broader class of superstring models, for example, in $SO(32)$ or in models of the heterotic string with direct compactification to 4-dimensional space-time.

In the latter case (Narain 1986; Kawai et al 1986) a wider class of GUT groups with the rank lower than

$$r \leqslant 22. \tag{11.2}$$

can be found.

Analysis of horizontal unification as the phenomenology of theories of everything is presented below based on cosmoparticle physics, may be useful for the proper selection of realistic models among the many possibilities that exist within models of superstrings.

Here we do not consider supersymmetric extensions of a model in which a number of new particles is predicted, extending the hidden sector of the theory. The properties of these particles strongly depend on the details of supersymmetry breaking and require special consideration.

To construct a realistic model of horizontal symmetry breaking, we must introduce a fairly wide set of parameters.

However:

1. number of these parameters is less than in a realistic model without horizontal symmetry;

2. most of these parameters are fixed by the experimental data on the properties of quarks and leptons;

3. a set of non-trivial physical phenomena predicted by the model, in principle, provides full verification of the model and determination of all of its parameters.

These new phenomena occur at very high energy scale of horizontal symmetry breaking F which is of the order of magnitude

$$F > 10^5 \div 10^6 \,\mathrm{GeV}, \tag{11.3}$$

which makes them inaccessible to the study in accelerators even in the distant future.

However, the combination of experimental searches for indirect effects in rare processes of the known particles with an analysis of their cosmological and astrophysical effects let us study physics, predicted on these scales, as well as the cosmological scenario based on this physics.

The model proposed in (Berezhiani, Khlopov 1990a, b, c; 1991; Berezhiani et al 1990a, b, 1992; Sakharov, Khlopov 1993; 1994a, b; 1995; 1996) satisfies the *conditions of naturalness*.

1. Natural suppression of flavour changing neutral currents (FCNC) (Glashow, Weinberg 1977)

In the model there are no duplicates of light scalars $SU(2)\times U(1)$ that transform under a non-trivial representation G_H (vertical–horizontal field) and lead to unacceptably strong effects of FCNC.

Yukawa couplings, responsible for generating the mass of quarks and leptons, include only the standard Higgs doublet $SU(2)\times U(1)$, while the left-handed polarized and right-handed polarized fermionic components transform as the adjoint representation of the horizontal symmetry group.

The price for this is the introduction of additional superheavy fermions that make up the hidden sector of the theory. Mixing with these heavy fermions gives masses to quarks and leptons. The so-called 'see-saw' mechanism (Gell-Mann et al 1979) is implemented not only for the neutrinos but also for all the quarks and leptons.

In addition, this approach provides a natural explanation for the hierarchy of the electroweak scales and GUT scale based on the mechanisms proposed for the GUT models with a single generation, for example, on the basis of supersymmetric extensions $SU(5)$, $SU(10)$, etc. (Nilles 1983; Vysotsky 1985).

The mass of the quark and lepton generations, generated by vertical–horizontal fields, would need unnaturally fine tuning of its parameters, even in the supersymmetric case.

2. Natural horizontal hierarchy

The observed mass hierarchy of generations, say, charged leptons

$$m_e : m_\mu : m_\tau = 1:200:4000 \tag{11.4}$$

is explained by the much more moderate hierarchy of horizontal symmetry breaking. Parameters of such a breaking are proportional to the square root of the masses of quarks and leptons. So it does not require special mechanisms to maintain a hierarchy with allowance for radiative loop corrections.

3. The natural solution to QCD problem of *CP*-violation (Peccei, Quinn 1977)

Although the model assumes the existence of only one Higgs doublet, it provides a natural inclusion of the $U(1)$ Peccei–Quinn symmetry (Peccei, Quinn 1977a,b), related to the heavy Higgs fields, which break the horizontal symmetry G_H on scale F. Breaking of this global $U(1)$ symmetry leads to the existence of pseudo-Goldstone boson α, like the invisible axion, with the scale of interaction F (Zhitnitsky 1980; Dine et al 1981; Wise et al 1981).

The boson α has a coupling with both diagonal and non-diagonal transitions of quarks and leptons in respect of the flavour, that is, it is also a singlet-type familon (Wilczek 1982, Anselm, Uraltsev 1983).

Finally, this boson is associated with the mechanism of generating Majorana neutrino masses, thus becoming the Majoron of a singlet type (Chikashige et al 1980).

Inevitable consequences of the model include:

a) flavour changing neutral currents associated with the axion and interactions of horizontal bosons;

b) the existence of Majorana neutrino mass and the mass hierarchy of neutrinos of different generations;

c) instability of heavy neutrinos with respect to axion decay into lighter neutrinos;

d) the existence of metastable superheavy fermions.

The model can be checked by a combination of laboratory tests, such as:

- search for neutrino mass
- search for neutrino oscillations,
- study of transitions $K^0 \leftrightarrow \bar{K}^0$ and $B^0 \leftrightarrow \bar{B}^0$,
- search for axion decays $\mu \to e\alpha, K \to \pi\alpha$, and so on

together with an analysis of cosmological and astrophysical effects.

The latter include the study of the effect of radiation of axions on the processes of stellar evolution, the study of the impact of the effects of primordial axion fields and massive unstable neutrino on the dynamics of formation of the large-scale structure of the Universe, as well as analysis of the mechanisms of inflation and baryosynthesis based on the physics of the hidden sector of the model.

1.2. Gauge model of broken symmetry between the generations

Consider the extension $SU(2)\otimes U(1)$ model of electroweak interactions (Sakharov, Khlopov 1994b; Khlopov, Sakharov 1994, 1995), suggesting a local chiral symmetry between generations of fermions (Chkareuli 1980; Berezhiani, Khlopov 1990a, b, c, 1991; Berezhiani et al 1990a, b).

Quarks and leptons are arranged in the following representation of the $SU(2)\otimes U(1)\otimes SU(3)_H$ group

$$\begin{gathered} f_{L\alpha}: \begin{pmatrix} u \\ d \end{pmatrix}_{L\alpha} \begin{pmatrix} 2 & \frac{1}{3} & 3 \end{pmatrix}; \begin{pmatrix} \nu \\ l \end{pmatrix}_{L\alpha} \begin{pmatrix} 2 & -1 & 3 \end{pmatrix}; \\ f_R^\alpha : u_R^\alpha \begin{pmatrix} 1 & \frac{4}{3} & \bar{3} \end{pmatrix}; \; d_R^\alpha \begin{pmatrix} 1 & -\frac{2}{3} & \bar{3} \end{pmatrix}; \; e_R^\alpha \begin{pmatrix} 1 & -2 & \bar{3} \end{pmatrix}, \end{gathered} \tag{11.5}$$

where $SU(3)_H$, the index $\alpha = 1, 2, 3$ indicates the generation.

Scalars which break the horizontal symmetry can be transformed as $SU(3)_H$ sextets or triplets. They all have to be singlets for the group

$SU(2)\otimes U(1)$ to prevent the electroweak symmetry breaking at the scale of the $SU(3)_H$ breaking scale.

Generation of the realistic mass matrix for quarks and leptons requires at least three such 'horizontal' scalars.

At least one of them with the greatest vacuum expectation value should be a sextet:

$$\xi^{(0)}_{\{\alpha,\beta\}};\ \alpha,\beta=1,2,3, \tag{11.6}$$

where the curly brackets denote symmetrization with respect to group indices. Only the triplet fields can not generate realistic mass matrices. For the other two scalars

$$\xi^{(1)}_{\{\alpha,\beta\}};\ \alpha,\beta=1,2,3 \tag{11.7}$$

and

$$\xi^{(2)}_{\{\alpha,\beta\}};\ \alpha,\beta=1,2,3 \tag{11.8}$$

we do not specify their representation in the $SU(3)_H$ group, pointing to it only in cases where the sextet and triplet representations lead to different consequences.

We introduce additional fermions in the form of

$$\begin{gathered} F^{\alpha}_{L}:U^{\alpha}_{L}\left(1\ \ \frac{4}{3}\ \ \overline{3}\right);\ D^{\alpha}_{L}\left(1\ \ -\frac{2}{3}\ \ \overline{3}\right);\ E^{\alpha}_{L}\left(1\ \ -2\ \ \overline{3}\right)\\ F_{R\alpha}:U_{R\alpha}\left(1\ \ \frac{4}{3}\ \ 3\right);\ D_{R\alpha}\left(1\ \ -\frac{2}{3}\ \ 3\right);\\ E_{R\alpha}\left(1\ \ -2\ \ 3\right);\ N_{R\alpha}\left(1\ \ 0\ \ 3\right) \end{gathered} \tag{11.9}$$

(Berezhiani 1983; Berezhiani, Khlopov 1990a, b, c, 1991; Berezhiani et al 1990a, b).

These fermions cancel the $SU(3)_H$ anomaly of ordinary quarks and leptons (11.5), which requires the existence of a partner also for the neutrino, a neutral lepton N.

In the most general case, Yukawa couplings for quarks and leptons allowed from the symmetry of the model are given by the Lagrangian in the form

$$L_{\text{Yuk}}=g_f\overline{f}_{L\alpha}F_{R\alpha}\phi^0+G^{(n)}_F\overline{F}^{\alpha}_{L}F_{R\beta}\overline{\xi}^{(n)\alpha\beta}+G_\eta\overline{F}^{\alpha}_{L}f^{\alpha}_{R}\eta+\text{h.c.}, \tag{11.10}$$

where $n = 0, 1, 2$; $f = u, d, e$; $F = U, D, E$ (Berezhiani 1983; Berezhiani, Khlopov 1990a, b, c, 1991; Berezhiani et al 1990a, b; Khlopov, Sakharov 1994, 1995, Sakharov, Khlopov 1994b). For neutrinos such links have the form

$$L_{\text{Yuk},\nu} = g_\nu \bar{\nu}_{L\alpha} N_{R\alpha} \phi^0 + G_N^{(n)} N_{R\alpha} C N_{R\beta} \xi^{(n)\alpha\beta} + \text{h.c.} \tag{11.11}$$

(Berezhiani, Chkareuli 1983; Berezhiani, Khlopov 1990a, b, c, 1991; Berezhiani et al 1990a, b; Khlopov, Sakharov 1994a, b; Sakharov, Khlopov 1994b).

Here ϕ^0 is the neutral component of the standard Higgs $SU(2)\otimes U(1)$ doublet. It corresponds to the representation of $SU(2)\otimes U(1)\otimes SU(3)_H$ group

$$\phi \; (2 \;\; \text{-1} \;\; 1) \tag{11.12}$$

and has a vacuum expectation

$$\langle \phi^0 \rangle = \vartheta = \left(\sqrt{8} G_F\right)^{-1/2} = 250 \text{ GeV}. \tag{11.13}$$

In Lagrangian (11.10) Higgs field η is a real $SU(2)\otimes U(1)\otimes SU(3)_H$ singlet scalar, having the expectation value

$$\langle \eta \rangle = \frac{\mu}{G_\eta}. \tag{11.14}$$

Yukawa couplings (11.10)–(11.11) are invariant under the global axial transformations $U(1)_H$:

$$\begin{gathered} f_L \to f_L \exp\{i\omega\};\; f_R \to f_R \exp\{i\omega\}; \\ F_L \to F_L \exp\{i\omega\};\; F_R \to F_R \exp\{i\omega\}; \\ \phi \to \phi;\; \xi^{(n)} \to \xi^{(n)} \exp\{2i\omega\}, \end{gathered} \tag{11.15}$$

where $n = 0, 1, 2$.

The Higgs potential also possess this $U(1)$ symmetry, which ensures the absence of triple couplings of the type

$$L_{\text{tril}} = \Lambda \xi^{(0)}_{\alpha\beta} \xi^{(1)\alpha} \xi^{(2)\beta} + \text{h.c.} \tag{11.16}$$

These couplings do not appear as radiative effects of any other (gauge or Yukawa) interaction. Thus, their absence in the Lagrangian is quite natural (Berezhiani, Chkareuli 1983; Berezhiani, Khlopov 1990a, b, c, 1991; Berezhiani et al 1990a, b; Khlopov, Sakharov 1994.1995, Sakharov, Khlopov 1994b) .

Analysis of the Higgs potential (Berezhiani, Chkareuli 1983; Berezhiani, Khlopov 1990a, b, c, 1991; Berezhiani et al 1990a, b; Khlopov, Sakharov 1994a, b) shows that the matrix of expectation values is as follows:

$$V_H = \left\langle \xi^{(0)} + \xi^{(1)} + \xi^{(2)} \right\rangle = \begin{pmatrix} r_1 & p_1 & p_3 \\ \pm p_1 & r_2 & p_2 \\ \pm p_3 & \pm p_2 & r_3 \end{pmatrix}, \tag{11.17}$$

where the plus and minus signs correspond to the sextet and triplet representations of scalars $\xi^{(1)}$ and $\xi^{(2)}$ respectively. Expectation values have a natural (5–10)-fold hierarchy

$$r_1 > p_1 > r_2 > p_2 > p_3 > r_3. \tag{11.18}$$

Substituting the vacuum expectation values of scalar fields into the expressions (11.10)–(11.11) for the Yukawa couplings, we obtain a complete 6×6 mass matrix for charged fermions:

$$\begin{matrix} & f_R & F_R \\ \overline{f}_L & \multicolumn{2}{c}{\multirow{2}{*}{}} \end{matrix} \quad \begin{array}{c} \\ \overline{f}_L \\ \overline{F}_L \end{array} \begin{array}{c} \begin{matrix} f_R & F_R \end{matrix} \\ \begin{pmatrix} 0 & g_f \vartheta \\ \mu & M_F \end{pmatrix} \end{array} \tag{11.19}$$

and the corresponding mass matrix for neutrinos

$$\begin{array}{c} \\ \overline{\nu}_L \\ \overline{N}_L \end{array} \begin{array}{c} \begin{matrix} \nu_R & N_R \end{matrix} \\ \begin{pmatrix} 0 & g_\nu \vartheta \\ g_\nu \vartheta & M_N \end{pmatrix} \end{array}. \tag{11.20}$$

Here the mass matrix of 3×3 is given by

$$M_F = \sum \left\langle \xi^{(n)} \right\rangle G_F^{(n)} \tag{11.21}$$

with $F = U, D, E$; a neutrino states are defined as

$$N_L = C\overline{N}_R;\ \nu_R = C\overline{\nu}_L.$$

It should be noted that only the sextet scalars contribute to the Majorana mass matrix M_N.

Thus, we have a Dirac 'see-saw' mechanism of mass generation of quarks and leptons and usual Majorana 'see-saw' mechanism for neutrino masses. In the latter case, the neutral lepton state N_R plays the role of right-handed neutrino.

The mass matrices of the quarks and charged leptons, obtained after block-diagonalization of matrices (11.19), have the form

$$m_f = \frac{\mu}{M_F} g_f \vartheta, \tag{11.22}$$

where $f = u, d, e$; and the mass term for the neutrino after reduction to the block-diagonal matrix (11.20) equals

$$m_\nu = \frac{(g_\nu \vartheta)^2}{M_N}. \tag{11.23}$$

The mass hierarchy of the generations is inverted with respect to the hierarchy of symmetry breaking $SU(3)_H \otimes U(1)_H$;

$$SU(3)_H \otimes U(1)_H \Rightarrow SU(2)_H \otimes U'(1)_H \Rightarrow U''(1)_H \Rightarrow I, \tag{11.24}$$

where I is the trivial group of identity transformations.

Here, the intermediate $SU(2)_H \otimes U'(1)_H$ horizontal symmetry refers to the second and third generations of quarks and leptons. Global $U''(1)_H$ symmetry remains unbroken after the second stage of symmetry breaking and correponds to the third generation only.

This case is called a model with inverse hierarchy in direct contrast to the model of direct hierarchy, where the mass hierarchy of quarks and leptons is broken parallel to $SU(3)_H \otimes U(1)_H$ symmetry breaking (Berezhiani 1985; Berezhiani, Khlopov 1990a, b, c, 1991; Berezhiani et al 1990a, b; Khlopov Sakharov 1994, Sakharov, Khlopov 1994b).

Breaking of the global $U(1)_H (U''(1)_H)$ symmetry leads to the existence of the massless Goldstone boson α, called the archion, which has both diagonal and non-diagonal (with respect to flavour) couplings with the quarks and leptons and thus represents the familon of the singlet type (Anselm, Uraltsev 1983; Anselm, Berezhiani 1985). The archion model was described in Chapter 2.

In the minimal $SU(2)_H \otimes SU(3)_H$ model the archion α is a particle of the arion type (Anselm 1982; Anselm, Uraltsev 1982), having extremely weak interaction with photons and gluons due to strong compensation of the appropriate triangle diagrams.

Due to the fact that the interaction of the archion with ordinary matter is strongly suppressed, there are no strong astrophysical constraints on the ϑ''_H scale (see review in Kim 1987; Cheng 1988).

At the level of the lowest acceptable scale of symmetry breaking of the generations processes with the change of flavour can take place with a high probability, such as

$$\mu \to e\alpha,\ \tau \to \mu\alpha,\ K \to \pi\alpha,\ B \to K\left(K^*\right)\alpha,\ D \to \pi\left(\rho\right)\alpha. \quad (11.25)$$

The search for such meson decays (Bereziani et al 1990b), archion decays (Sakharov et al 1996b; Khlopov et al 1996) and FCNC effects caused by gauge 'horizontal' bosons in the systems of K-, D- and B- neutral mesons (Sakharov et al 1996, Khlopov et al 1996e) could provide experimental verification of the investigated model (Belotsky et al 1998).

In any realistic extension of this scheme of horizontal unification, for example, in the gauge model based on the $SU(5)\otimes SU(3)_H$ symmetry the triangular anomalies arise due to the inevitable presence in GUT of extra heavy fermions, which leads to interaction of the archion with photons and gluons. In this case, the $U''(1)_H$ symmetry can be identified with the Peccei–Quinn symmetry (Peccei, Quinn 1977) and the archion is a special type of the invisible axion (Chapter 2).

Then, on the scale

$$\vartheta''_H = \vartheta_{PQ} \quad (11.26)$$

constraints are imposed by the astrophysical estimates of the energy loss by stars through radiation of archions:

$$\vartheta_{PQ} > 10^6 \text{ GeV}. \quad (11.27)$$

Constraint (11.27) is based on observational data about the Sun and red giants and follows from the theory of stellar evolution.

At each stage in the evolution of stars, the time scale of evolution is determined by the rate of energy loss L at a fixed energy Q released in thermonuclear reactions in the considered stage, and is equal to

$$t_{\text{ev}} \sim \frac{Q}{L}. \quad (11.28)$$

The additional energy losses L_α due to radiation of axions lead to a decrease in the time evolution t'_{ev}

$$t'_{\text{ev}} \sim \frac{Q}{L + L_\alpha}. \quad (11.29)$$

The observational data on the Sun, the stars of the main sequence and red giants give an upper limit on the rate of emission of axions, and thus a lower limit on the scale of the axion (archion) interaction.

On the other hand, the maximum allowable effect of the radiation of the archions accelerates the evolution of dense and hot central parts of stars that may be important for interpreting data on the SN1987A Supernova. The appropriate presupernova did not exhibit properties expected for a star at the end of its evolution. This can be explained by the high rate of evolution of the central nucleus, which can not be observed directly.

The emission of the archions from collapsing stars can reduce the total energy and characteristic emission time of neutrinos. The claims of the observation of a neutrino signal preceding the SN1987A Supernova explosion restrict the possible axion luminosity and, therefore, can be used to limit the permissible scale of the interaction of axions

$$\vartheta_{PQ} > 10^{10}\,\mathrm{GeV}. \tag{11.30}$$

Because of the complexities associated with the calculations of neutrino transport in a collapsing star in the neutrino opacity conditions, the constraints (11.30) should be used very carefully.

This constraint includes free propagation of the archions in a collapsing star which is true when the archions interact so weakly that it corresponds to the scale (11.30). At a smaller scale interaction of the archion, the collapsing star is opaque to the archions and this weakens the effect of archion radiation. It is essential to carry out a very detailed and accurate analysis of the transfer of the archions in dense matter to obtain reliable constraints on the allowable values of the scale ϑ_{PQ}. In any case, for the scale

$$\vartheta_{PQ} \sim 10^{6}\ \mathrm{GeV} \tag{11.31}$$

the energy loss due to axion emission decreases to the extent that the loss becomes compatible with the presence of a neutrino signal from SN1987A.

2. Early Universe in the model of horizontally unification

2.1. Inflation dynamics and the energy scale

The inflationary scenario may find its physical base in the model of horizontal unification (MHU), as the Higgs field η (see (11.14)), which

determines the mass term independent of flavour, can play the role of the inflaton.

In general, the expectation value $\langle\eta\rangle$ determines in the MHU the suppression of the masses of neutral fermions (light neutrinos) with respect to the charged fermions (see Section 1)

$$\frac{m_\nu}{m_f} \propto \frac{g_f\langle\phi^0\rangle}{G_\eta\langle\eta\rangle}. \tag{11.32}$$

It is assumed that the global symmetry is spontaneously broken at the energy scale f, and that the real field has the potential

$$V(\eta) = \lambda\left(\eta^2 - \frac{f^2}{2}\right)^2. \tag{11.33}$$

Thus, the consideration is reduced to the inflationary model with a single slow rolling scalar field with potential (11.33). This means that the simplest implementation of the horizontal symmetry corresponds to the simplest scenario of chaotic inflation.

To fix the parameters of the inflaton potential, we can use observational constraints on the energy density of the inflaton in the period when the observed fluctuations of the microwave background were generated.

This happened as soon as the appropriate scale crossed the Hubble radius during inflation, as a rule, when the scale factor was of the order of exp $\{-60\}$ of its size at the end of inflation. This value is commonly referred to as the 60th e-folding from the end of inflation.

Using the constraint for the 60th e-folding from the end of inflation, we can analyze the evolution of the system and limit the energy density of the inflaton during the end of inflation.

In such an approach to the reconstruction of the inflaton potential the equation of motion must be rewritten in the Hamilton–Jacobi form (see Khlopov, Sakharov 1999; Khlopov 1999 and references therein)

$$\left(\frac{d^2H}{d\eta^2}\right)^2 - \frac{12\pi}{m_{\text{Pl}}^2}H^2 = -\frac{32\pi^2}{m_{\text{Pl}}^4}V(\eta), \; \dot{\eta} = -\frac{m_{\text{Pl}}^2}{4\pi}\frac{dH}{d\eta}. \tag{11.34}$$

In principle, the Hamilton–Jacobi formalism allows us to treat the dynamics of the evolution of the scalar field exactly, at least at the classical level.

Using the potential of the inflaton, we can calculate the amplitude of density fluctuations $\delta_H(k)$. It is most convenient to perform calculations for 60th e-folding so that the COBE conditions can be satisfied.

For models with fairly flat spectra and negligible gravitational waves, as is the case in models of horizontal unification (Khlopov, Sakharov 1999; Khlopov 1999), the corresponding value δ_H reproduces the COBE result

$$\delta_H \cong 1.7 \cdot 10^{-5}. \tag{11.35}$$

We can also estimate the value of the Hubble constant during the end of inflation H_{end}, which in a broad range of vacuum expectation values f will be equal to

$$H_{\text{end}} \cong 1.8 \cdot 10^{-7} m_{\text{Pl}} \tag{11.36}$$

(Khlopov, Sakharov 1998).

Interestingly, the spectrum of density fluctuations grows toward the red. This is supported by the results of the WMAP experiment (Spergel et al 2003 and 2010).

2.2. Formation of primordial black holes in a model of horizontal unification

The ultraviolet behaviour of the spectrum of density fluctuations can be effectively limited by the analysis of the primordial black holes (PBH) in the early Universe.

Following the discussion in Chapter 4, we consider a spherically symmetric Gaussian distribution with amplitude $\delta(M)$ and the background equation of state $p = \gamma\varepsilon$, where $0 < \gamma < 1$. The probability to form a PBH with mass M is given by the tail of the Gaussian distribution of density fluctuations (compare with (4.8))

$$\beta_0(M) \approx \delta(M)\exp\left\{-\frac{\gamma^2}{2\delta^2(M)}\right\}. \tag{11.37}$$

This probability determines the fraction of the total density which turns into PBHs with mass M. The mass of the PBH, formed at the time t, must be at least $\gamma^{3/2}$ of the mass of the horizon, so that

$$M \approx \gamma^{3/2}\frac{t}{t_{\text{Pl}}} m_{\text{Pl}}. \tag{11.38}$$

Typically, the value γ in the early Universe is taken to be equal to 1/3, which corresponds to the equation of state for the RD stage, considered for the early Universe in the old Big Bang model.

In the case $\gamma = 0$, both equation (11.37) and equation (11.38) are not valid (Chapter 4).

During the dust stage, which corresponds to the equation of state $p = 0$, the density fluctuations grow and form gravitationally bound objects.

The fraction of the total density of turning into PBHs depends on the probability that these objects are contracted within their Schwarzschild radius. The minimum probability corresponds to the direct collapse of primordial black holes directly in the period of formation (Polnarev, Khlopov 1982, compare with (4.44)) and is

$$\beta(M) \approx 2 \cdot 10^{-2} \delta(M)^{13/2}. \quad (11.39)$$

If the dust stage occurs during the period

$$t_1 < t < t_2, \quad (11.40)$$

then the probability of (11.39) holds for the formation of a PBH with the mass in the range (compare with (4.41))

$$M_1 \le M \le M_{\max}, \quad (11.41)$$

where M_1 is the mass inside the cosmological horizon at the moment t_1 and $M_{\max}$ is the mass of an object separated from the expansion directly at t_2. The latter is given implicitly by (see (4.43))

$$M_{\max} = \left[\delta(M_{\max})\right]^{3/2} \frac{t_2}{t_{Pl}} m_{Pl}. \quad (11.42)$$

The product of the last two factors on the right hand side of equation (11.42) is the mass within the cosmological horizon at the moment t_2. In order to determine $M_{\max}$ we must know the form of $\delta(M)$.

The soft equation of state can be derived in the heating period at the end of chaotic inflation (Khlopov et al 1985), which is realized in an inflationary model based on the model of horizontal unification (MHU). In this model, there are inflationary field oscillations around the potential minimum starting at

$$t_1 = H_{end}^{-1}, \quad (11.43)$$

while heating takes place.

Friction due to the interaction of the scalar inflaton field with other particles converts the kinetic energy of the field oscillations in the background radiation. Heating is completed in a period determined by the decay width Γ_η. In the present case Γ_η is given by

$$\Gamma_\eta\left(\eta \to \bar{F}f\right) = \frac{G_\eta^2 m_\eta}{8\pi}. \tag{11.44}$$

We can calculate G_η from the potential (11.33), using the expression for the self-interaction of the field η and mass of η. Generating minimum self-interaction coupling of the inflaton as the radiative effect from fermion loops according to (11.10), we obtain a constant λ_η

$$\lambda_\eta \cong \frac{G_\eta^4}{8\pi^2}. \tag{11.45}$$

On the other hand, Γ_η can be estimated by reconstructing the inflaton potential and we get for a wide range of scales f

$$\Gamma_\eta \cong 10^{-14} m_{\rm Pl}. \tag{11.46}$$

From equations (11.36) and (11.46) it is clear that

$$\Gamma_\eta^{-1} \gg H_{\rm end}^{-1} \tag{11.47}$$

and that a relatively long stage of oscillation of a coherent scalar field with the dust equation of state should take place.

The dust stage, starting with a period of the end of inflation at t_1, continuing until the end of heating of the Universe at

$$t_2 = \Gamma_\eta^{-1}, \tag{11.48}$$

when the scalar field rapidly decays to relativistic particles. However, at the selected numerical parameters the PBHs do not have an extended spectrum and the peak of their mass distribution is close to M_1, which coincides with the mass (4.42). In the most general case of non-minimal self-interacting inflaton its self-interaction coupling constant is fixed by the condition (11.35), but weaker couplings with other fields lead to an extended range of PBH masses.

Substituting the value (11.35) into (11.39), we find the probability of PBH formation, resulting in a small effect of their existence and evaporation. Nevertheless, the need for internal consistency of the model of horizontal unification makes the picture more complicated, and constraints on the formation of primordial black holes play an important role in its analysis.

Indeed, in a model of horizontal unification it is necessary to impose the condition

$$G_\eta f \leqslant G_F^{(n)} \left\langle \xi^{(n)} \right\rangle, \tag{11.49}$$

which ensures the correct structure of the mass matrix of fermions, generated by the Dirac 'see-saw' mechanism (11.22). This condition requires that

$$f \leqslant 10^{-6} m_{\rm Pl} \tag{11.50}$$

for the minimum of self-interaction of the inflaton (11.45).

However, an inevitable consequence of such a low value of f is the problem of domain walls formed due to fluctuations of a real inflaton field in the post-inflationary dust stage (see Khlopov, Sakharov 1999 and references therein).

To resolve this problem, we must either take

$$f \cong m_{\rm Pl} \tag{11.51}$$

and eliminate the condition of minimal coupling of the inflaton, or introduce the complex inflaton field. In the first case, the duration of post-inflationary dust stage becomes abnormally large, while in the latter case the non-minimal model of the axion is used.

Thus, even at this level of study of the problem, we find that the decision should lead to a more complex multi-parameter cosmological scenario.

2.3. Early 'horizontal' phase transitions

The interaction of scalar fields with the inflaton leads to phase transitions in the inflationary stage. The reason for this is that in this interaction the Higgs fields acquire a positive mass. When the amplitude of the slow rolling inflaton field falls below a certain critical value η_c, the mass term in the Higgs potential changes its sign and then a phase transition occurs.

As stated in (Sakharov, Khlopov 1993), 'horizontal' phase transitions give rise to characteristic peaks in the spectrum of density fluctuations. Kofman and Linde (1987) first pointed out the existence of such peaks in the scenario of chaotic inflation.

These peaks arise for fluctuations on the scale leaving the horizon during the period of the order of (40÷1) e-folding before the end of inflation, re-enter the horizon in the radiation or dust stage and can, in principle, be contracted to form a PBH.

Let us now calculate the mass and size of a typical PBH, formed from such a fluctuation.

Suppose that after the phase transition the Universe expanded $\exp\{N_e\}$ times. Then, at the end of inflation, the physical scale, which left the horizon during the phase transition is

$$l_e \sim H_0^{-1} \exp(N_e), \tag{11.52}$$

where H_0 is the Hubble constant during inflation.

If the equation of state $p = \varepsilon/3$, then the scale factor in the Universe is growing after inflation as follows

$$a(t) \propto \sqrt{tH_0}. \tag{11.53}$$

The scale

$$l = H_0 \sqrt{tH_0} \exp(N_e) \tag{11.54}$$

is equal to the scale of the cosmological horizon in the period

$$t_k = H_0^{-1} \exp(2N_e). \tag{11.55}$$

When the fluctuations reach the density contrast $\delta{\sim}1$, they form black holes with the size of

$$r_{bh} \sim H_0^{-1} \exp(2N_e) \tag{11.56}$$

and with mass

$$M \cong \frac{m_{\mathrm{Pl}}^2}{H_0} \exp(2N_e). \tag{11.57}$$

If PBHs are formed in the dust stage, the time of their formation is

$$t_k = H_0^{-1} \exp(3N_e) \tag{11.58}$$

and the masses of the PBHs are

$$M \cong \frac{m_{\mathrm{Pl}}^2}{H_0} \exp(3N_e). \tag{11.59}$$

We can include horizontal scalars in the effective inflationary potential, which we take here in the form

$$V\left(\eta,\xi^{(0)},\xi^{(1)},\xi^{(2)}\right)=-\frac{m_\eta^2}{2}\eta^2+\frac{\lambda_\eta}{4}\eta^4-\sum_{i=0}^{2}\frac{m_i^2}{2}\left(\xi^{(i)}\right)^2+$$

$$+\sum_{i=0}^{2}\frac{\lambda_i^2}{4}\left(\xi^{(i)}\right)^4+\sum_{i=0}^{2}\frac{\nu_i^2}{2}\eta^2\left(\xi^{(i)}\right)^2. \qquad (11.60)$$

In the model of horizontal unification the masses m_i of the bosons are associated with the scale of horizontal symmetry breaking, which can be given by the expressions

$$m_i^2 \approx 10^{-3} V_i^2, \qquad (11.61)$$

where the vacuum expectation values of Higgs fields

$$V_i=\left\langle \xi^{(i)}\right\rangle,\ i=0,1,2 \qquad (11.62)$$

follow the hierarchy

$$V_2 : V_1 : V_0 \cong 1:30:200 \qquad (11.63)$$

and

$$\lambda_i=\lambda_\xi \cong 10^{-3}. \qquad (11.64)$$

The minimal interaction of the inflaton with the horizontal scalars is determined by the fermion loop corrections due to Yukawa coupling (11.10), and is given by

$$\nu_i=\nu_\xi \cong \frac{G_\eta^2\left(G_F^{(i)}\right)^2}{8\pi^2}. \qquad (11.65)$$

Because of the interaction (11.60) between the inflaton and the horizontal scalars a phase transition occurs when the amplitude of the field η drops below

$$\eta_{ci}=\frac{m_i}{\sqrt{\nu_i}}. \qquad (11.66)$$

Such a phase transition gives rise to characteristic peaks in the spectrum of adiabatic perturbations with the density contrast

$$\delta \cong \frac{4}{9s};\ s=\left(\sqrt{\frac{9}{4}+k\cdot 10^5\left(\frac{V_i}{m_{\rm Pl}}\right)^2}-\frac{3}{2}\right), \qquad (11.67)$$

where $k \sim 1$ (Sakharov, Khlopov 1993).

We can link the scale of horizontal symmetry breaking with *e*-folding and the mass of primordial black holes formed by these peaks. From the condition that all three horizontal phase transitions do not lead to the formation of PBHs with mass greater than 1000 grams, we obtain the limit

$$V_2 \leqslant 10^{13} \text{ GeV}. \tag{11.68}$$

2.4. Large-scale modulation in the distribution of the primordial axion field

Analysis of the model of cold dark matter (CDM), based on a model of the invisible axion (Sakharov, Khlopov 1994a; Sakharov et al 1996) shows that there is a large-scale inhomogeneity in the distribution of the energy density of coherent oscillations of the primordial axion field. These inhomogeneities, characteristic for all other models of axion cold dark matter, have been named archioles (Sakharov, Khlopov 1994a).

The archioles represent the formations, reflecting the Brownian percolation vacuum structure of axion walls bounded by strings, fixed in a strongly inhomogeneous initial distribution of cold dark matter. They seem to arise in any cosmological scenario with an invisible axion as its essential element.

In general, the axion field arises in the early Universe after the phase transition that breaks the Peccei–Quinn (PQ) symmetry. The axion corresponds to the phase θ of the complex Higgs boson that breaks the PQ-symmetry and remains almost massless until the QCD phase transition in the early Universe. The true vacuum state is degenerate for all values θ in this period.

Due to instanton effects, the PQ-symmetry is not free from the colour anomaly, which is effectively described by a potential

$$V(\theta) = \Lambda_1^4 \left(1 - \cos(\theta N)\right). \tag{11.69}$$

This potential eliminates the degeneracy of the vacuum state and leads to a non-zero axion mass described by the equation

$$m_\alpha = A_c \frac{m_\pi f_\pi}{F_\alpha}, \tag{11.70}$$

where F_α is the scale of PQ-symmetry violation and the constant $A_c \sim 1$ depends on the choice of the axion models (Chapter 2).

During the period of QCD phase transition the axion mass depends on temperature (Ipser, Sikivie 1983), until the Compton wavelength of the axion does not compare within the cosmological horizon and the condition

$$m(\tilde{t})\cdot\tilde{t}\approx 0.75 \qquad (11.71)$$

is satisfied.

The axion mass and coherent oscillations of the axion field are then 'switched on'. The coherent oscillations of the axion field correspond to a condensate with zero momentum and with energy density (see for review Berezhiani et al 1992) given by

$$\rho_a(T)=\left(\frac{39.14}{2}\right)\left(\frac{T_1^2 m_a}{m_{\text{Pl}}}\right)\left(\frac{T}{T_1}\right)^3\theta^2 F_a^2, \qquad (11.72)$$

where T is the temperature of the thermal background radiation and $T_1=\Lambda_1$.

In cosmology of the invisible axion, it is generally assumed that the average energy density of coherent oscillations is distributed uniformly and that it corresponds to the average phase values of the amplitude of the axion field

$$\overline{\theta}=1.$$

However, the local value of the energy density of coherent oscillations of the axion field depends on the local phase θ, which determines the local amplitude of coherent oscillations.

As shown in (Sakharov, Khlopov 1994a), the initial large-scale inhomogeneity of the distribution of θ should be reflected in the distribution of the energy density of coherent axion field oscillations. Large-scale modulation of the phase θ distribution arise from the formation and transformation of topological defects in a sequence of two (PQ and QCD) phase transitions.

When the temperature of the Universe falls to F_a in the course of expansion, PQ-phase transition takes place and is accompanied by the formation of axion strings (Harari, Sikivie 1987). These strings may be infinite or closed. The numerical calculation of string formation shows (Vilenkin 1982) that about 80% of the total length of strings corresponds to infinite Brownian lines. This means that 80% of the total length of strings within any finite area corresponds to strings that extend beyond this area. The remaining 20% of this length is attributed to the closed strings. Infinite strings form a random Brownian network with the step

$$L(t) = \lambda t, \tag{11.73}$$

where $\lambda \cong 1$.

When the temperature of the Universe falls to $T \cong T_1$, the potential (11.69) begins to make a significant contribution to the total potential. This removes the degeneracy of the vacuum θ so that the minimum energy for the number of generations $N = 1$ corresponds to a vacuum

$$\theta = 2\pi k, \tag{11.74}$$

where k is an integer, for example, zero.

The vacuum value of the phase θ can not always be zero, since the phase must vary by $\Delta\theta = 2\pi$ along the closed path around the string. Consequently, at this pass we transfer from the vacuum state with $\theta = 0$ to the vacuum state with $\theta = 2\pi$.

The vacuum state θ is fixed at all points around the string. An exception is the surface in the direction $\theta = \pi$ from the string. Crossing this surface, we get from one vacuum to another, and thus axion walls form along this surface. The thickness of such a wall, bounded by the strings, is of the order

$$\delta \cong m_a^{-1}. \tag{11.75}$$

The initial value of the phase θ should be close to π near the wall, and the amplitude of coherent oscillations in (11.72) is determined by the difference between the initial local phase θ and the vacuum value. Thus, the maximum oscillation energy of the axion field is given by

$$\rho_a(\theta \sim \pi) = \pi^2 \rho_a(\overline{\theta} \sim 1). \tag{11.76}$$

This means that the distribution of the coherent oscillations of the axion field is modulated by the non-linear inhomogeneities in which the relative density contrast is

$$\frac{\delta\rho}{\rho} > 1.$$

Sakharov and Khlopov (1994a) called such inhomogeneities archioles.

The scale of this modulation in the density distribution is greater than the cosmological horizon due to the presence of the component of the infinite strings in the structure of axion strings.

In other words, the large-scale structure of the archioles copies the vacuum structure formed by the axion walls bounded by strings, in a period when the mass of the axion is 'switched on'. This structure of the archioles is preserved in the RD stage when the vacuum structure itself breaks up and disappears.

The spectrum of the density inhomogeneities, reflecting the large-scale Brownian modulation of the distribution of the energy density of coherent axion field oscillations was studied in (Sakharov et al 1996)

Density inhomogeneities, associated with the Brownian structure of the archioles, have been described in terms of a two-point autocorrelation function which has the form

$$\left\langle \frac{\delta\rho}{\rho}(\vec{k}), \frac{\delta\rho}{\rho}(\vec{k}') \right\rangle = 12\rho_a\mu_a k^{-2}\delta^3(\vec{k}+\vec{k}')\tilde{t}^{-1}f^{-2}G^2t^4, \tag{11.77}$$

where ρ is the background density, G is the gravitational constant,

$$f = \begin{cases} \dfrac{1}{6\pi}, & p = \dfrac{1}{3}\varepsilon, \\ \dfrac{3}{32\pi}, & p = 0, \end{cases},$$

ρ_a is the total energy density of the Brownian lines, μ_a is the linear density of the structure. The mean-square of the fluctuations of the mass of the archioles is given by

$$\left(\frac{\delta M}{M}\right)^2 (k,t) = 12\rho_a\mu_a\tilde{t}^{-1}f^{-2}G^2kt^4. \tag{11.78}$$

Using equations (11.77) and (11.78), we can calculate the quadrupole anisotropy of the cosmic microwave background radiation caused by the archioles structure

$$\frac{\delta T}{T} \cong 2.3\cdot 10^{-6}\left(\frac{F_a}{10^{10}\,\mathrm{GeV}}\right)^2. \tag{11.79}$$

This leads to a model-independent constraint on the scale of PQ-symmetry violation in the invisible axion model

$$F_a \leqslant 1.5\cdot 10^{10}\,\mathrm{GeV}. \tag{11.80}$$

This upper limit on F_a greatly differs from the upper limit which is quantitatively close to (11.80) and was obtained in (Davis 1986) when comparing the density of axions from the decay of axion strings with a critical density.

The difference is that the density of the axions, formed by the decay of axion strings, strongly depends on assumptions about the spectrum of axions. The estimate of the density of the axions from the decay of strings has the uncertainty factor of 70. The corresponding uncertainty also appears in the upper bound on F_a. Arguments that lead to the constraint (11.80) are free from these ambiguities because they are model-independent.

The small-scale evolution of the archioles can provide a physical basis for the RMS fluctuation of the mass distribution within spheres with the radius $8h^{-1}$ Mpc. Such fluctuations are used for the normalization of the amplitude of the spectrum of density fluctuations in the standard model of CDM.

At the smallest scales corresponding to the horizon at a time when the axion mass was 'switched on', the evolution of archioles in the beginning of the matter dominance in the Universe should lead to the formation of the smallest gravitationally bound axion objects.

Such objects are formed at redshifts $z_{\rm MD} = 4 \cdot 10^4$ with a minimum mass

$$M \cong \rho_a \tilde{t}^3 \cong 10^{-6} M_\odot \qquad (11.81)$$

and have a typical minimum

$$R \cong \tilde{t}\left(\frac{1+z_\Lambda}{1+z_{\rm MD}}\right) \cong 10^{13}\,\text{cm}, \qquad (11.82)$$

where z_Λ is the redshift corresponding to the period when the axion mass was 'switched on'.

We can expect that the distribution of the mass of the axion objects at small scales will have a peak around the lowest mass, so it is very likely that the galactic halo will contain objects with the mass

$$M \sim \left(10^{-6} \div 10^{-1}\right) M_\odot \qquad (11.83)$$

and size

$$R \sim 10^{13} \div 10^{15}\,\text{cm}. \qquad (11.84)$$

This may have interesting implications for the theoretical interpretation of microlensing events in the MACHO experiment.

Even in its simplest form the model of horizontal unification provides a mechanism of baryosynthesis with the baryon number non-conservation not related to the GUT model. This mechanism combines non-conservation of (B+L) in electroweak transitions at high temperatures and non-equilibrium transitions with $\Delta L = 2$ due to the physics of Majorana neutrino masses.

At the correct order of magnitude and sign of the average baryon asymmetry the spatial variation of the baryonic charge can take place. Such a change may be due to modulation $\theta(x)$, if the *CP*-violation by axions plays an important role in the baryosynthesis processes. Baryon asymmetry in this case may depend on $\theta(x)$ (see, e.g., Yoshimura 1983), which is the sum of constant and spatially dependent quantities

$$\Delta(\vec{x}) = \Delta_0 + \Delta_1 \sin\theta(\vec{x}). \tag{11.85}$$

If the spatially-dependent amplitude of the axion contribution exceeds the constant baryon excess $\Delta_1 > \Delta_0$, an excess of antibaryons forms in the region corresponding to the condition

$$\left|\theta(x) - \frac{3}{2}\pi + 2\pi k\right| > \arccos\left(\frac{\Delta_0}{\Delta_1}\right). \tag{11.86}$$

Baryosynthesis with the dominance of the axions can give a non-trivial picture of the evolution of domains of antimatter (Khlopov 1998). Small domains in such a scenario annihilate before the first second of the expansion, thus avoiding the severe constraints on the annihilation of antimatter (see Chapters 6 and 7).

Large domains, which satisfy these constraints, can form during their evolution objects of antimatter in the Universe. The minimum mass of such objects is determined by the condition of survival relative to the annihilation with surrounding matter. The upper limit on the total mass of these domains comes from the observed gamma background.

Current data do not exclude the existence of globular clusters of anti-stars in the halo of our galaxy (Khlopov 1998). These clusters should be a source of an anti-nuclear component of galactic cosmic rays, which can be tested in the AMS experiment on the International Space Station.

Therefore, in the cosmology based on a model of horizontal unification the concept of the antiworld in baryon-asymmetrical Universe can be implemented.

3. Model of horizontal unification and cosmology of dark matter

3.1. Unified description of various forms of dark matter on the basis of the hidden sector of models of horizontal unification

In the case of broken symmetry of generations, the 'see-saw' mechanism of generation of neutrino mass gives the mass hierarchy inverse to the hierarchy of horizontal symmetry breaking.

The mass of the ordinary light neutrino is inversely proportional to the Majorana mass of the heavy partner (see (11.23))

$$m_\nu \propto M_N^{-1}. \tag{11.87}$$

In the case of the inverse hierarchy model, the same inverse proportionality holds for the heavy partners of quarks and charged leptons. According to (11.22) and (11.23), the hierarchy of the Majorana neutrino masses is similar to the hierarchy of masses of ordinary quarks and leptons:

$$m_{\nu_e} : m_{\nu_\mu} : m_{\nu_\tau} \sim m_e : m_\mu : m_\tau. \tag{11.88}$$

In the model of horizontal unification, the mixing angles that characterize the amplitude of neutrino oscillations are determined by the relative rotation in the horizontal group space of the mass matrices of neutrinos and charged leptons. Therefore, the experimental data on neutrino mixing angles in the studies of neutrino oscillations impose upper limits on the deviation of the mass ratios of neutrinos from the strict proportionality of the corresponding ratio of the masses of charged leptons. At large mixing angles, for example, close to the case of maximum mixing, the ratio of proportionality (11.88) is strongly distorted.

For the sextet representation of the horizontal Higgs fields $\xi^{(1)}$ and $\xi^{(2)}$, the neutrino mass matrices given by expression (11.23) are non-diagonal, and heavy neutrinos ν_H can decay into light neutrinos ν_L and archions with a lifetime

$$\tau(\nu_H \to \nu_L + \alpha) = \frac{16\pi}{g_{HL}^2 m_H}, \tag{11.89}$$

where

$$g_{HL} = g_{\nu_H \nu_L} = \frac{m_{HL}}{\vartheta_{PQ}} \tag{11.90}$$

and m_{HL} is the corresponding off-diagonal element of the mass matrix of neutrinos.

For the 'low' scale of symmetry breaking of generations

$$\vartheta_{PQ} \sim 10^6 \text{ GeV} \quad (11.91)$$

the predicted effect of the Majorana neutrino mass is very close to present-day sensitivities of the search of the neutrinoless double decay. In particular, these predictions correspond to the evaluation of the electron neutrino mass based on the results of the Heidelberg–Moscow experiments (2002). However, these results are not confirmed by other groups.

On the other hand, predicted in a model with the 'low' scale (11.91) the values of the difference of the squares of neutrino masses are in sharp contrast with these parameters determined from the experimental data on neutrino oscillations in SNO experiments (SNO collaboration, Ahmad et al 2002), KAMLAND (KAMLAND collaboration, Eguchi et al 2003), K2K (K2K collaboration, Ahn et al 2003) and SuperKamiokande (SuperKamiokande collaboration, Kajita, Totsuka 2001). Therefore, these data exclude the model with the 'low' scale, making this cosmological scenario, based on this model, illustrative only.

Because the interactions of the archion with fermions of the lightest generation (u, d, e, ν_e) are suppressed, the existing constraints on the proper scale of the invisible axion weaken to

$$\vartheta_{PQ} \geqslant 10^6 \text{ GeV}, \quad (11.92)$$

making the archion model pretty close to the hadronic axion model (Kim 1979; Shifman et al 1980, see Chapter 2).

It turns out (Berezhiani et al 1992) that the archion model avoids the serious problems associated with the relic stable supermassive quark Q predicted by the model of the hadronic axion.

We can estimate the frozen-out concentration of such quarks and the corresponding hadrons in the Universe and find a contradiction between these estimates and the upper limit on their concentration, as follows from searches for anomalous isotopes of nuclei. Consequently, the theory of the hadronic axion must be supplemented by a mechanism of instability of the supermassive quarks.

The inclusion of the hadronic axion models in the GUT model leads to the inevitable existence of a supermassive lepton associated with the axion. A mixture of the superheavy quark with the ordinary light quark, which leads to instability Q, leads to the existence of the axion interaction with leptons, so that the axion will not be hadronic.

Because of these difficulties in the model of the hadronic axion, the archion model is of particular interest because the supermassive quark becomes unstable in the model and the interaction of the axion with the leptons is suppressed.

Since in the model of horizontal unification the scale of symmetry breaking determines:

the neutrino mass

$$m_\nu \propto \vartheta_{PQ}^{-1}, \tag{11.93}$$

neutrino lifetime

$$\tau(\nu_H \to \nu_L a) \propto \vartheta_{PQ}^5 \tag{11.94}$$

and the average density of the primordial axion field

$$\rho_a = \rho_a(\bar{\theta} = 1) \propto \vartheta_{PQ}, \tag{11.95}$$

changing the parameter ϑ_{PQ} we reproduce all the main types of cosmological models of structure formation in the Universe.

For large ϑ_{PQ} the axion field must dominate the Universe. Dominance of stable massive neutrinos corresponds to lower ϑ_{PQ}. And finally, the lowest allowed values of this parameter correspond to cosmological models with massive unstable neutrinos (Chapter 9).

In a model of horizontal unification (MHU), continuously changing the fundamental physical parameter ϑ_{PQ}, we obtain a smooth transition from the dominance of one form of dark matter to the dominance of other such forms in the Universe. The model allows us to make definite predictions for each type of dark matter, based on the combination of cosmological, astrophysical and physical limitations.

For fixed total dark matter density ρ_{DM} the equation

$$\rho_a(\vartheta_{PQ}) + \sum_i \rho_{\nu i} + \rho_a^{\rm dec}(\vartheta_{PQ}) + \sum_i \rho_{\nu i}^{\rm dec}(\vartheta_{PQ}) = \rho_{DM} \tag{11.96}$$

becomes an equation with respect to ϑ_{PQ}.

Equation (11.96) takes into account the primordial contribution of axions and neutrinos with the lifetime exceeding the lifetime of the Universe, as well as the contribution of axions and neutrinos from the decay of unstable neutrinos.

This equation defines a discrete set of cosmological models with different types of dark matter that forms the structure of the Universe.

In general, there are six different scenarios of dark matter that can be realized within the model of horizontal unification.

3.2. Models of dark matter resulting in horizontal unification

1. Scenario of cold dark matter (CDM)
The cosmological evolution of the axion field after PQ-symmetry breaking follows in MHU general features of cosmology of the 'invisible' axion.

As a result of the hierarchy of symmetry $SU(3)_H \otimes U(1)_H$ breaking, the $U(1)''_H$ group appears to be PQ-symmetry of only one generation of quarks and leptons. This allows us to automatically solve the problem of the cosmological θ-domains in the theory of the axion (Ipser, Sikivie 1983).

In the axion theory with $N > 1$ fermion generations, discrete vacuum degeneracy with various N after PQ-phase transition leads to overproduction of the vacuum walls at the boundary of θ-domains. This problem does not exist in the axion theory with only one generation, to which the MHU is reduced

As discussed in Section 2, the stochastic distribution of the axion field

$$\theta = \frac{\alpha}{\vartheta_{PQ}} \tag{11.97}$$

can lead to a change of the field by 2π when going along a closed circuit, which also leads to the formation of the structure of axion strings, rapidly decaying when the axion mass is switched on

$$T \leqslant 800 \text{ MeV}. \tag{11.98}$$

The axion field varies with the local amplitude $\theta(\vec{x}) - \theta_{\text{vac}}$. The oscillation energy density decreases during the expansion process as (see (11.72))

$$\rho_a \propto a(t)^{-3}. \tag{11.99}$$

Here $a(t)$ is the scale factor. The present-day density of the axions is

$$\rho_a = \left(\frac{\vartheta_{PQ}}{4 \cdot 10^{12} \text{GeV}} \right) \cdot \rho_{\text{cr}}, \tag{11.100}$$

where ρ_{cr} is the critical density (Abbott, Sikivie 1983; Preskill et al 1983; Dine, Fishler 1983).

According to (Davis 1986), intense radiation of axions by the decaying axion cosmic strings increases the present-day cosmological axion density up to

$$\rho_a = \left(\frac{\vartheta_{PQ}}{2 \cdot 10^{10}\,\text{GeV}}\right) \cdot \rho_{\text{cr}}. \tag{11.101}$$

Comparing these predictions with the total cosmological density, taking it to be equal ρ_{cr}, we obtain an upper limit on ϑ_{PQ}. If compared with the dark matter density, the constraint is four times stronger.

However, in the inflationary cosmology there may be cases in which the axion field in the observable Universe has an amplitude

$$\left|\theta(\vec{x}) - \theta_{\text{vac}}\right| \ll 1 \tag{11.102}$$

(Linde 1988b).

The exponential expansion of areas with the axion fields of type (11.102) also leads to the absence of axion strings, and hence to the fact that the axion density will not increase due to the decay of these strings.

It would seem that in this case there is no cosmological upper limit on the scale ϑ_{PQ}, so that this scale can reach even the Planck scale $\vartheta_{PQ} \sim m_{\text{Pl}}$. However, the scale ϑ_{PQ} in MHU may be limited even in this case.

In cosmology of horizontal unification, both axions and inflationary cosmology are derived from the same physical model. This allows self-consistent interpretation of the axions in inflationary cosmology.

Following the discussion started in Section 2, we can 'fix' the potential of the inflaton in relation to the COBE, WMAP and PLANCK data and take into account the condition of absence of the phase transition in the inflationary stage.

Based on the scenario of chaotic inflation which arises in the MHU, to avoid peaks in the density of fluctuations, we must eliminate the possibility of horizontal transitions in the inflationary stage, which leads to the constraint

$$\vartheta_{PQ} < 2 \cdot 10^{10}\ \text{GeV} \tag{11.103}$$

(Sakharov, Khlopov 1993).

Under the condition (11.103), the horizontal phase transitions can occur only in the post-inflationary stage or in the RD stage after heating of the Universe.

In both cases, PQ-symmetry breaking is followed by the formation of the structure of the axion strings. When the axion mass is switched on, the vacuum walls bound by the strings form and decay and this

is reflected in the structure of the archioles (see Section 2). The observational data on the quadrupole anisotropy and estimates (11.79) of the effect determined by the archioles give the model-independent limit (11.80).

According to (Berezhiani et al 1992) at

$$\vartheta_{PQ} < 10^8 \text{ GeV} \tag{11.104}$$

the coherent oscillations of the axion field thermalize after the QCD phase transition. The rate of interaction of the axions with the equilibrium nucleon–antinucleon pairs in the reactions

$$a + N(\bar{N}) \to \pi + N(\bar{N}) \tag{11.105}$$

under the condition (11.104) is higher than the rate of expansion. As a result, the axion condensate is evaporated and the structure of the archioles decay.

At low values of ϑ_{PQ} corresponding to (11.104), the evaporation of the axion condensate leads to the formation of a thermal axion background. The gas of primordial axions in this case is a thermal gas of particles with zero spin and mass m_a, which is decoupled from the plasma and radiation at a temperature of 60–160 MeV, depending on the value ϑ_{PQ}.

During decoupling, the relative concentration of thermal axions and neutrinos is given by the equilibrium value (see Chapter 3)

$$\frac{n_a}{n_{\nu\bar{\nu}}} = \frac{2}{3}. \tag{11.106}$$

Depending on the exact values of ϑ_{PQ} the present-day relative concentration of the gas of the thermal axions should be corrected with the account for the effect of heating of the neutrino by pairs of muons and pions present in equilibrium during the decoupling of thermal axions. Say, for $\vartheta_{PQ} \sim 10^6$ GeV the present-day relative concentration of thermal axions is

$$\frac{n_a}{n_{\nu\bar{\nu}}} = 0.57. \tag{11.107}$$

The existence of the distribution of the axion field before the QCD phase transition and evaporation of the condensate could have led to inhomogeneous baryosynthesis considered in the previous section.

Thus, at the 'low' values of the scale ϑ_{PQ} (11.104) the primordial axions represent not cold but hot dark matter. However, this form of hot dark matter is not dominant in the MHU cosmology with the 'low' scale. On the other hand, as mentioned in section 3.1, the data on neutrino oscillations exclude the 'low scale' value and all its corresponding cosmological scenarios.

2. Scenario of hot dark matter (HDM)
Dominance of τ-neutrinos with the concentration of (3.118), predicted by the standard Big Bang model, corresponds to the mass ν_τ

$$m_{\nu_\tau} = 20 \text{ eV} \tag{11.108}$$

and the lifetime exceeding the age of the Universe

$$\tau(\nu_\tau \to \nu_\mu + a) > t_U. \tag{11.109}$$

The cosmological density of such ν_τ is expressed through the parameters of the model of horizontal unification as

$$\rho_{\nu_\tau} = \frac{6.6 \cdot 10^{13} \text{GeV}}{\vartheta_{PQ}} \cdot \left(\frac{g_\nu^2}{G_N} \right) \cdot \rho_{\text{cr}}. \tag{11.110}$$

For a combination of the Yukawa constants used in the model of horizontal unification in the range

$$10^{-6} < \frac{g_\nu^2}{G_N} < 10^{-4}, \tag{11.111}$$

the HDM scenario with dominance of ν_τ is realized in the MHU in the appropriate range of scales

$$10^8 \text{ GeV} < \vartheta_{PQ} < 10^{10} \text{ GeV}. \tag{11.112}$$

The range (11.112) practically coincides with the interval corresponding to the axion cold dark matter and is defined by constraints (11.80) and (11.104). This allowed to consider in (Khlopov, Sakharov 1999) a model of a mixed dark matter CDM + HDM as a possible solution in the cosmology of horizontal unification. The data on neutrino oscillations exclude this possibility.

3. Relativistic unstable dark matter (UDM)
In the model of horizontal unification, this solution corresponds to the

dominance in the Universe of relativistic archions and ν_μ which arise from the decay of ν_τ with the mass

$$m_{\nu_\tau} \cong 50 \div 100 \text{ eV} \tag{11.113}$$

and lifetime

$$\tau\left(\nu_\tau \to \nu_\mu + a\right) = 4 \cdot 10^{15} \div 10^{16} \text{ s.} \tag{11.114}$$

The muon neutrinos from the decays (11.114) are relativistic in the present-day Universe due to the fact that their mass does not exceed 5 eV.

The considered solution of (11.96) follows from the expression for the energy density of relativistic decay products of the τ-neutrino

$$\rho_{\nu_\mu}^{\text{dec}} + \rho_a^{\text{dec}} = \left(\frac{\vartheta_{PQ}}{10^{10}\,\text{GeV}}\right)^{3/2} \cdot \frac{1}{x} \cdot \left(\frac{g_V^2}{G_N}\right) \cdot \rho_{\text{cr}}, \tag{11.115}$$

where $x \sim 1$ is a parameter of neutrino mixing and takes into account the redshift of the relativistic decay products.

4. Scenario of non-relativistic UDM

The dominance of the non-relativistic ν_μ with a mass of about 10 eV, both primordial and from the decay of ν_τ with the mass of the order of 100 eV and a lifetime ~10^{15} s, gives that

$$\tau\left(\nu_\mu \to \nu_e + a\right) > t_U. \tag{11.116}$$

The density of non-relativistic muon neutrinos dominating the Universe in this solution of (11.96) can be expressed through the parameters of horizontal unification as

$$\rho_{\nu_\mu} = \frac{6.6 \cdot 10^{12}\,\text{GeV}}{\vartheta_{PQ}} \cdot \left(\frac{g_V^2}{G_N}\right) \cdot \rho_{\text{cr}}. \tag{11.117}$$

5. Scenario of the relativistic hierarchical neutrino decay (HND)

The present-day Universe is characterised by the dominance of relativistic archions and electron neutrinos from the decay of muon neutrinos with the mass

$$m_{\nu_\mu} \cong 50 \div 100 \text{ eV} \tag{11.118}$$

and the lifetime

$$\tau\left(\nu_{\mu} \rightarrow \nu_{e} + a\right) = 4 \cdot 10^{15} \div 10^{16}\,\mathrm{s} \tag{11.119}$$

subject to rapid decay of the τ-neutrino with the mass

$$m_{\nu_\tau} \cong 1 \div 10\ \mathrm{keV} \tag{11.120}$$

and the lifetime

$$\tau\left(\nu_{\tau} \rightarrow \nu_{\mu} + a\right) \leqslant 10^{8} \div 10^{10}\,\mathrm{s}. \tag{11.121}$$

The present-day density is in this case

$$\rho_{\nu_e}^{\mathrm{dec}} + \rho_{a}^{\mathrm{dec}} = \left(\frac{\vartheta_{PQ}}{10^{8}\,\mathrm{GeV}}\right)^{3/2} \cdot \frac{1}{x} \cdot \left(\frac{g_{\nu}^{2}}{G_{N}}\right) \cdot \rho_{\mathrm{cr}}. \tag{11.122}$$

6. Scenario of non-relativistic HND

Dominance of the non-relativistic or semi-relativistic archions arising in the early decay of ν_τ, which satisfy the conditions (11.120) and (11.121), and subsequent decay of ν_μ satisfying the conditions (11.118) and (11.119), is realized in the case

$$m_a > m_{\nu_e}. \tag{11.123}$$

The main contribution to the inhomogeneous dark matter (in rich clusters of galaxies and the halos of galaxies) is provided by both the background of the primordial thermal archions and non-relativistic archions from early decays of ν_τ. The total density also takes into account the dominant contribution of the homogeneous background of the archions from the recent decay

$$\rho_a = \frac{5 \cdot 10^{5}\,\mathrm{GeV}}{\vartheta_{PQ}} \cdot \rho_{\mathrm{cr}}. \tag{11.124}$$

In the opposite case,

$$m_a < m_{\nu_e}, \tag{11.125}$$

nonrelativistic ν_e, both primordial and from the decay of ν_μ and ν_τ contribute to the present-day cosmological density

$$\rho_{\nu_e} = \frac{3.3 \cdot 10^{10}\,\mathrm{GeV}}{\vartheta_{PQ}} \cdot \left(\frac{g_{\nu}^{2}}{G_{N}}\right) \cdot \rho_{\mathrm{cr}}. \tag{11.126}$$

Therefore, the model of horizontal unification provides a unified physical basis for cold dark matter (primordial axion condensate), HDM (stable τ-neutrino), the relativistic and non-relativistic unstable dark matter and hierarchical neutrino decay scenarios regarded as the solution of (11.96).

A complete set of solutions to (11.96) can be realized in dependence on ϑ_{PQ} for the combination of Yukawa couplings in the range

$$1.5\cdot 10^{-6} < \frac{g_\nu^2}{G_N} < 4\cdot 10^{-5}. \tag{11.127}$$

The graphical solution of equation (11.96) is shown in Fig. 11.1.

On the other hand, the set of cosmological and astrophysical constraints, discussed in section 2, leaves only two small intervals: near

$$\vartheta_{PQ} \sim 10^{10}\ \text{GeV}, \tag{11.128}$$

where the scenarios of CDM and HDM, or a mixture thereof with a possible admixture of archioles may take place; and close to

$$\vartheta_{PQ} \sim 10^{6}\ \text{GeV} \tag{11.129}$$

in which the HND scenarios 5 and 6 are implemented.

The UDM scenarios 3 and 4 are completely eliminated by the constraints on ϑ_{PQ} obtained from observations of the neutrino pulse from SN1987A.

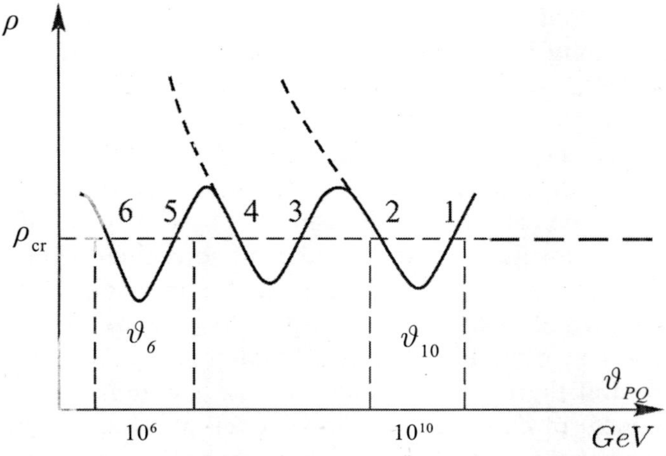

Fig. 11.1. Set of dark matter scenarios in a model of horizontal unification, determined by the scale of horizontal symmetry breaking.

The first possibility, corresponding to (11.128), relates to the more conservative 'standard' CDM scenario of formation of the large-scale structure. The non-trivial effect predicted by the MHU is the percolation of the structure of archioles, which may be desirable in the CDM model to improve its agreement with the observational data on the large-scale structure.

The second solution offers a non-trivial possibility for the formation of the large-scale structure. The HND scenarios 5 and 6 combine the attractive features of HDM, CDM and UDM models. This makes the HND scenario an attractive non-trivial example for the development of detailed models of formation of the cosmological large scale structure and for comparing the predictions of such models with astronomical data.

Indeed, in HND scenarios the dominance of ν_τ the with the mass of the order of 1–10 keV in the $10^8 \div 10^{10}$ s period, generates short-wavelength fluctuations in the spectrum of density fluctuations ν_μ with a mass of 50–100 eV. These short-wave fluctuations correspond to the scale of galaxies, since they provide the growth of baryonic fluctuations with the masses

$$10^7 \leqslant M \leqslant 10^{10} M_\odot. \tag{11.130}$$

Muon neutrinos from the decay ν_τ are relativistic during the decay period t_d, but then at

$$t_{nr} \sim t_d \cdot \left(\frac{m_{\nu_\tau}}{2m_{\nu_\mu}} \right)^2 \tag{11.131}$$

they undergo the red shift to non-relativistic energies and double the spectrum of the long-range components of the spectrum of fluctuations inherent in HDM models and stimulating the formation of a distinctive cellular structure of voids and superclusters.

Although only one type of dark matter particles – the muon neutrino – ensure the formation of the structure, the model is effectively of the two-component type. The structure is formed by a mixture of the cold component of primordial ν_μ that have the range typical of the thermal gas of the particles with mass 1–10 keV, and the hot component ν_μ from the decay of ν_τ, which behaves with respect to gravitational instability as a thermal gas of particles with masses of several eV.

The primordial thermal background of the archions which are the coldest component of dark matter in this model, plays an important role in the evolution of the very shortwave part of the density perturbations, including, in particular, the formation of massive halos outside the visible part of the galaxies. It should be noted that (Madsen 1990)

constraints from the density in the phase space on the mass of particles in the halo (Tremaine, Gunn 1979) may be weakend or non-existent in the case of the Bose gas.

The physical basis of the HND (MHU with the low scale (11.129)) could be verified by experiments in accelerators in searches for the axion decays of mesons and baryons and the effects of 'horizontal' gauge bosons in the systems of the neutral K-, D-, and B-mesons. This possibility of efficient search for the effects resulting from the same physical fundamentals as the investigated cosmological model and is also predicted with the probability determined by the correspondence of the cosmological model to the observations, illustrates the prospects for investigating the physical fundamentals of present-day cosmology by the methods of experimental cosmoparticle physics (Belotsky et al 1998).

Both solutions of the MHU for dark matter, corresponding to the relations (11.128) and (11.129), provided a natural basis for mixed CDM + HDM models of formation of the large-scale structure.

As already mentioned, the main problem of the discussed simplest implementation of the proposed cosmological scenarios is associated with the evidence for neutrino oscillations and constraints on the mass of all neutrino types m_ν <1 eV, arising from these data. It eliminates the dominant role of the three known types of neutrinos, and, consequently, the role of the known particles in the cosmological evolution, opening up new spaces of the Universe to new not yet known forms of matter. For example, HND or CDM+HDM scenarios can be implemented in cosmology of sterile neutrinos.

This is perhaps deeply symbolic. Extension of the forms of dark matter, corresponding to an increasingly wide hidden sector of the modern theory of elementary particles, respectively, reduces the cosmological value of the neutrinos, while enhancing the historic role of neutrinos in this process. In the same extent as the cosmology of massive neutrinos has revealed the need for cosmological non-baryonic dark matter, the cosmological scenarios based on the physics of neutrino masses in a model of horizontal unification can stimulate the development of physically non-trivial multi-component scenarios of dark matter and energy determining the evolution of the Universe.

3.3. History of the Universe based on the model of horizontal unification

Thus, the above scenarios determine the history of the Universe based on the model of horizontal symmetry. Those are two physically self-consistent inflationary scenarios of dark matter in the baryon-asymmetric

Universe.

In these scenarios, all steps of the cosmological evolution correspond quantitatively to the parameters of particle theory.

The physics of the inflaton corresponds to the Dirac 'see-saw' mechanism of generation of the mass of the quarks and charged leptons, lepton processes with non-conservation of the lepton number in the combined ($\Delta L = 2$) + (electroweak processes of baryon number non-conservation) baryosynthesis are based on the physics of Majorana neutrino masses. The parameters of axion CDM in the solution (11.128) with a high energy scale, as well as the masses and lifetimes of neutrinos in HND physically correspond to the scale and structure of breaking of the symmetry of generations.

Even in the minimal realization of the inflationary CDM scenario, based on the solution of (11.128), we find additional non-trivial features, such as the post-inflationary dust stage, the horizontal phase transitions, axions strings, the structure of archioles, the possibility of PBH formation and evaporation, as well as possible spatial dependence of the baryonic charge and the formation of domains of the antibaryons in the baryon-dominated Universe.

The qualitatively similar pattern of the very early Universe, based on the MHU with a low energy scale is supplemented by alternating stages of the dominance of massive unstable neutrinos and their decay products in which the large-scale structure of the Universe forms.

Two-photon decays of archions in the HND scenario leads to the effect of small ionization during recombination, to complete ionization of the baryonic matter during the recent decay of the muonic neutrinos, and should lead to the existence of non-thermal electromagnetic background in the ultraviolet, optical and infrared ranges.

From a physical point of view, the models of horizontal unification with high and low energy scales are still far from complete versions of the phenomenology of 'theory of everything' (TOE).

It is necessary to add supersymmetry to solve the problem of the quadratic divergence of Higgs mass, Grand Unification – to ensure the axion solution of strong *CP*-violation, shadow and (or) mirror matter – to restore the equivalence between the left-handed and right-handed coordinate systems, and so on.

Therefore, even if the presented cosmological scenarios do not contradict the experimental and observational data, in principle *they might not give* the best agreement with the observations.

For example, the addition of archioles and hot dark matter can not substantially eliminate the difficulties of CDM models for structure formation.

Conversely, the cold component in the HND may be insufficient to explain the effects of dark matter on scales of small galaxies. The dominance of the relativistic decay products of the neutrino may find difficulty in comparing the observed large-scale structure with the observations of anisotropy of cosmic microwave background radiation.

All these drawbacks are easily addressed in more complex scenarios with more complex physical bases. These scenarios reveal a non-trivial cosmological picture, and present-day cosmoparticle physics is strong enough to confront them and analyze them in all their physical variety. Cosmoparticle analysis of the MHU is proof of that.

Anyway, MHU provides a single basis for a theoretical description of both the structure of elementary particles and the structure of the Universe.

Such an approach that combines cosmology and elementary particle physics is an essential feature of cosmoparticle physics. Combining the individual results of the study of special problems of cosmology and particle physics, a model of horizontal unification is only a first step toward a realistic description of a single unified foundation of the structure of the macro- and micro-world.

A method for identifying the main elements of the theory, based on the detailed design of its 'low-energy' basis, can provide valuable recommendations for the selection of realistic options for a complete 'theory of everything' (the superstring theory, for example), which is obviously important from the standpoint of the existing theoretical uncertainties in the search for fundamental foundations of physics and cosmology.

It is important to note the important epistemological aspect of the investigations. We have demonstrated the fundamental possibility of a detailed study of the multi-parameter 'hidden' sector of particle theory.

An example of a model of horizontal unification gives hope to test any realistic multi-parameter scenario of superhigh energy physics. Developed in every detail, such a scenario could lead to indirect effects, available for experimental and observational tests. If the number of such effects exceeds the number of independent parameters of the model, then there is an 'overdetermined' system of equations for the unknown parameters.

Thus, a general approach to experimental verification of the theory, based on the overdetermined system of equations for the parameters of the theory, can be implemented.

In the framework of cosmoparticle physics the indirect effects, predicted by theory, can be used for a detailed analysis in the case where a direct experimental verification is not possible.

References

[1] Abbot, L., Farhi, E., Wise, M. (1982) *Phys. Lett.*, **117B**, 29.
[2] Abbot, L.F., Sikivie, P. (1983) *Phys. Lett.*, **B120**, 133.
[3] Abbot, L.F., Wise, M.B. (1984) *Nucl. Phys.*, **B244**, 541.
[4] Adams, F.C., Freeze, K. (1991) *Phys. Rev.,* **D43**, 353.
[5] Adler, S.L. (1969) *Phys. Rev.,* **177**, 2426.
[6] Affeck, I., Dine, M. (1985) *Nucl. Phys.,* **B249**, 361.
[7] Aharonian, F.A., Vardanian, V.V. (1985) Preprint EFI-827-54-85-YEREVAN.
[8] Akhmedov E.Kh. and Khlopov M.Yu (1988a), *Yadernaya Fizika*, **47**, 1679.
[9] Akhmedov, E.Kh., Khlopov, M.Yu. (1988b) *Mod. Phys. Lett.,* **A3**, 451.
[10] Albrecht, A., Steinhardt, P., Turner, M., Wilczek, F. (1982) *Phys. Rev. Lett.,* **48**, 1437.
[11] Albrecht, A., Steinhardt, P.J. (1982) *Phys. Rev. Lett.*, **48**, 1220.
[12] Albrecht, A., Turok, N. (1985) *Phys. Rev. Lett.*, **54**, 1868.
[13] Albrecht, A., Turok, N. (1989) *Phys. Rev.,* **D40**, 973.
[14] Alcock, C., Farhi, E. (1985) *Phys. Rev. Lett.,* **D32**, 1273.
[15] Alcock, C., Olinto, A. (1988) *Ann. Rev. Nucl. Part. Phys.,* **38**, 161.
[16] Alcock, C.R., Fuller, G.M., Mathews, C.J. (1987) *Astrophys. J.,* **320**, 439.
[17] Alexeyev, S.O., Pomazanov, M.V. (1997) *Phys. Rev.,* **D55**, 2110.
[18] Alford, M.G., Wilczek, F. (1989) *Phys. Rev. Lett.,* **62**,1071.
[19] Alichanian, A.I., Alihanov, A.I. and Dzelepov, B.S. (1938) *Phys. Rev.,* **53**, 766.
[20] Alichanian, A.I., Nikitin, S.J. (1938) *Phys. Rev.,* **53**, 767.
[21] Aliev, T.M., Vysotsky, M.I. (1981) *Usp. Fiz. Nauk,* **135**, 709 [*Sov. Phys. Usp.*, **24**, 1008].
[22] Alpher, R.A., Herman, R. (1988) *Phys. Today* **41**, 24.
[23] Anselm, A.A. (1982) *Pis'ma ZhETF*, **36**, 46.
[24] Anselm, A.A. and Uraltsev, N.G. (1983) *Pis'ma ZhETF,* **84**, 1961.
[25] Anselm, A.A., Berezhiani, Z.G. (1985) *Phys. Lett.,* **162B**, 349.
[26] Anselm, A.A., Uraltsev, N.G., Khlopov M.Yu (1985), *Yadernaya Fizika,* **41**, 1678.
[27] Anselm, A.A., Uraltsev, N.G. (1982) *Phys. Lett.,* **116B**, 161.
[28] Appelquist, T., Chodos, A. (1983) *Phys. Rev.,* **D28**, 772.
[29] Applegate, J.H., Hogan, C.J. (1985) *Phys. Rev.,* **D31**, 3037.
[30] Applegate, J.H., Hogan, C.J., Scherrer, R.J. (1987) *Phys. Rev.,* **D35**, 1151.
[31] Arnold, P., McLerran, L. (1987) *Phys. Rev.,* **D36**, 581.
[32] Arnold, V.I., Shandarin, S.F., Zeldovich, Ya.B. (1982) *Geophys. Astrophys. Fluid Dynamics,* **20**, 111.
[33] Aryal, M., Everett, A.E. (1987) *Phys. Rev.,* **D35**, 3105.
[34] Azimov, Ya.E., Dokshitzer, Yu.P., Khoze, V.A. (1983) *Yadernaya Fizika,* **37**, 703.
[35] Babul, A., Paczynsky, B., Spergel, D. (1987) *Astrophys. J., Lett.,* **316**, L49.
[36] Bahcall, J.N. (1984) *Astrophys. J.,* **276**, 156; **276**, 169; **287**, 926.
[37] Bahcall, J.N., Flynn, C., Gould, A. (1992) *Astrophys. J.,* **389**, 234.

[38] Balakirev, A.N. (1976) *Astrofiz. Zh.*, **53**, 119.
[39] Balestra, F., Guaraldo, K., Piragino, G., Sapozhnikov, M.G., Falomkin, I.V., Khlopov, M.Yu. et al.,(1985) *JINR Rapid Comm.*, **6**, 11.
[40] Balestra, F., Piragino, G., Pontecorvo, D.B., Sapozhnikov, M.G., Falomkin, I.V. and Khlopov, M.Yu. (1984) *Yadernaya Fizika,* **39**, 990 [*Sov. J. Nucl. Phys.*, **39**, 626].
[41] Bardeen, J., Steinhardt P., Turner, M. (1983) *Phys. Rev.,* **D28**, 679.
[42] Bardeen, J.M. et al. (1986) *Astrophys. J.,* **304**, 15.
[43] Barrow, J.D., Copeland, E.J., Liddle, A.R. (1992) *Phys. Rev.,* **D46**, 645.
[44] Batlen, A.H., Fletcher, J.M., Mann, P.J. (1978) *Publ. Pominion. Astrophys. Obs.*, **15**, 121.
[45] Batusov, Yu.A., et al (1984) *Lett. Nuovo Cim.,* **41**, 223.
[46] Belavin, A.A., Polyakov, A.M., Schwartz, A.S., Tyupkin, Yu.S. (1975) *Phys. Lett.*, **59B**, 85.
[47] Belinsky V.A., Grishchuk, L.P., Khalatnikov, I.M. and Zeldovich, Ya.B. (1985) *Phys. Lett.,* **155B**, 232.
[48] Bell, J.S., Jackiw, R. (1969) *Nuovo Cim.,* **AGO**, 47.
[49] Bellido, J.C., Linde, A., Wands, D. (1996) *Phys. Rev.,* **54D**, 6040.
[50] Belotsky, K., et al (2008) *Phys. Atom. Nucl.*, **71**, 147.
[51] Belotsky, K.M., Damour, T., Khlopov, M.Yu. (2002b) *Phys. Lett.,* **B529**, 10.
[52] Belotsky, K.M., Fargion, D., Khlopov, M.Yu., Konoplich, R., Shibaev, K. (2003) *Phys. Rev.,* **D68**, 0540271; hep-ph\0210153.
[53] Belotsky, K.M., Golubkov, Yu.A., Konoplich, R.V., Sakharov, A.S., Khlopov, M.Yu. (2000) *Yadernaya Fizika,* **63**, 290 [*Phys. Atom. Nucl.* **63**, 233].
[54] Belotsky, K.M., Khlopov, M.Yu. (2001) *Gravitation and Cosmology,* **7**, 189.
[55] Belotsky, K.M., Khlopov, M.Yu. (2003) in: Proc. 15th Rencontres de Blois: Physical Cosmology (Chateau de Blois, France, 16–22 Jun 2003).
[56] Belotsky, K.M., Khlopov, M.Yu., Sakharov, A.S., Shklyaev, A.A., Sudarikov, A.L. (1998) *Proc. Cosmion'96* Eds. KhlopovM.Yu., et al. (World Scientific).
[57] Belotsky, K.M., Khlopov, M.Yu., Shibaev, K. (2000) *Gravitation and Cosmology,* **6**, Suppl., 140.
[58] Belotsky, K.M., Khlopov, M.Yu., Shibaev, K. (2001) *Part. Nucl. Lett.,* **108**, 5.
[59] Belotsky, K.M., Khlopov, M.Yu., Shibaev, K. (2002a) *Yadernaya Fizika,* **65**, 407.
[60] Belotsky, K.M., Khlopov, M.Yu., Shibaev, K.I. (2009) in: *The Physics of quarks: New research* (Horizons in World Physics, V.265) eds. Watson, N.L., Grant, T. M., (NOVA Publishers), p.19.
[61] Belova, T.I., Kudryavtzev, A.E. (1988) *Physica,* **D32**,18.
[62] Bennet, D.P., Bouchet, F.R. (1988) *Phys. Rev., Lett.* **60**, 57.
[63] Bennett, C.L., et al. (1996) *Astrophys. J. Lett.,* **464**, L1.
[64] Bento, L., Berezhiani, Z. (2002) *Fortsch. Phys.,* **50**, 489.
[65] Berestetsky, V.B., Lifshits, E.M., Pitaevsky, L.P. (1968) *Relativistic Quantum Theory.* Part 1 (Moscow, Nauka).
[66] Berezhiani, Z.G. (1983) *Phys. Lett.,* **129B**, 99.
[67] Berezhiani, Z.G. (1985) *Phys. Lett.,* **150B**, 177.
[68] Berezhiani, Z.G., Chkareuli, J.K. (1982) *Pisma ZhETF,* **35**, 494.
[69] Berezhiani, Z.G., Chkareuli, J.K. (1983) *Pisma ZhETF,* **37**, 285.
[70] Berezhiani, Z.G., Chkareuli, J.K. (1983) *Yadernaya Fizika,* **37**, 1043.
[71] Berezhiani, Z.G., Comelli, D., Villante, F.L. (2001a) *Phys. Lett.,* **B503**, 362.
[72] Berezhiani, Z.G., Dolgov, A.D., Mohapatra, R.N. (1996) *Phys. Lett.,* **B375**, 26.
[73] Berezhiani, Z.G., Gianfagna, L., Giannotti, M. (2001b) *Phys. Lett.,* **B500**, 286.
[74] Berezhiani, Z.G., Khlopov, M.Yu. (1989) in: *Proc. XXIV Rencontres de*

Moriond, ed. Tran Thanh Van, J. (Editions Frontieres), p.179.

[75] Berezhiani, Z.G., Khlopov, M.Yu. (1990a) *Yadernaya Fizika,* **51**, 1157 [*Sov. J. Nucl. Phys.*, **51**, 739].

[76] Berezhiani, Z.G., Khlopov, M.Yu. (1990b) *Yadernaya Fizika*, **51**, 1479 [*Sov. J. Nucl. Phys.,* **51**, 935].

[77] Berezhiani, Z.G., Khlopov, M.Yu. (1990c) *Yadernaya Fizika,* **52**, 96 [*Sov. J. Nucl. Phys.,* **52**, 60].

[78] Berezhiani, Z.G., Khlopov, M.Yu. (1991) *Z. Phys. C-Particles and Fields*, **49**, 73.

[79] Berezhiani, Z.G., Khlopov, M.Yu., Khomeriki, R.R. (1990a) *Yadernaya Fizika,* **52**, 104 [*Sov. J. Nucl. Phys.,* **52**, 65].

[80] Berezhiani, Z.G., Khlopov, M.Yu., Khomeriki, R.R. (1990b) *Yadernaya Fizika,* **52**, 538 [*Sov. J. Nucl. Phys.,* **52**, 344].

[81] Berezhiani, Z.G., Sakharov, A.S., Khlopov, M.Yu. (1992) *Yadernaya Fizika*, **55**, 1918 [*Sov. J. Nucl. Phys.,* **55**, 1063].

[82] Berezhiani, Z.G., Vysotsky, M.I., Doroshkevich, A.G., Khlopov, M.Yu. and Yurov, V.P. (1990c) *Yadernaya Fizika,* **51**, 1614.

[83] Berezhiani, Z.G., Vysotsky, M.I., Khlopov, M.Yu. (1987) *Yadernaya Fizika,* **45**, 1719.

[84] Berkov, A.V., Nikitin, Yu.P., Sudarikov A.L., Khlopov M.Yu (1988) *Yadernaya Fizika,* **48**, 779.

[85] Berkov, A.V., Nikitin, Yu.P., Sudarikov A.L., Khlopov M.Yu (1987), *Yadernaya Fizika*, **46**, 1729.

[86] Bernabei, R., et al (2000) *Phys. Lett.*, **B480**, 23.

[87] Bernabei, R., et al (2003) *Riv. Nuovo Cim.*, **26**, 1.

[88] Bernabei, R., et al (2010) *Eur. Phys. J.,* **C67**, 39.

[89] Bernard, C.W. (1974) *Phys. Rev.,* **D9**, 3312.

[90] Bilenky, S. M., Hosek, J., Petcov, S.T. (1980) *Phys. Lett.,* **B94**, 495.

[91] Bilenky, S. M., Pontecorvo, B. (1978) *Phys. Rept.,* **41**, 225.

[92] Bisnovatyi-Kogan, G.S., and Novikov, I.D. (1980), *Astron. Zh.,* **57**, 890.

[93] Bisnovatyi-Kogan, G.S., Lukash V.N. and Novikov, I.D. (1980) *Proc IAU Symposium.*

[94] Bjorken, B.J., Drell, S.D. (1965) *Relativistic quantum mechanics* (McGraw-Hill, New York).

[95] Blau, S.K., Guth, A.H. (1987) in: *300 Years of Gravitation*, eds. Hawking, S.H., Israel, W. (Cambridge University Press).

[96] Blinnikov, S.I. (1996) A quest for weak objects and for invisible stars. astro-ph\9801015.

[97] Blinnikov, S.I., Khlopov M.Yu. (1980) Preprint ITEP N-126.

[98] Blinnikov, S.I., Khlopov M.Yu. (1982) *Yadernaya Fizika,* **36**, 809.

[99] Blinnikov, S.I., Khlopov, M.Yu. (1983a) *Astron. Zh.,* **60**, 623.

[100] Blinnikov, S.I., Khlopov, M.Yu. (1983b) *Solar Physics,* **82**, 383.

[101] Bond, J.R., Efstathiou G., Silk, J. (1980) *Phys. Rev. Lett.*, **45**, 300.

[102] Bond, J.R., Szalay, A.S., Turner, M.S. (1982) *Phys. Rev. Lett.*, **48**, 1636.

[103] Bond. J.R., Szalay, A.S. (1983) *Astrophys. J.,* **274**, 443.

[104] Boris, S., Golutvin, A., Laptin, L., Lubimov, V., Nagovizin, V., Nozik, V., Novikov, E., Soloshenko, V., Tihomirov, I., Tretyakov, E. (1987) *Phys. Rev. Lett.*, **58**, 2019.

[105] Bouchet, F., Bennet, D., Stebbins, A. (1988) *Nature,* **335**, 410.

[106] Brandenberger, R.H. (1987) *Nucl. Phys.*, **B293**, 812.

[107] Brown, R.W., Stecker, F.W. (1979) *Phys. Rev. Lett.*, **43**, 315.

[108] Bugrii, A.I., Trushevsky, A.A. (1977) *Astrophysica,* **13**, 361.
[109] Burns, M.L., Lovelace, R.V.E. (1982) *Astrophys. J.,* **202**, 87.
[110] Callan, C.G., Jr., (1982) Preprint LPTENS-82-20; *Nucl. Phys.,* **B212**, 391 (1983).
[111] Callan, C.G., Jr., Coleman, S. (1974) P\n\s.. Rei; D16, 1762.
[112] Candelas, P., Horovitz, G.T., Strominger, A., Witten, E. (1985) *Nucl. Phys.,* **B258**, 46.
[113] Carlson, E.D., Glashow, S.L. (1987) *Phys. Lett.,* **B193**, 168.
[114] Carr, B.J. (1975) *Astrophys. J.,* **205**, 1.
[115] Carr, B.J., Gilbert, J.H., Lidsey, J.E. (1994) *Phys. Rev.,* **50D**, 4853.
[116] Chechetkin V.M., Khlopov, M.Yu., Sapozhnikov, M.G., Zeldovich, Ya.B. (1982) *Phys. Lett.,* **118B**, 329.
[117] Chechetkin, V.M., Khlopov, M. Yu., Sapozhnikov, M.G. (1982) *Rivista Nouvo Cim.,* **5**, N-10, 1.
[118] Cheng, H.Y. (1988) *Phys. Rep.,* **158**, 1.
[119] Chikashige, Y., Mohapatra, R.N., Peccei, R.D. (1980) *Phys. Rev. Lett.* **5**,19 6.
[120] Chikashige, Y., Mohapatra, R.N., Peccei, R.D. (1981) *Phys. Lett.,* **98B**, 265.
[121] Chiu, H.Y. (1966) *Phys. Rev. Lett.,* **17**, 712.
[122] Chkareuli, Yu.K. (1980) *Pis'ma ZhETF,* **32**, 684.
[123] Chudnovsky, E.M., Field, G.B., Spergel, D.N., Vilenkin, A. (1986) *Phys. Rev.,* **D34**, 944.
[124] Ciarcellutti, P. (2010) *Int. J. Mod. Phys.,* **D19**, 2151, P.335 (2010)
[125] Cline, D. (2003) in Proc. of Physics Beyond the Standard Model, BEYOND'03 (Castel Ringberg, 9–15 June 2003). astro-ph\0310439.
[126] Cohen, A. G., De Rujula, A., Glashow S.L. (1997). Preprint CERN-TH-97-154; astro-ph\9707087; *Astrophys. J.,* **495**, 539 (1998).
[127] Cohen, A., Kaplan, D. (1987) *Phys. Lett.,* **199B**, 251.
[128] Cohen, A., Kaplan, D. (1988) *Nucl. Phys.,* **B308**, 913.
[129] Coleman, S. (1974) *Phys. Rev.,* **D15**, 2929.
[130] Coleman, S., Weinberg, E. (1973) *Phys. Rev.,* **D7**, 1888.
[131] Colpi, M., Shapiro, S.L., Wasserman, I. (1986) *Phys. Rev. Lett.,* **57**, 2485.
[132] Comelli, D., Pietroni, M., Riotto, A. (1994) *Nucl. Phys.,* **B412**, 441.
[133] Copeland, E.J., et al (1994) *Phys. Rev.,* **D49**, 6410.
[134] Cowie, L., Hu, E. (1987) *Astrophys. J.,* **318**, L33.
[135] Damour, T., Krauss, L.M. (1999) *Phys. Rev.,* **D59**, 063509.
[136] Dannenberg, A., Hall. L.J. (1988) *Phys. Lett.,* **198B**, 411.
[137] Davis, M., Lecar, M., Pryor, C. and Witten, E. (1981) *Astrophys. J.,* **250**, 423.
[138] Davis, M., Peebles, P.J.E. (1983) *Astrophys. J.,* **267**, 465.
[139] Davis, R.L. (1986) *Phys. Lett.,* **180B**, 225.
[140] De Vaucouleurs, J. (1959) *Astron. Zh.,* **36**, 977.
[141] De Vaucouleurs, J. (1976) *Astrophys. J.,* **203**, 33.
[142] Deser, S., Zumino, B. (1976) *Phys. Lett.,* **62B**, 335.
[143] Dicus, D.S., Kolb, E.W., Teplitz, V.L., Wagoner, R.V. (1978) *Phys. Rēv.,* **D18**, 1829.
[144] Dimopoulos, S., Esmailzadeh, R., Hall, L. J., Starkman, G.D. (1988) Preprint SLAC-PUB-4532; *Nucl. Phys.,* **B311**, 699 (1989)
[145] Dimopoulos, S., Esmailzadeh, R., Hall, L.J., Starkman, G.D. (1987) Preprint SLAC-PUB-4356; *Astrophys. J.,* **330**, 545 (1988)
[146] Dimopoulos. S., Hall, L.J. (1987) *Phys. Lett.,* **196B**, 135.
[147] Dine, M. and Fischler, W. (1983) *Phys. Lett.,* **B120**, 137.
[148] Dine, M., Fischler, W., Srednicky, M. (1981) *Phys. Lett.,* **104B**, 199.
[149] Dirac, P.A.M. (1934) *Proc. Phys. Soc.,* **A133**, 60.
[150] Dirac, P.A.M. (1948) *Phys. Rev.,* **74**, 817.

[151] Dolan, L., Jackiw, R. (1974) *Phys. Rev.,* **D9**, 3320.
[152] Dolgov, A.D. (1980) *Yadernaya Fizika,* **32**, 1606.
[153] Dolgov, A.D. (1980) *ZhETF,* **79**, 337.
[154] Dolgov, A.D. (1981) *Yadernaya Fizika,* **33**, 1309 [*Sov. J. Nucl. Phys.,* **33**, 700].
[155] Dolgov, A.D. (1983) in: *The very early Universe*, eds. Gibbons, G., Hawking, S.W., Siklos, S.T., Cambridge University Press, Cambridge, 1983, 449.
[156] Dolgov, A.D. (1992) *Phys. Rept.*, **222**, 309.
[157] Dolgov, A.D. (1996) in Proc. Oak Ridge 1996, Future prospects of baryon instability search (Oak Ridge, TN, 28-30 Mar 1996) p.101; hep- ph\9605280
[158] Dolgov, A.D. (2001) *Nucl. Phys., Proc. Suppl.,* **95**, 42.
[159] Dolgov, A.D. (2002a) *Nucl. Phys., Proc. Suppl.* **113**, 40.
[160] Dolgov, A.D. (2002b) *Phys. Rept.,* **370**, 333.
[161] Dolgov, A.D., Illarionov, A.F., Kardashev, N.S., Novikov, I.D. (1987) *ZhETF,* **94**, 1.
[162] Dolgov, A.D., Kirilova, D.P. (1988) *Int. J. Mod. Phys.*, **A3**, 267.
[163] Dolgov, A.D., Zeldovich, Ya.B. (1981) *Rev. Mod. Phys.*, **53**, 1.
[164] Dolgov. A.D., Linde. A.D. (1982) *Phys. Lett.,* **116B**, 329.
[165] Domogatsky, G.V., Zheleznykh, I.M. (1969), *Yadernaya Fizika,* **10**, 1238.
[166] Doroshkevich, A.G. (1980) *Astron. Zh.*, **57**, 259.
[167] Doroshkevich, A.G. (1983a) *Sov. Astron. J. Lett.,* **9**, 238.
[168] Doroshkevich, A.G. (1983b) *Sov. Astron. J. Lett.,* **9**, 271.
[169] Doroshkevich, A.G. and Khlopov, M.Yu. (1984a) *Mon. Not. Roy. Astr. Soc.,* **211**, 277.
[170] Doroshkevich, A.G. Klypin, A.A., Khlopov, M.Yu. (1985) *Pism'a Astron. Zh.,* **11**, 483.
[171] Doroshkevich, A.G., Khlopov, M.Yu. (1981), *Astron. Zh.,* **58**, 913.
[172] Doroshkevich, A.G., Khlopov, M.Yu. (1983) *Pism'a Astron. Zh.,* **9**, 323.
[173] Doroshkevich, A.G., Khlopov, M.Yu. (1984b) *Yadernaya Fizika,* **39**, 869 [*Sov. J. Nucl. Phys.*, **39**, 551].
[174] Doroshkevich, A.G., Khlopov, M.Yu. (1985) *Pism'a Astron. Zh.,* **11**, 563.
[175] Doroshkevich, A.G., Khlopov, M.Yu., Klypin, A.A. (1989) *Mon. Not. Roy. Astr., Soc.,* **239**, 923.
[176] Doroshkevich, A.G., Klypin, A.A., Khlopov, M.Yu. (1980a) *Astron. Zh.,* **65**, 248.
[177] Doroshkevich, A.G., Klypin, A.A., Khlopov, M.Yu. (1988b) in: *Large scale structures of the Universe,* eds. Au-douze, J., Szalay, A.S. (Dordrecht, Reidel), p.293.
[178] Doroshkevich, A.G., Saar, E.M., Shandarin, S.F. (1978) *Mon. Not. Roy. Astr. Soc.,* **184**, 643.
[179] Doroshkevich, A.G., Zeldovich, Ya.B. (1981) *ZhETF*, **80**, 801.
[180] Doroshkevich, A.G., Zeldovich, Ya.B., Sunyaev, R.A., Szalay, A.S., Khlopov, M.Yu. (1980) *Proc. Neutrino-80* (Erice, 1980).
[181] Doroshkevich, A.G., Zeldovich, Ya.B., Sunyaev, R.A., Szalay, A.S., Khlopov, M.Yu. (1981) *Ann. N.Y. Acad. Sci.,* **375**, 32.
[182] Doroshkevich, A.G., Zeldovich, Ya.B., Sunyaev, R.A., Khlopov, M.Yu. (1980a) *Pis'ma Astron. Zh.,* **6**, 456.
[183] Doroshkevich, A.G., Zeldovich, Ya.B., Sunyaev, R.A., Khlopov, M.Yu. (1980a) *Pis'ma Astron. Zh.,* **6**, 465.
[184] Dover, C.B., Gaisser, T.K., Steigman, G. (1979) *Phys. Rev. Lett.,* **42**, 1117.
[185] Dubrovich, V.K., Khlopov, M.Yu. (1989) *Astron. Zh.,* **66**, 232.
[186] Duff, M.J., Nilsson, B.E.W., Pope, C.N. (1976) *Phys. Rep.,* **130**, 1.

[187] Dymnikova, I.G. (1996) *Int. J. Mod. Phys.*, **D5**, 529.
[188] Dymnikova, I.G., Khlopov, M.Yu. (1998) Proc. Cosmion'96 Eds. Khlopov M.Yu. et al. (World Scientic).
[189] Dymnikova, I.G., Khlopov, M.Yu. (2000) *Mod. Phys. Lett.*, **A15**, 2305.
[190] Dymnikova, I.G., Khlopov, M.Yu. (2001) *Eur. Phys. J.*, **C20**, 139.
[191] Dymnikova, I.G., Krawczyk, M. (1995) *Mod. Phys. Lett.*, **A10**, 3069.
[192] Einasto, J. et al.,(1980) *Mon. Not. Roy. Astr. Soc.*,**193**, 353.
[193] Ellis, J. R., Nanopoulos, D.V., Sarkar S. (1984) Preprint CERN-TH-4057-84, *Nucl. Phys.*, **B259**, 175 (1985).
[194] Ellis, J., Enqvist, K., Nanopoulos, D.V., Olive, K.A. (1987) *Phys. Lett.*, **191B**, 343.
[195] Ellis, J., Enqvist, K., Nanopoulos, D.V., Zwirner, F. (1986) *Nucl. Phys.*, **B276**, 14.
[196] Ellis, J., Nanopoulos, D., Olive, K., Tamvakis, K. (1983) *Nucl. Phys.*, **B221**, 524.
[197] Ellis, J., Nanopoulos, D.V., Quiros, M. (1986) *Phys. Lett.*, **174B**, 176.
[198] Ellis, J., Olive, K.A. (1987) *Phys. Lett.*, **193B**, 525.
[199] Englert, F., Brout, R. (1964) *Phys. Rev. Lett.* **13**, 321.
[200] Enqvist, K., Kainulainen, K., Maalampi, J. (1990) *Phys. Lett.*, **B244**, 186.
[201] Enqvist, K., Rez, A.I., Semikoz, V.B. (1995) *Nucl. Phys.*, **B436**, 49.
[202] Enrico, N., Esteban, R. (1990) *Phys. Lett.*, **B245**, 109.
[203] Ershov, A.A. Smirnov, G.T., Sorochenko, R.L., Tsivilev, A.P. (1988), *Pis'ma Astron Zh.*, **14**, 209.
[204] Everett, E. (1981) *Phys. Rev.*, **D24**, 858.
[205] Fabian, A.C., Pringle, J.E., Rees, M.J. (1975) *Not. Roy. Astr. Soc.*, **172**, 15p.
[206] Falk, S.W., Schramm, D.N. (1978) *Phys. Lett.*, B79, 511.
[207] Fargion, D., Khlopov, M.Yu. (2003) *Astropart. Phys.*, **19**, 441.
[208] Fargion, D., Khlopov, M.Yu., Konoplich, R.V., Mignani R. (1996) *Phys. Rev.*, **D54**, 4684.
[209] Fargion, D., Khlopov, M.Yu., Konoplich, R.V., Mignani R. (1998) *JETP Letters*, **68**, 685.
[210] Fargion, D., Khlopov, M.Yu., Konoplich, R.V., Mignani, R. (1995) *Phys. Rev.*, **D52**, 1828.
[211] Fargion, D., Konoplich, R.V., Grossi, M., Khlopov, M.Yu. (2000) *Astropart. Phys.*, **12**, 307.
[212] Feynman, R.P., Hibbs, A.R. (1965) *Quantum mechanics and path integrals* (McGraw-Hill, New York).
[213] Field, G.B. (1965) *Astrophys. J.*, **142**, 531.
[214] Filippov, S.S., Khlopov, M.Yu., Levitan, Yu.L., Sedelnikov, E.V. and Sobol, I.M. (1996) in: *Proc. 1st International conference on cosmopartcle physics, Cosmion-94 dedicated to 80th Anniversary of Ya.B. Zeldovich and 5th Memorial of A.D. Sakharov, (Moscow, December 5–14, 1994)*, eds.Khlopov, M.Yu., Prokhorov, M.E., Starobinsky, A.A., Tran Thanh Van, J. (Editions Frontieres), p.93.
[215] Foot, R. (2003) Experimental implications of mirror matter-type dark matter. astro-ph\0309330.
[216] Foot, R., Ignatiev, A.Yu., Volkas, R.R. (2001) *Phys. Lett.*, **B503**, 355.
[217] Foot, R., Ignatiev, A.Yu., Volkas, R.R. (2002) *Astropart. Phys.*, **17**, 195.
[218] Ford, L.H. (1987) *Phys. Rev.*, D35, 2339.
[219] Freedman, D.Z., van Nieuwenhuizen, P., Ferrara, S. (1976) *Phys. Rev.*, **D13**, 3214.

[220] Frenk, C.S., White, S.D.M., Davis, M. (1983) *Astrophys. J.*, **271**, 87.
[221] Freund. P.G.O., Rubin, M.A. (1980) *Phys. Lett.*, **B97**, 233.
[222] Fukugita, M., Watamura S. (1982) *Phys. Rev. Lett.*, **48**, 1522.
[223] Fuller, G.M., Mathews, C.J., Alcock, C.R. (1988) *Phys. Rev.*, **D37**, 1380.
[224] Gamov, G.A. (1948) *Phys. Rev.*, **74**, 505.
[225] Gamow, G.A. (1946) *Phys. Rev.*, **70**, 572.
[226] Garfinkle. D., Vachaspati, T. (1987) *Phys. Rev.*, **D36**, 2229.
[227] Gasilov, V.A., Maslyankin, V.I., Khlopov, M.Yu (1985) *Astrofizika*, **23**, 191.
[228] Gell-Mann, M., Ramond, P., Slansky, R. (1979) in: *Supergravity*, eds Van Nieuwenhuizen, P., Freedman, D. (North Holland), p.186.
[229] Gelmini, G., Schramm, D.N., Valle, J.W.F. (1984) *Phys. Lett.*, **14GB**, 311.
[230] Gelmini, G.B., Roncadelli, M. (1981) *Phys. Lett.*, **99B**, 411.
[231] Georgi, H., Glashow, S.L. (1974) *Phys. Rev. Lett.*, **32**, 438.
[232] Gershtein, S.S., Zeldovich, Ya.B. (1966), *Pis'ma ZhETF*, **4**, 174.
[233] Gibbons, G.W., Hawking, S. (1977) *Phys. Rev.*, **D15**, 2738.
[234] Ginzburg, V.L. (ed.), Dogiel, V.A., Berezinsky. V.S., Bulanov, S.V., Ptuskin, V.S. (1990) *Astrophysics of Cosmic rays* (Amsterdam, Netherlands, North-Holland).
[235] Glashow. S.L., Manohar, A. (1985) *Phys. Rev., Lett.* 54, 2306.
[236] Glashow. S.L., Weinberg, S. (1977) *Phys. Rev.*, **D15**, 1958.
[237] Gliese, W. (1977) *Highlights Astron,* **4**, P.II, 11.
[238] Glinner, E.B. (1965) *ZhETF*, **49**, 542.
[239] Glinner, E.B. (1970) *Doklady Akad. Nauk,* **192**, 771.
[240] Glinner, E.B., Dymnikova, I.F. (1975) *Pis'ma Astron. Zh.*, **1**, 7.
[241] Goddard. P., Mansfeld, P. (1986) *Rep. Progr. Phys.*, **49**, 725.
[242] Goldstone, J. (1961) *Nuovo Cim.*, **19**, 154.
[243] Golubkov, D.Yu., Golubkov, Yu.A., Khlopov, M.Yu. (2000) *Gravitation and Cosmology,* **6**, Suppl., 101.
[244] Golubkov, Yu.A., Fargion, D., Khlopov, M.Yu., Konoplich, R.V., Mignani R. (1999) *Pisma ZhETF*, **69**, 402 [*JETP Letters*, **69**, 434]
[245] Golubkov, Yu.A., Khlopov, M.Yu. (2001) *Yadernaya Fizika,* **64**, 1904 [*Phys. Atom. Nucl.*, **64**, 1821].
[246] Goncharov, A.S., Linder, A.D. (1986) *Fizika Elementarnykh Chastits i Atomnogo Yadra,* **17**, 837.
[247] Goncharov, S.A., Linde, A.D., Muhanov, V.F. (1987) *Int. J. Theor. Phys.*, **A2**, 561.
[248] Gorbunov, D.S., Rubakov, V.A. (2011a) *Introduction to the theory of the early Universe Hot Big Bang theory* (World Sci.).
[249] Gorbunov, D.S., Rubakov, V.A. (2011b) *Introduction to the theory of the early Universe. Cosmological perturbations and inflationary theory* (World Sci.)
[250] Gratton, P.G. (1982) *Astrophys. J.*, **257**, 640.
[251] Grawert, G., Luders, G., Rollnik, H. (1959) *Fortschritte Der Phys.*, **7**, 291.
[252] Green, M., Schwartz G., Witten, E. *Superstring theory*, Moscow, Mir, 1990, vol. 1 and 2.
[253] Green, M.B., Schwarz, J.H. (1984) *Phys. Lett.*, **B149**, 117.
[254] Grigoriev, D.Yu., Rubakov, V.A. (1988) *Nucl. Phys.*, **B299**, 67.
[255] Gross, D.J., Harvey, J.A., Martinec, E., Rohm, R. (1985) *Phys. Rev. Lett.*, **54**, 502.
[256] Gross, D.J., Pisarsky, R.D., Yaffe, L.G. (1981) *Rev. Mod. Phys.*, **53**, 43.
[257] Gross, D.J., Wilczek, F. (1973) *Phys. Rev. Lett.*, **30**, 1343.
[258] Gulyaev, S.A. Sorochenko, R.L., Tsivilev, A.P. (1987) *Pis'ma Astron. Zh.*, **23**, 191.
[259] Gunn, J.E., Lee, B.W., Lerche, I., Schramm, D.N., Steigman, G. (1978) *Astro-*

phys. J., **223**, 1015.
[260] Gunn, J.E., Petersen, B.A. (1965) *Astrophys. J.*, **142**, 1633.
[261] Guralnik, G.S., Hagen, C.R., Kibble, T.W.B. (1964) *Phys. Rev. Lett.* **13**, 585.
[262] Gurovich, V.Ts., Starobinsky, A. A. (1979) *Zh. Eksp. Teor. Fiz.*, **77**, 1683 [*Sov. Phys. JETP,* **50**, 844].
[263] Gurzadyan, V.G., Savvidy, G.K. (1986) *Astrophys. J.*, **160**, 203.
[264] Guth, A. (1981) *Phys. Rev.*, **D23**, 347.
[265] Guth, A., Pi, S.-Y. (1982) *Phys. Rev. Lett.*, **49**, 1110.
[266] Guth, A., Weinberg, E. (1981) *Phys. Rev.*, **D23**, 876.
[267] Guth, A., Weinberg, E. (1983) *Nucl. Phys.*, **B212**, 321.
[268] Haber, H.E., Kane, G.L. (1985) *Phys. Rep.*, **117**, 75.
[269] Hagiwara, K., et al (Particle Data Group) (2002) *Phys. Rev.*, **D66**, 010001.
[270] Harari, D., Sikivie, P. (1987) *Phys. Lett.*, **195B**, 361.
[271] Harrison, E.R. (1967) *Mem. Soc. Roy. Sci. Liege,* **15**, 15.
[272] Harrison, E.R. (1970) *Phys. Rev.*, **Dl**, 2726.
[273] Harvey, J.A., Kolb, E.W., Reiss, D.B., Wolfram, S. (1982) *Nucl. Phys.*, **B201**, 16.
[274] Hawking, S., Moss, I.G., Stewart, J.M. (1982) *Phys. Rev.*, **D26**, 2681.
[275] Hawking, S.W. (1974) *Nature,* **248**, 30.
[276] Hawking, S.W. (1975) *Comm. Math. Phys.*, **43**, 199.
[277] Hawking, S.W. (1976) *Phys. Rev.*, **D13**, 191.
[278] Hawking, S.W. (1982) *Phys. Lett.*, **115B**, 295.
[279] Hawking. S.W., Moss, I.G. (1982) *Phys. Lett.*, **HOB**, 35.
[280] Heidelberg–Moscow Collaboration, Klapdor-Kleingrothaus, H.V. (2002) *Yadernaya Fizika,* **65**, 2198 [*Phys. Atom. Nucl.*, **65**, 2135].
[281] Higgs, P.W. (1964) *Phys. Rev. Lett.*, **13**, 508.
[282] Hill, C.T., Schramm, D.N., Fry, J.N. (1989) *Comm. Nucl. Part. Phys.*, **19**, 25.
[283] Hill, C.T., Schramm, D.N., Walker. T.P. (1987) *Phys. Rev.*, **D36**, 1007.
[284] Hill, G., Hilditch, R.W., Barnes, J.W. (1979) *Mon. Not. Roy. Astr. Soc.*, **186**, 813.
[285] Hogan, C., Rees, M. (1984) *Nature,* **311**, 109.
[286] Holman, R., Kolb, E.W., Wang, Y. (1990) *Phys. Rev. Lett.*, **65**, 17.
[287] Holtzman, J.A. (1989) *Astrophys. J.*, Suppl., **71**, 1.
[288] Hosotani, Y. (1983) *Phys. Lett.*, **126B**, 309.
[289] House, F., Kilkenny, D. (1980) *Astron. Astrophys.*, **81**, 251.
[290] Huang, K. (1982) *Quarks, leptons, gauge fields* (World Scientific, Singapore).
[291] Hut, P. (1977) *Phys. Lett.*, **69B**, 85.
[292] Hut, P., White, S.D.M. (1984) *Nature,* **310**, 637.
[293] Ignatev, A.Yu., Volkas, R.R. (2000a) *Phys. Rev.*, **D62**, 023508.
[294] Ignatiev, A.Yu., Volkas, R.R. (2000b) *Phys. Lett.*, **B487**, 294.
[295] Ignatiev, A.Yu., Volkas, R.R. (2003) *Phys. Rev.*, **D68**, 023518.
[296] Ignatiev, A.Yu., Krasnikov, N.V., Kuzmin, V.A., Tavkhelidze, A.N. (1978) *Phys. Lett.*, **76B**, 436.
[297] Ipser, J., Sikivie, P. (1983) *Phys. Rev. Lett,* **50**, 925.
[298] Ipser, J., Sikivie, P. (1984) *Phys. Rev.*, **D30**, 712.
[299] Itoyama, H., McLerran, L., Taylor, T.R., van der Bij, J.J. (1987) *Nucl. Phys.*, **B279**, 380.
[300] Janes, K., Demarque, P. (1983) *Astrophys. J.*, **264**, 206.
[301] Jones, B.J.T. (1976) *Rev. Mod. Phys.*, **48**, 107.
[302] Jones, D.H.P. (1972) *Astrophys. J.*, **178**, 467.

[303] K2K collaboration, Ahn, M.H., et al (2003) *Phys. Rev. Lett.*, **90**, 04180.
[304] Kadnikov, A.F. Maslyankin, V.I., Khlopov, M.Yu. (1989) *Astrofizika*, **31**, 177.
[305] Kajante, K., Kurkio-Suonio, H. (1986) *Phys. Rev.*, **D34**, 1719.
[306] Kalashnikov, O.K., Khlopov, M.Yu. (1983) *Phys. Lett.*, **127B**, 407.
[307] Kaluza, Th. (1921) *Sitzungsber. Preuss. Akad. Wiss. Berlin. Math. Phys.*, **Kl**, 966.
[308] KamLAND collaboration, Eguchi, K., et al (2003) *Phys. Rev.*, **D64**, 021802.
[309] Kane, G.L., Kani, I. (1986) *Nucl. Phys.*, **B277**, 525.
[310] Kaplan, D.B., Manohar, A.V. (1986) *Phys. Rev. Lett.*, **56**, 2004.
[311] Kardashev, N.S., Blome, H.J., Priester, W. (1987) *Corn. Astrophys.*, **13**, 87.
[312] Kawai, H., Lewellen, D., Tye, S.H.H. (1986) *Phys. Rev. Lett.*, **57**, 1832.
[313] Kazama, Y., Yang, C.N., Goldhaber, A.S. (1977) *Phys. Rev.*, **D15**, 2287.
[314] Kazanas, D. (1980) *Astrophys. J.*, **241**, L59.
[315] Khlopov, M.Yu. (1979) *Priroda*, **12**, 90.
[316] Khlopov, M.Yu. (1980) *Yadernaya Fizika*, **32**, 1600.
[317] Khlopov, M.Yu. (1981) *Pis'ma ZhETF*, **33**, 100.
[318] Khlopov, M.Yu. (1982) *EChAYa*, **55**, 765.
[319] Khlopov, M.Yu. (1985) *Universe as a laboratory for elementary particles*, *Priroda*, **5**, 20.
[320] Khlopov, M.Yu. (1987) *Universe as a giant accelerator, Ser. Fizika*, Znanie, Moscow, 1/87.
[321] Khlopov, M.Yu. (1988) in: *Magnetic stars*, eds. Glagolevsky, Yu.V., Kopylov, M., (Leningrad, Nauka), 201.
[322] Khlopov, M.Yu. (1989) *Cosmoparticle physics, Ser. Fizika*, Znanie, Moscow, 89/3.
[323] Khlopov, M.Yu. (1990) *Astronuclear experiment ASTROBELIX, ser. Fizika*, Znanie, Moscow, 7/90.
[324] Khlopov, M.Yu. (1992) *Sov.* J. *Nucl. Phys.*, **55**, 765.
[325] Khlopov, M.Yu. (1995) in: *Dark matter in cosmology, clocks and tests of fundamental laws*, eds. Guiderdoni, B., Greene, G., Hinds, D., Tran Thanh Van, J. (Editions Frontieres), p.133.
[326] Khlopov, M.Yu. (1996a) in: Dark matter in cosmology. Quantum measurement. Experimental Gravitation, eds. Tran Thanh Van, J. et al. (Editions Frontieres), p.255.
[327] Khlopov, M.Yu. (1996b) in: *Proc. 1st International conference on Cosmoparticle physics Cosmion-94, dedicated to 80th Anniversary of Ya.B. Zeldovich and 5th Memorial of A.D. Sakharov,* (Moscow, December 5–14, 1994), eds. Khlopov, M.Yu., Prokhorov, M.E., Starobinsky, A.A., Tran Thanh Van, J. (Editions Frontieres), p.1.
[328] Khlopov, M.Yu. (1996c) in: *Proc. 1st International conference on Cosmoparticle physics Cosmion-94, dedicated to 80th Anniversary of Ya.B. Zeldovich and 5th Memorial of A.D. Sakharov,* (Moscow, December 5–14, 1994), eds. Khlopov, M.Yu., Prokhorov, M.E., Starobinsky, A.A., Tran Thanh Van, J. (Editions Frontieres), p.67.
[329] Khlopov, M.Yu. (1998) *Gravitation and Cosmology,* **4**, 69.
[330] Khlopov, M.Yu. (1999) *Cosmoparticle physics* (World Scientific).
[331] Khlopov, M.Yu. (2006) *JETP Letters,* **83**, 1.
[332] Khlopov, M.Yu. (2010) *Res. Astron. Astrophys*., **10**, 495.
[333] Khlopov, M.Yu. et al. Astrodamus collaboration (1996a) in: *Proc. 1st International conference on Cosmoparticle physics Cosmion-94, dedicated to 80th Anniversary of Ya.B. Zeldovich and 5th Memorial of A.D. Sakharov,* (Moscow,

December 5–14, 1994), eds. Khlopov, M.Yu., Prokhorov, M.E., Starobinsky, A.A., Tran Thanh Van, J. (Editions Frontieres), p.99.

[334] Khlopov, M.Yu. et al. Astrodamus collaboration (1996b) in: *Proc. 1st International conference on Cosmoparticle physics Cosmion-94, dedicated to 80th Anniversary of Ya.B. Zeldovich and 5th Memorial of A.D. Sakharov,* (Moscow, December 5–14, 1994), eds. Khlopov, M.Yu., Prokhorov, M.E., Starobinsky, A.A., Tran Thanh Van, J. (Editions Frontieres) p.107.

[335] Khlopov, M.Yu. et al. Astrodamus collaboration (1996c) in: The Dark side of the Universe eds. Bernabei, R., Incicchitti, A. (World Scientific), p.242.

[336] Khlopov, M.Yu., Beskin, G.M., Bochkarev, N.G., Pustilnik, L.A., Pustilnik, S.A. (1989) Preprint FERMI-LAB-Pub-89/193A.

[337] Khlopov, M.Yu., Beskin, G.M., Bochkarev, N.G., Pustilnik, L.A., Pustilnik, S.A. (1991) *Astron. Zh.,* **69**, 121.

[338] Khlopov, M.Yu., Chechetkin, V.M. (1985) *Test of GUTs in DUMAND. KIAM* preprint N-41.

[339] Khlopov, M.Yu., Chechetkin, V.M. (1987) *Fiz. Elementarn. Chastits Atomn. Yadra,* **18**, 627 [*Sov. J. Part. Nucl.*, **18**, 267].

[340] Khlopov, M.Yu., Konoplich, R.V., Mignani, R., Rubin, S.G., Sakharov, A.S. (2000c) *Astropart. Phys.,* **12**, 367.

[341] Khlopov, M.Yu., Konoplich, R.V., Rubin, S.G. (1996) in: *Proc. 1st International conference on Cosmoparticle physics Cosmion-94, dedicated to 80th Anniversary of Ya.B. Zeldovich and 5th Memorial of A.D. Sakharov,* (Moscow, December 5–14, 1994), eds. Khlopov, M.Yu., Prokhorov, M.E., Starobinsky, A.A., Tran Thanh Van, J. (Editions Frontieres), p.307.

[342] Khlopov, M.Yu., Malomed, B.A., Zeldovich, Ya.B. (1985) *Mon. Not. Roy. Astr. Soc.*, **214**, 100.

[343] Khlopov, M.Yu., Mayorov, A.G., Soldatov, E.Yu. (2010) *Int. J. Mod. Phys.*, **D19**, 1385.

[344] Khlopov, M.Yu., Naselsky, P.D., Polnarev, A.G. (1985) *Proc. III Int. Sem. Quantum gravity* (Moscow, 1984), eds. Markov, M.A., Berezin, V.A., Frolov, V.P. (Singapore, World Scientific), p.690.

[345] Khlopov, M.Yu., Petcov, S.T. (1981) *Phys. Lett.,* **B99**, 117.

[346] Khlopov, M.Yu., Polnarev, A.G. (1980) *Phys. Lett.,* **B97**, 383.

[347] Khlopov, M.Yu., Polnarev, A.G. (1983) in: *Proc. of the Nuffield Workshop – The very early Universe*, eds. Gibbons, G.W., Hawking, S.W., Siclos, S.T.C. (Cambridge University Press), p.407.

[348] Khlopov, M.Yu., Rubin S.G. (2004) *Cosmological pattern of microphysics in the inflationary Universe* (Kluwer Academic).

[349] Khlopov, M.Yu., Rubin, S.G., Sakharov, A.S. (1999) *Gravitation and Cosmology,* **5**, 246.

[350] Khlopov, M.Yu., Rubin, S.G., Sakharov, A.S. (2000a) *Phys. Rev.,* **D63**, 0835051.

[351] Khlopov, M.Yu., Rubin, S.G., Sakharov, A.S. (2000b) *Gravitation and Cosmology,* **6**, Suppl., 51.

[352] Khlopov, M.Yu., Rubin, S.G., Sakharov, A.S. (2002) *Gravitation and Cosmology,* **8**, Suppl., 57; astro-ph\0202505.

[353] Khlopov, M.Yu., Sakharov, A.S. (1994) in: *Particle astrophysics, atomic physics and gravitation*, eds. Tran Thanh Van, J., Fontaine, G., Hinds, E. (Editions Frontieres), p.197.

[354] Khlopov, M.Yu., Sakharov, A.S. (1995) *Gravitation and Cosmology,* **3**, 219.

[355] Khlopov, M.Yu., Sakharov, A.S. (1996) in: *Proc. 1st International conference on Cosmoparticle physics Cosmion-94, dedicated to 80th Anniversary of Ya.B.*

Zeldovich and 5th Memorial of A.D. Sakharov, (Moscow, December 5–14, 1994), eds. Khlopov, M.Yu., Prokhorov, M.E., Starobinsky, A.A., Tran Thanh Van, J. (Editions Frontieres), p.273.
[356] Khlopov, M.Yu., Sakharov, A.S. (1999) *Gravitation and Cosmology,* **5**, Suppl., 107.
[357] Khlopov, M.Yu., Sakharov, A.S., Shklyaev, A.A., Sudarikov, A.L. (1996) in: *Proc. 1st International conference on Cosmoparticle physics Cosmion-94, dedicated to 80th Anniversary of Ya.B. Zeldovich and 5th Memorial of A.D. Sakharov,* (Moscow, December 5–14, 1994), eds. Khlopov, M.Yu., Prokhorov, M.E., Starobinsky, A.A., Tran Thanh Van, J. (Editions Frontieres), p.289.
[358] Khlopov, M.Yu., Sakharov, A.S., SokoloV, D.D. (1999) *Nucl. Phys.,* (Proc. Suppl.) **B72**, 105.
[359] Khlopov, M.Yu., Shibaev, K. (2002) *Gravitation and Cosmology,* **8**, Suppl., 45.
[360] Khlopov. M.Yu., Linde, A.D. (1984) *Phys. Lett.,* **138B,** 265.
[361] Kibble, T.W.B. (1976) *J. Phys.,* **A9**, 1387.
[362] Kibble, T.W.B. (1980) *Phys. Rep.,* **67**, 183.
[363] Kibble, T.W.B. (1985) *Nucl. Phys.,* **B252**, 227.
[364] Kibble, T.W.B., Lazarides, G., Shaft, Q. (1982a) *Phys. Rev. Lett.,* **113B**, 237.
[365] Kibble, T.W.B., Lazarides, G., Shaft, Q. (1982b) *Phys. Rev.,* **D26**, 435.
[366] Kim, J.E. (1979) *Phys. Rev. Lett.,* **43**, 103.
[367] Kim, J.E. (1987) *Phys. Rep.,* **150**, 1.
[368] Kirzhnits, D.A. (1972) *Zh. Eksp. Teor. Fiz.,* **15**, 745.
[369] Kirzhnits, D.A., Linde. A.D. (1972) *Phys. Lett.,* **42B**, 471.
[370] Klein, O. (1926) *Z. Phys.,* **37**, 895.
[371] Klinkhammer, F.R., Norman, C.A. (1981) *Astrophys. J.,* **243**, LI.
[372] Klypin, A., Holtzman, J.A., Primack, J.R., Regos, E. (1993) *Astrophys. J.,* **416**, 1.
[373] Kobzarev, I.Yu., Martem'yanov, B.V., Okun, L.B. Shchepkin, M.G. (1980) *Yadernaya Fizika,* **32**, 1590 [*Sov. J. Nucl. Phys*., **32**, 823].
[374] Kobzarev, I.Yu., Okun, L.B., Pomeranchuk, I.Ya. (1966) *Yadernaya Fizika,* **3**, 1154.
[375] Kofman, L., Linde, A., Starobinsky, A.A. (1996) *Phys. Rev. Lett.,* **76**, 1011.
[376] Kofman, L., Linde, A.D., Starobinsky, A.A. (1987) *Phys. Lett.,* **157B**, 361.
[377] Kofman, L.A., Linde, A.D. (1987) *Nucl. Phys.,* **B282**, 555.
[378] Kofman, L.A., Linde, A.D., Einasto, J. (1987) *Nature,* **326**, 48.
[379] Kofman, L.A., Pogosyan, Yu.D. (1988) *Phys. Lett.,* **214B**, 508.
[380] Kogan, Ya.I., Khlopov, M.Yu. (1987) *Yadernaya Fizika,* **44**, 1344.
[381] Kogan, Ya.I., Khlopov, M.Yu. (1987) *Yadernaya Fizika*, **46**, 314,
[382] Koikawa, T. (1991) *Prog. Theor. Phys.,* **85**, 1049.
[383] Kolb, E.W., Perry, M.J., Walker, T.P. (1986) *Phys. Rev.,* **D33**, 869.
[384] Kolb, E.W., Seckel, D., Turner, M. S. (1985) *Nature,* **314**, 415.
[385] Kolb, E.W., Turner, M.S. (1983) *Ann. Rev. Nucl. Part. Sci*., **33**, 645.
[386] Kolb, E.W., Turner, M.S. (1990) Front.Phys. 69, 1
[387] Kondratyuk, L.A., Shmatikov, M.Zh. (1983) *Yadernaya Fizika,* **38**, 361.
[388] Konoplich, R.V. (1980) *Yadernaya Fizika,* **32**, 1132.
[389] Konoplich, R.V., Khlopov, M.Yu. (1988) *Yadernaya Fizika,* **47**, 891.
[390] Konoplich, R.V., Khlopov, M.Yu. (1988) *Yadernaya Fizika,* **57**, 452 [*Phys. Atomn. Nucl.,* **57**, 425].
[391] Konoplich, R.V., Rubin, S.G., Sakharov, A.S., Khlopov, M.Yu. (1994) *Pis'ma Astron. Zh.,* **24**, 1.
[392] Konoplich, R.V., Rubin, S.G., Sakharov, A.S., Khlopov, M.Yu. (1999) *Yadernaya*

Fizika, **62**, 1705 [*Phys. Atomn. Nucl.*, **62**, 1593].
[393] Konoplich, R.V., Sorokina, E.V., Khlopov, M.Yu. (1994) *Yadernaya Fizika,* **57**, 1338 [*Phys. Atomn. Nucl.*, **57**, 1268].
[394] Kosover, D, A., Hall, L.J., Krauss, L.M. (1985) *Phys. Lett.,* **150B**, 436.
[395] Krauss, L.M. (1983) *Phys. Lett.,* **128V**, 37.
[396] Kuijken, K., Gilmore, G. (1989) *Mon. Not. Roy. Astr. Soc.,* **239**, 571; **239**, 605; **239**, 651.
[397] Kuijken, K., Gilmore, G. (1991) *Astrophys. J. Lett.,* **367**, L9.
[398] Kumachenko, Yu.Ya., Khlopov, M.Yu. (1990) *Yadernaya Fizika,* **51**, 1081.
[399] Kurki-Suonio, H., Matzner, R.A., Centrello, J.M., Rothman, T., Wilson, J.R. (1988) *Phys. Rev.,* **D38**,1091.
[400] Kuzmin, V.A. (1970) *Pis'ma ZhETF,* **12**, 335.
[401] Kuzmin, V.A., Rubakov, V.A., Shaposhnikov, M.E. (1985) *Phys. Lett.,* **B155**, 36.
[402] Kuzmin, V.A., Shaposhnikov, M.E. and Tkachev, I.I. (1992) *Phys. Rev.,* **45D**, 466.
[403] Kuzmin, V.A., Shaposhnikov, M.E., Tkachev, I.I. (1981) *Phys. Lett.,* **B105**, 167.
[404] La, D., Steinhardt, P.J. (1989) *Phys. Rev. Lett.,* **62**, 376.
[405] Landau, L.D., Lifshits, E.M. (1964) *Statistical physics* (Moscow, Nauka).
[406] Landau, L.D., Lifshits, E.M. (1973) *Field theory* (Moscow, Nauka).
[407] Landsman, N.P., van Weert, Ch.G. (1987) *Phys. Rep.,* **45**, 141.
[408] Langacker, P., Pi, S.-Y. (1980) *Phys. Rev. Lett.,* **45**, 1.
[409] Lazarides, G., Panagiotakopoulos, C., Shaft, Q. (1986) *Phys. Rev. Lett.,* **56**, 557.
[410] Lazarides, G., Panagiotakopoulos, C., Shaft, Q. (1987) *Phys. Lett.,* **192B**, 323.
[411] Lazarides, G., Shaft, Q. (1982) *Phys. Lett.,* **115B**, 21.
[412] Lazarides, G., Shaft, Q., Walsh, T. (1982) *Nucl. Phys.,* **B195**, 157.
[413] Lee, T.D. (1973) *Phys. Rev.,* **D8**, 1226.
[414] Lee, T.D., Yang, C.N. (1956) *Phys. Rev.,* **104**, 254.
[415] Lee, W., Weinberg, S. (1977) *Phys. Rev. Lett.* **39**, 165.
[416] Levitan, Yu.L., E.V., Sobol, I.M., Khlopov, M.Yu., Chechetkin, V.M. (1988) *Yadernaya Fizika,* **47**, 168.
[417] Levitan, Yu.L., Sedelnikov, E.V., Sobol, I.M., Khlopov, M.Yu. (1994), *Yadernaya Fizika,* **57**, 1466.
[418] Liddle, A.R. (1993) *Phys. Rev.,* **D49**, 739.
[419] Liddle, A.R., Lyth, D.H. (1993) *Phys. Rep.,* **231**, 1.
[420] Lidsey, J.E. et al (1995) Preprint FERMILAB-Pub-95/80-a astrophi9508078.
[421] Linde, A.D. (1979) *Repts. on Progress in Physics,* **42**, 389.
[422] Linde, A.D. (1982a) *Phys. Lett.,* **108B**, 389.
[423] Linde, A.D. (1982b) *Phys. Lett.,* **116B**, 335.
[424] Linde, A.D. (1983a) *Pisma ZhETF,* **38**, 149.
[425] Linde, A.D. (1983b) *Phys. Lett.,* **129V**, 177.
[426] Linde, A.D. (1984) *Usp. Fiz. Nauk,* **114**, 177.
[427] Linde, A.D. (1985) *Phys. Lett.,* **160B**, 243.
[428] Linde, A.D. (1986) *Mod. Phys. Lett.,* **A2**, 81.
[429] Linde, A.D. (1988) *Phys. Lett.,* **200B**, 272.
[430] Linde, A.D. (1988) *Phys. Lett.,* **B201**, 437.
[431] Linde, A.D. (1990) *Particle physics and inflationary cosmology* (Harwood, Chur, Switzerland).
[432] Lindley, D. (1980) *Mon. Not. Roy. Astr. Soc.,* **193**, 593.
[433] Luty, M.A. (1992) *Phys. Rev.,* **D45**, 455.
[434] Lubimov, V.A., Novikov, E.G., Nozhik, V.Z., Tretyakov, E.F., Kozik, V.S. (1980)

Yadernaya Fizika, **32**, 310 [*Sov. J. Nucl. Phys.*, **32**, 2800].
[435] Ma, C.P. (1996) *Astrophys. J.,* **471**, 13.
[436] MacGibbon, J.H., Carr, B.J. (1991) *Astrophys. J.,* **371**, 447.
[437] Madsen, J. (1990) in: *Particle astrophysics. The early Unierse and cosmic structures,* eds. Alimi, J.M., Blanchard, A., Bouquet, A., Martin de Volnay, F. and Tran Thanh Van, J. (Editions Frontieres), p.205.
[438] Malaney, R.A., Fowler, W.A. (1989) *Annals Phys.*, **192**, 45.
[439] Malaney, R.A., Mathews, G.J. (1993) *Phys. Rept.,* **229**, 145.
[440] Maltoni, M., Novikov, V.A., Okun, L.B., Rozanov, A. N., Vysotsky, M.I. (2000) *Phys. Lett.,* **B476**, 107.
[441] Mao, S., Paczynski, B. (1996), *Astrophys. J.,* **473**, 57.
[442] Markov, M.A. (1964) *Phys. Lett.,* **10**, 122.
[443] Markov, M.A. (1978) *On the stability of elementary black holes*, Preprint ICTP, Trieste, IC\78\41.
[444] Markov, M.A. (1993) *Phys. Lett.,* **A172**, 331.
[445] Martemyanov, B.V., Khlopov, M.Yu., Shchepkin, M.G. (1980) *Pis'ma ZhETF,* **32**, 484.
[446] Mather, J. C. et al (1990) *Astrophys. J.,* **354**, L37.
[447] Mather, J. C. et al (1994) *Astrophys. J.,* **420**, 439.
[448] Matzner, M. (1988) *Computer in Physics,* **2**, 51.
[449] Melott, A.L., et al (1983) *Phys. Rev. Lett.,* **51**, 935.
[450] Mikheyev, S.P., Smirnov, A.Yu. (1985) *Yadernaya Fizika,* **42**, 1441.
[451] Mikheyev, S.P., Smirnov, A.Yu. (1986) *Zh. ETF,* **91**, 7 [*Sov. Phys. JETP,* **64**, 4].
[452] Miyama, S., Sato, K. (1978) *Prog. Theor. Phys.,* **59**, 1012.
[453] Mohapatra, C.N. (1987) *Phys. Lett.,* **186B**, 303.
[454] Morgan, D.L., Hughes, V.W. (1970) *Phys. Rev.,* **D2**, 1389.
[455] Moss, I.G. (1994) *Phys. Rev.,* **D50**, 676.
[456] Mukhanov, V. (2005) *Physical foundations of cosmology* (Cambridge Univ. Press).
[457] Mukhanov, V.F., Feldman, H.A., Brandenberger, R.H. (1992) *Phys. Rept.,* **215**, 203.
[458] Nakamura, K. et al (2010) *J. Phys. G,* **G37**, 075021.
[459] Nambu, Y. (1970) *Proc. Int. Conf. on Symmetries and Quark Models,* p.269.
[460] Narain, K.S. (1986) *Phys. Lett.,* **169B**, 41.
[461] Naselsky, P.D. (1978) *Sov. Astron. J. Lett.,* **4**, 387.
[462] Naselsky, D., Polnarev, A.G. (1986) *Astron. Zh.*, **62**, 833.
[463] Nielsen, H.B., Olesen, P. (1973) *Nucl. Phys.,* B61, 45.
[464] Nilles, H.P. (1983) *Phys. Rep.,* **86**, 200.
[465] Nilles, H.P. (1984) *Phys. Rep.,* **110**, 1.
[466] Nissen, P.E., Twarog, B.A., Crawford, D.L. (1987) *Astron. J.,* **93**, 634.
[467] Novikov, I.D., Polnarev, A.G., Starobinsky, A.A., Zeldovich, Ya.B. (1979) *Astron. Astrophys.*, 80, 104.
[468] Novikov, V.A., Okun, L.B., Rozanov, A. N., Vysotsky, M.I. (2002) *Phys. Lett.* **B529**, 111.
[469] Novikov, V.A., Shifman, M.A., Vainshtein, A.I., Zakharov, V.I. (1981) *Nucl. Phys.,* **B191**, 301.
[470] Occhionero, F., Amendola, L. (1994) *Phys. Rev.,* **D50**, 4846.
[471] Oknyansky, V. (1999) *Gravitation and Cosmology,* **5**, Suppl., 97.
[472] Okun, L.B. (1983) On a search for mirror particles. Preprint ITEP-149-1983.
[473] Okun, L.B. (1984) *Physics of elementary particles* (Moscow, Nauka).

[474] Okun, L.B. (2007) *Phys. Usp.*, **50**, 380.
[475] Okun, L.B., Zeldovich, Ya.B. (1976) *Comm. Nucl. And Part. Phys.*, **6**, 69.
[476] Okun, L.B. (1984) *Leptons and quarks* (Moscow, Nauka).
[477] Olshanetsky, M.A. (1982) *Usp. Fiz. Nauk,* **136**, 421,
[478] Oort, J.H. (1965) in: *Stellar Dynamics. Galactic Structure* (Chicago–London), p.455.
[479] Ostriker, J., Thompson, C., Witten, E. (1986) *Phys. Lett.,* **B180**, 231.
[480] Ozernoi, L.M., Chernin, A.Zh. (1981) *Astron. Zh.,* **44**, 1131.
[481] Parker, L. (1968) *Phys. Rev. Lett.,* **21**, 562.
[482] Peacock, J.A., Dodds, S.J. (1994) *Mon. Not. Roy. Astr. Soc.,* **267**, 1020.
[483] Peccei, R.D., Quinn, H.R. (1977a) *Phys. Rev. Lett.,* **38**, 1440.
[484] Peccei, R.D., Quinn, H.R. (1977b) *Phys. Rev.,* **D16**, 1791.
[485] Pedoussaut, A., Ginestat, N., Carquillat, J.M. (1984) *Astron. Astrophys.,* Suppl. Ser. **58**, 601.
[486] Pedoussaut, A., Ginestat, N., Carquillat, J.M., Vigneau, J. (1988) *Astron. Astrophys.*, Suppl. Ser. **75**, 441.
[487] Penzias. A.A., Wilson, R.W. (1965) *Astrophys. J.,* **142**, 419.
[488] Petcov, S.T. (1977) *Sov. J. Nucl. Phys.,* **25**, 340.
[489] Pi, S.Y. (1984) *Phys. Rev. Lett.,* **52**, 1725.
[490] Pogosyan, Yu.D., and Starobinsky A.A. (1995) *Astrophys. J.,* **447**, 465.
[491] Politzer, H.D. (1973) *Phys. Rev. Lett.,* **30**, 1346.
[492] Polnarev, A.G., Khlopov, M.Yu. (1982a) *Astron. Zh.,* **59**, 15.
[493] Polnarev, A.G., Khlopov, M.Yu. (1982a) *Astron. Zh.,* **59**, 639.
[494] Polnarev, A.G., Khlopov, M.Yu. (1982a) *Astron. Zh.,* **59**, 639.
[495] Polnarev, A.G., Khlopov, M.Yu. (1982a) *Usp. Fiz. Nauk,* **145**, 369,
[496] Polyakov, A.M. (1974) *Pisma ZhETF,* **20**, 430.
[497] Preskill, J., Wise, M., Wilczek, F. (1983) *Phys. Lett.,* **B120**, 127.
[498] Preskill, J.R. (1979) *Phys. Rev. Lett.,* **43**, 1365.
[499] Press, W.H., Schechter, P. (1974) *Astrophys. J.,* **187**, 425.
[500] Press, W.H., Teukolsky, S. (1977) *Astrophys. J.,* **213**, 183.
[501] Raffelt, G., Seckel, D. (1988) *Phys. Rev. Lett.*, **60**, 1793.
[502] Raffelt, G.G. (1996) *Stars as laboratories for fundamental physics: The astrophysics of neutrinos, axions, and other weakly interacting particles* (Chicago Univ. Press).
[503] Raffelt, G.G. (2000) *Phys. Rept.*, **333**, 593.
[504] Rajaraman, R. (1982) *Solitons and instantons. An Introduction to solitons and instantons in quantum field theory* (North-Holland Publishing Company. Amsterdam–New York–Oxford).
[505] Reiss, D. (1982) *Phys. Lett.,* **115B**, 217.
[506] Rodionov, B.U. (1996) in: *1st International conference on Cosmoparticle physics Cosmion-94, dedicated to 80th Anniversary of Ya.B. Zeldovich and 5th Memorial of A.D. Sakharov,* (Moscow, December 5–14, 1994), eds. Khlopov, M.Yu., Prokhorov, M.E., Starobinsky, A.A., Tran Thanh Van, J. (Editions Frontieres), p.473.
[507] Rothman, T., Matzner, R. (1981) *Astrophys. Space Sci.,* **75**, 229.
[508] Rubakov, V.A. (1981) *Pis'ma ZhETF,* 33, 658.
[509] Rubakov, V.A. (1982) *Nucl. Phys.,* **B203**, 311.
[510] Rubakov, V.A., Sazhin, M.V., and Veryaskin, A.V. (1982) *Phys. Lett.,* **115B**, 189.
[511] Rubakov, V.A., Shaposhnikov, M.E. (1996) *Usp. Fiz. Nauk*, **166**, 493 [*Usp. Fiz. Nauk*, **39**, 461].
[512] Rubin, S.G., Sakharov, A.S., Khlopov, M.Yu. (2001) *ZhETF,* **92**, 921.

[513] Ruffini, R., Bonazzola, S. (1969) *Phys. Rev.,* **187**, 1767.
[514] Ruiz, F.R., Alvarez-Estrado, R.F. (1987) *Z. Phys.*, **C34**, 131.
[515] Sachs, R.K., Wolf, A.M. (1967) *Astrophys. J.,* **147**, 73.
[516] Sakharov, A.D. (1948) *ZhETT,* **18**, 631.
[517] Sakharov, A.D. (1967) *Pis'ma ZhETF,* **5**, 32.
[518] Sakharov, A.D. (1979) *ZhETF*, **76**, 1172.
[519] Sakharov, A.D. (1989) *Vestnik AN SSSR*, **4**, 39.
[520] Sakharov, A.S., Khlopov, M.Yu. (1993) *Yadernaya Fizika*, **56**, 220 [*Sov. J. Nucl. Phys.*, **56**, 412].
[521] Sakharov, A.S., Khlopov, M.Yu. (1994b) *Yadernaya Fizika*, **57**, 690 [*Phys. Atomn. Nucl.* **57**, 651].
[522] Sakharov, A.S., Sokolov, D.D., Khlopov, M.Yu. (1996) *Yadernaya Fizika*, **59**, 1050.
[523] Sakharov, A.S., Sudarnikov, A.L., Khlopov, M.Yu., Shklyaev, A.A. (1996b) *Yadernaya Fizika*, **59**, 295.
[524] Sakharov, A.S., Sudarnikov, A.L., Khlopov, M.Yu., Shklyaev, A.A. (1996b) *Yadernaya Fizika*, **59**, 2041.
[525] Sakharov, A.S.,Khlopov, M.Yu. (199a) *Yadernaya Fizika*, **57**, 514 [*Sov. J. Nucl. Phys.*, **57**, 485].
[526] Salopek, D.S., Bond, J.R. (1990) *Phys. Rev.,* **D42**, 3936.
[527] Salucci, P. (2003) in: *Proc. of Physics Beyond the Standard Model*, BEYOND'03 ed. H.V. Klapdor-Kleingrothaus (Springer, Heidelberg).
[528] Sarkar, S., Cooper, A.M. (1984) *Phys. Lett.,* **B148**, 347.
[529] Sato, H., Takahahara, F. (1980) *Progr. Theor. Phys.*, **54**, 2029.
[530] Sato, K. (1981a) *Phys. Lett.,* **99B**, 66.
[531] Sato, K. (1981b) *Mon. Not. Roy. Astr. Soc.,* **195**, 467.
[532] Sato, K., Kobayashi, H. (1987) *Progr. Theor. Phys.*, **58**, 1775.
[533] Sato, N., Kodama, H., Sato, K. (1985) *Progr. Theor. Phys.*, **74**, 405.
[534] Sazhin, M.V., Khlopov, M.Yu. (1989) *Astron. Zh.*, **66**, 191.
[535] Schechter, J., Valle, J.W.F. (1982) *Phys. Rev.,* **D25**. 774.
[536] Scherk, J. (1979) *Phys. Lett.,* **88B**, 265.
[537] Scherrer, R.J., Turner, M.S. (1985) *Phys. Rev.,* **D31**, 681.
[538] Schmidt, M., Schneider, D.P., Gunn, J.E. (1986) *Astrophys. J.,* **306**, 411.
[539] Schramm, D. N., Steigman, G. (1979) *Phys. Lett.,* **B87**, 141.
[540] Schramm, D.N., Copi, C. (1996) in: *1st International conference on Cosmoparticle physics Cosmion-94, dedicated to 80th Anniversary of Ya.B. Zeldovich and 5th Memorial of A.D. Sakharov,* (Moscow, December 5–14, 1994), eds. Khlopov, M.Yu., Prokhorov, M.E., Starobinsky, A.A., Tran Thanh Van, J. (Editions Frontieres), p.77
[541] Schramm, D.N., Steigman, G. (1981) *Astrophys. J.,* **151**, 459.
[542] Schwartz, A.S. (1976) *Nucl. Phys.,* **B112**, 358.
[543] Schwartz, A.S. (1982) *Nucl. Phys.,* **B208**, 141.
[544] Schwartz, A.S., Tyupkin, Yu.A. (1982) *Nucl. Phys.,* **B209**, 427.
[545] Schwinger, J. (1966) *Phys. Rev.,* **144**, 1087.
[546] Sedelnikov, E.V. (1999) *Pisma Zh. Eksp. Teor. Fiz.*, **69**, 609.
[547] Sedelnikov, E.V. (2000) *Pisma Zh. Eksp. Teor. Fiz.,* **72**, 513.
[548] Sedelnikov, E.V., Fillippov, S.S., Khlopov, M.Yu. (1995) *Yadernaya Fizika*, **58**, 280.
[549] Sedelnikov, E.V., Khlopov, M.Yu. (1996) *Yadernaya Fizika*, **59**, 1045.
[550] Senjanovic, G., Stecker, F.W. (1980) *Phys. Lett.,* **B96**, 285.
[551] Senjanovic, G., Wilczek, F., Zee, A. (1984) *Phys. Lett.,* **B141**, 389.
[552] Severny, A.B., Kotov, V.A., Tsap, T.T. (1976) *Nature,* **259**, 87.

[553] Severny, A.B., Kotov, V.A., Tsap, T.T. (1979) *Astron. Zh.*, **56**, 1137.
[554] Shabalin, E.P. (1980) *Yadernaya Fizika,* 32, 249 [*Sov. J. Nucl. Phys.*, **32**, 129].
[555] Shafi, Q., Wetterich, C. (1983) *Phys. Lett.*, **129B**, 387.
[556] Shandarin, S.F., Doroshkevich, A.G., Zeldovich, Ya.B. (1983) *Sov. Phys. Uspekhi,* **139**, 83.
[557] Shaposhnikov M.E. (1987) *Nucl. Phys.*, **B287**, 757.
[558] Shaposhnikov M.E. (1988) *Nucl. Phys.*, **B299**, 797.
[559] Shaposhnikov M.E. (1996) in: *Proc. 2nd International Sakharov Conference on Physics* (Moscow, Russia, 20–23 May 1996) p.320.
[560] Shifman, A.I., Vainshtein, A.I., Zakharov, V.I. (1979) *Nucl. Phys.*, **B147**, 385.
[561] Shifman, M.A., Vainshtein, A.I., Zakharov, V.I. (1980) *Nucl. Phys.*, **B166**, 493.
[562] Shuryak, E.V. (1980) *Phys. Rep.*, **61**, 71.
[563] Shvartman, V.F., Braginsky, V.B., Gershtein, S.S., Zeldovich, Ya.B., Khlopov, M.Yu. (1982) *Pis'ma ZhETF,* **36**, 224.
[564] Shvartsman, V.F. (1969) *Pis'ma ZhETF,* **9**, 315.
[565] Shvartsman, V.F. (1977) *Comm. of SAO*, **19**, 1.
[566] Sikivie, P. (1982) *Phys. Rev. Lett.*, **48**, 1156.
[567] Sikivie, P. (1983) *Phys. Rev. Lett.*, **51**, 1415.
[568] Sikivie, P. (2010) *Int. J. Mod.Phys.*, **A25**, 554.
[569] Silk, J. (1968) *Astrophys. J.*, **151**, 459.
[570] Silk, J., Srednicki, M. (1984) *Phys. Rev. Lett.*, **53**, 624.
[571] Silk, J., Turner, M. (1987) *Phys. Rev.*, **D35**, 419.
[572] Smith, A.G., Vilenkin, A. (1987) *Phys. Rev.*, **D36**, 987.
[573] Smoot, G.F. et al. (1992) *Astrophys. J. Lett.*, **300**, 90.
[574] SNO collaboration, Ahmad, Q.R., et al (2002) *Phys. Rev. Lett.*, **89**, 011301.
[575] Sohnius, M.F. (1985) *Phys. Rep.*, **128**, 39.
[576] Sommerfeld, A. (1931) *Annalen der Physik,* **403**, 257.
[577] Souslin, V.M., Khlopov, M.Yu., Chechetkin, V.M., Chouyanov, V.A. (1982) KIAM Preprint N-30.
[578] Spergel, D.N., et al (2003) *Astrophys. J.*, Suppl. **148**, 175; astroph\0302209.
[579] Starobinsky, A.A. (1979) *Pis'ma ZhETF,* **30**, 719.
[580] Starobinsky, A.A. (1983) *Pis'ma Astron. Zh.*, **9**, 1983.
[581] Starobinsky, A.A. (1980) *Phys. Lett.*, **91**, 99.
[582] Starobinsky, A.A. (1982) *Phys. Lett.*, **117B**, 175.
[583] Stauffer, D. (1979) *Phys. Rep.*, **54**, 1.
[584] Stecker, F.W. (1978) *Nature,* **273**, 493.
[585] Stecker, F.W., Shaft, Q. (1983) *Phys. Rev. Lett.*, **50**, 928.
[586] Steigman, G. (1976) *Ann. Rev. Astron. Astrophys.*, **14**, 339.
[587] Steigman, G., Olive, K. A., Schramm, D.N., Turner, M. S. (1986) *Phys. Lett.*, **B176**, 33.
[588] Steigman, G., Olive, K.A., Schramm, D.N. (1979) *Phys. Rev. Lett.*, **43**, 239.
[589] Steigman, G., Schramm, D.N., Gunn, J.E. (1977) *Phys. Lett.*, **B66**, 202.
[590] Steigman, G., Turner, M.S. (1985) *Nucl. Phys.*, **B253**, 375.
[591] Sunyaev, R.A., Zeldovich, Ya.B. (1969) *Astrophys. Space Sci.*, **4**, 285.
[592] Sunyaev, R.A., Zeldovich, Ya.B. (1970) *Astrophys. Space Sci.*, **7**, 0.
[593] SuperKamiokande collaboration, Kajita, T., Totsuka, Y. (2001) *Rev. Mod. Phys.*, **73**, 85.
[594] Szalay, A.S., Marx, G. (1976) *Astron. Astrophys.* **49**, 437.
[595] t'Hooft, G. (1974) *Nucl. Phys.*, **B79**, 276.
[596] t'Hooft, G. (1976) *Phys. Rev. Lett.*, **37**, 8.
[597] Terasawa, N., Sato, K. (1989) *Phys. Rev.*, **D39**, 2893.

[598] Toomre, A., Toomre, J. (1972) *Astrophys. J.,* **178**, 623.
[599] Toussaint, D., Treiman, S.B., Wilczek, F., Zee, A. (1979) *Phys. Rev.,* **D19**, 1036.
[600] Tremaine, S., Gunn, J.E. (1979) *Phys. Rev. Lett.,* **42**, 407.
[601] Tryon, E.P. (1973) *Nature*, **246**, 396.
[602] Tsivilev, A.P. (1991) *Pis'ma Astron. Zh.*, **17**, 3.
[603] Tsivilev, A.P., et al (2002) *Gravitation and Cosmology,* **8**, Suppl., 126.
[604] Tully, R.B. (1982) *Astrophys. J.,* **257**, 389.
[605] Turner, M.S. (1985) *Phys. Rev.,* **D31**, 1212.
[606] Turner, M.S. (1990) *Phys. Rep.,* **197**, 67.
[607] Turner, M.S., Cohen, A.G., Kaplan, D.B. (1988) *Phys. Lett.,* **B216**, 20.
[608] Turner, M.S., Steigman, G., Krauss, L.M. (1984) *Phys. Rev. Lett.*, **92**, 2090.
[609] Turner, M.S., Weinberg, E.J., Widrow, L.M. (1992) *Phys. Rev.,* **D46**, 2384.
[610] Turner. M.S., Widrow, L.M. (1988) *Phys. Rev.,* **D37**, 2743.
[611] Turner. M.S., Widrow,L. (1986) *Phys. Rev. Lett.,* **57**, 2237.
[612] Turok, N., Bhattacharjee, P. (1984) *Phys. Rev.,* **D29**, 1557.
[613] Twarog, B.A. (1983) *Astrophys. J.,* **267**, 207.
[614] Vachaspati, T., Vilenkin, A. (1985) *Phys. Rev.,* **D31**, 3052.
[615] Vachaspati. T., Vilenkin, A. (1984) *Phys. Rev.,* **D30**, 2936.
[616] Vainer, B.V., Naselsky, P.D. (1977) *Pisma Astron. Zh.,* **3**, 147.
[617] Vainshtein, A.I., Zakharov, V.I., Novikov, V.I., Shifman, A, I. (1982) *Usp. Fiz. Nauk,* **136**, 553.
[618] Valle, J.W.F. (1983) *Phys. Lett.,* **131B**, 87.
[619] Vandenberg, P.A. (1983) *Astrophys. J.,* Suppl. Ser. **51**, 29.
[620] Vilenkin, A. (1981a) *Phys. Rev. Lett.,* **46**, 1169; 1496 (E).
[621] Vilenkin, A. (1981b) *Phys. Rev.,* **D24**, 2082.
[622] Vilenkin, A. (1981c) *Phys. Lett.,* **B107**, 47.
[623] Vilenkin, A. (1981d) *Phys. Rev.,* **D23**, 852.
[624] Vilenkin, A. (1982) *Nucl. Phys.,* **B196**, 240.
[625] Vilenkin, A. (1983a) *Phys. Rev.,* **D27**, 2848.
[626] Vilenkin, A. (1985) *Phys. Rep.,* **121**, 263.
[627] Vilenkin, A., Everett, A.E. (1982) *Phys. Rev. Lett.,* **48**, 1867.
[628] Vishniac, E.T., Olive, K., Seckel, D. (1987) *Nucl. Phys.,* **B289**, 717.
[629] Voloshin, M.B., Dolgov, A.D. (1982) *Yadernaya Fizika,* **35**, 213.
[630] Voloshin, M.B., Okun, L.B. (1978) *Pis'ma ZhETF,* **28**, 156.
[631] Voloshin, M.B., Ter-Martirosyan, K.A. (1984), *Theory of gauge interactions of elementary particles* (Moscow, Energoatomizdat).
[632] Vysotsky, M.I. (1985) *Sov.Phys. Usp.* **28**, 667.
[633] Vysotsky, M.I., Dolgov, A.D., Zeldovich, Ya.B. (1987) *Pis'ma ZhETF,* **26**, 200.
[634] Vysotsky, M.I., Zeldovich, Ya.B., Khlopov M.Yu and Chechetkin, V.M. (1978) *Pis'ma ZhETF,* **27**, 533.
[635] Wagoner, R. V. (1973) *Astrophys. J.,* **179**, 343.
[636] Wagoner, R.V., Fowler, W.A., Hoyle, F. (1967) *Astrophys. J.,* **148**, 3.
[637] Warren, S.J., Hewlett, P.O., Irwin, M.J., McMahou, R.J., Bridgeland, M.T., Bunclark, P.S., Kibblewhite, E.J. (1987) *Nature,* **325**, 131.
[638] Watkins. R., Widrow, L.M. (1992) *Nucl. Phys.,* **B374**, 446.
[639] Weinberg, S. (1972) *Gravitation and cosmology,* (John Wiley and Sons, Inc. New York–London–Sydney–Toronto).
[640] Weinberg, S. (1974) *Phys. Rev.,* **D9**, 3357.
[641] Weinberg, S. (1978) *Phys. Rev. Lett.,* **40**, 223.
[642] Weinberg, S. (1979) *Phys. Rev. Lett.* **42**, 850.
[643] Weinberg, S. (1982) *Phys. Rev. Lett.,* **48**, 1303.
[644] White, S.D.M., Frenk, C.S., Davis, M. (1983) *Astrophys. J.,* **274**, L1.

[645] Widrow, L.M. (1988) *Phys. Rev.*, **D39**, 3571.

[646] Wielen, R. (1971) *Astrophys. Space Sci.*, **13**, 300.

[647] Wilczek, F. (1978) *Phys. Rev. Lett.*, **40**, 279.

[648] Wilczek, F. (1982) *Phys. Rev. Lett.*, **49**, 1549.

[649] Wilczek, F. (1983) in: *The Very early Universe* eds. Gibbons, G., Hawking, S.W., Siklos, S.T. (Cambridge University Press, Cambridge) p.9.

[650] Wise, M.B., Georgi, H., Glashow, S.L. (1981) *Phys. Rev. Lett.*, **47**, 402.

[651] Witten, E. (1981) *Nucl. Phys.*, **B177**, 477.

[652] Witten, E. (1984) *Phys. Rev.*, **D30**, 272.

[653] Witten, E. (1985) *Nucl. Phys.*, **B258**, 75.

[654] Wolfenstein, L. (1978) *Phys. Rev.*, **D17**, 2369.

[655] Wright, E.L. et al. (1992) *Astrophys. J. Lett.*, **300**, 100.

[656] Yahil, A., Beaudet, G. (1976)) *Astrophys. J.*, **206**, 26.

[657] Yamamoto, K. (1986) *Phys. Lett.*, **168B**, 341.

[658] Yamamoto, K. (1987) *Phys. Lett.*, **194B**, 390.

[659] Yang, C.N., Mills, R.L. (1954) *Phys. Rev.*, **96**, 191.

[660] Yang, J.-M., Schramm, D.N., Steigman, G., Rood, R. T. (1979) *Astrophys. J.*, **227**, 697.

[661] Yokoyama, J., Kodama, H., Sato, K., Sato, N. (1987) *Int. J. Mod. Phys.*, **A2**, 1808.

[662] Yoshimura M. (1983) *Phys. Rev. Lett.*, **51**, 439.

[663] Yoshimura, M. (1978) *Phys. Rev. Lett.*, **41**, 281; **42**, 746.

[664] Zavelsky, A.S. (1939) *Phys. Rev.*, **55**, 317.

[665] Zeldovich, Ya.B. (1961) *ZhETF*, **41**, 1609.

[666] Zeldovich, Ya.B. (1968) *Usp. Fiz. Nauk*, **95**, 209.

[667] Zeldovich, Ya.B. (1972) *Mon. Not. Roy. Astr. Soc.*, **160**, 1.

[668] Zeldovich, Ya.B. (1976) *Pis'ma ZhETF*, **24**, 29.

[669] Zeldovich, Ya.B. (19777) *Pis'ma Astron. Zh.*, **1**, 8.

[670] Zeldovich, Ya.B. (1980) *Mon. Not. Roy. Astr. Soc.*, **192**, 63.

[671] Zeldovich, Ya.B. (1981) in: *Quantum gravity*, Proc. of the 2nd Seminar, eds. M.A.Markov, P.C.West (Plenum), p.253.

[672] Zeldovich, Ya.B., Illarionov, A.F., Sunyaev, R.A. (1972) *ZhETF*, **62**, 1217.

[673] Zeldovich, Ya.B., Khlopov, M.Yu. (1978) *Phys. Lett.*, **79B**, 239.

[674] Zeldovich, Ya.B., Khlopov, M.Yu. (1981) *Pism'a ZhETF*, **34**, 148.

[675] Zeldovich, Ya.B., Khlopov, M.Yu. (1981) *Usp. Fiz. Nauk*, **135**, 45.

[676] Zeldovich, Ya.B., Khlopov, M.Yu. (1984) *Astrophysics and elementary particles*, Lectures MIFI, Moscow.

[677] Zeldovich, Ya.B., Khlopov, M.Yu. (1988) *Drama of ideas in understanding nature*, (Moscow, Nauka).

[678] Zeldovich, Ya.B., Klypin, A.A., Khlopov, M.Yu., Chechetkin, V.M. (1980) *Yadernaya Fizika*, **31**, 1286 [*Sov. J. Nucl. Phys.*, **31**, 6640].

[679] Zeldovich, Ya.B., Kobzarev I.Yu., Okun, L.B. (1974) *ZhETF*, **67**, 3.

[680] Zeldovich, Ya.B., Novikov, I.D. (1975) *Structure and evolution of the Universe*, (Moscow, Nauka).

[681] Zeldovich, Ya.B., Okun, L.B., Pikelner, S.B. (1967) *Usp. Fiz. Nauk*, **84**, 113.

[682] Zeldovich, Ya.B., Podurets, M.A. (1965) *Astron Zh.*, **42**, 983.

[683] Zeldovich, Ya.B., Starobinsky, A.A., Khlopov, M.Yu., Chechetkin, N.M. (1977) *Pis'ma Astron. Zh.*, **3**, 208.

[684] Zeldovich, Ya.B., Starobinsky, A.A. (1976) *JETP Lett.*, **24**, 616.

[685] Zeldovich, Ya.B., Sunyaev, R.A. (1980) *Pis'ma Astron. Zh.*, **6**, 451,

[686] Zhitnitsky, A.R. (1980) *Yadernaya Fizika*, **31**, 197.

[687] Zumino, B. (1975) *Nucl. Phys.*, **B89**, 535.

Index

D

E

F

G